Opportunistic Networks

Opportunistic Networks

Fundamentals, Applications and Emerging Trends

Edited by Anshul Verma, Pradeepika Verma, Sanjay Kumar Dhurandher and Isaac Woungang

CRC Press
Taylor & Francis Group
Boca Raton London New York

CRC Press is an imprint of the
Taylor & Francis Group, an **informa** business

First edition published 2022
by CRC Press
6000 Broken Sound Parkway NW, Suite 300, Boca Raton, FL 33487-2742

and by CRC Press
2 Park Square, Milton Park, Abingdon, Oxon, OX14 4RN

CRC Press is an imprint of Taylor & Francis Group, LLC

ISBN: 978-0-367-67730-5 (hbk)
ISBN: 978-0-367-67732-9 (pbk)
ISBN: 978-1-003-13258-5 (ebk)

Typeset in Sabon
by Straive, India

Contents

Preface

OVERVIEW AND GOALS

The opportunistic network is a result of intensive research of so many years on Mobile ad hoc networks (MANETs), therefore it is called an extension of MANET. Like MANET, an opportunistic network is also an autonomous, dynamic, decentralized, and infrastructure-less network that is made of wireless mobile devices. It is also ad hoc because the network is formed dynamically "on the fly" and does not have the pre-existing infrastructure, unlike infrastructure-full networks where routers and access points are used to manage network information. Instead, each node acts as an end node (source or destination) as well as a router (forward data of other nodes). The next forwarder of the data is determined dynamically based on network connectivity and the routing protocol in use. Unlike MANETs, opportunistic networks consider several wireless networks' properties, such as nodes mobility, disconnection of nodes, network partitions, and links' instability, as advantages rather than exceptions. This often results in the design of opportunistic networks being significantly more complex and challenging than that of MANETs.

In opportunistic networks, mobility of the nodes is used as a way for the communication between disconnected clusters of nodes rather than an exception that should be overcome. It does not create a complete path between a source and destination during a transmission. There is no need for source and destination to be connected to the same network at the same time. Despite this, messages can be exchanged between source and destination by using the *store-carry-and-forward* approach in which intermediate nodes store messages in the absence of better forwarding nodes, carry messages from one location to other, and forward to the other nodes those that have better opportunity to bring the messages closer to the destination. This approach of message transmission in highly intermittent connected ad hoc wireless networks became more favorable than the traditional MANET paradigm.

However, the opportunistic network is an emerging and recent area of research, and many industrial applications have been developed based on this network. The opportunistic network is made of heterogeneous mobile devices that exist everywhere, due to the widespread use of cellular and pervasive mobile devices by the people, and are emerging day by day. Hence, many more concrete futuristic mobile applications can be developed by adopting the opportunistic networking paradigm. To make this research area more adaptable for practical and industrial use, there is a need to further investigate several research challenges in all aspects of opportunistic networks. Therefore, this book will provide theoretical, algorithmic, simulation, and implementation-based research developments related to fundamentals, applications, and emerging research trends in opportunistic networks.

The following are some of the important features of this book, which, we believe, would make it a valuable resource for our readers:

- This book is designed, in structure and content, with the intention of making the book useful at all learning levels.
- Most of the chapters of the book are authored by prominent academicians, researchers, and practitioners, with solid experience and exposure to research on the area of routing in opportunistic networks. These contributors have been working in this area for many years and have a thorough understanding of the concepts and practical applications.
- The authors of this book are distributed in a large number of countries, and most of them are affiliated with institutions of worldwide reputation. This gives this book an international flavor.
- The authors of each chapter have attempted to provide a comprehensive bibliography section, which should greatly help interested readers to dig further into the aforementioned research area.
- Throughout the chapters of this book, most of the core research topics have been covered, making the book particularly useful for industry practitioners working directly with the practical aspects behind enabling the technologies in the field.

We have attempted to make the different chapters of the book look as coherent and synchronized as possible.

TARGET AUDIENCE

This book will be beneficial for academicians, researchers, developers, engineers, and practitioners working in or interested in the fields related to opportunistic networks, delay tolerant networks, intermittently connected ad hoc networks, vehicular ad hoc networks, and mobile ad hoc networks. This book is expected to serve as a reference book for developers and engineers working in the mobile application development and telecommunications industries, and for a graduate/postgraduate course in computer science and engineering/information technology.

Acknowledgments

We are extremely thankful to the 28 authors of the 14 chapters of this book, who have worked very hard to bring this unique resource forward for helping the students, researchers, and community practitioners. We feel that it is contextual to mention that as the individual chapters of this book are written by different authors, the responsibility of the contents of each of the chapters lies with the concerned authors.

We like to thank Randi Cohen, Publisher – Computer Science & IT, and Gabriella Williams, Editor, who worked with us ohe beginning, for their professionalism. We also thank Daniel Kershaw, Editorial Assistant at Taylor & Francis Group, and his team members who tirelessly worked with us and helped us in the publication process.

This book is a part of the research work funded by "Seed Grant to Faculty Members under IoE Scheme (under Dev. Scheme No. 6031)"[1] and "DST-Science and Engineering Research Board (SERB), Government of India (File no. PDF/2020/001646)"[2].

NOTES

1 Awarded to Anshul Verma at Banaras Hindu University, Varanasi, India.
2 Awarded to Pradeepika Verma at Indian Institute of Technology (BHU), Varanasi, India.

Contributors

Soumaia A. Al Ayyat
The American University in Cairo
New Cairo, Egypt

Sherif G. Aly
The American University in Cairo
New Cairo, Egypt

Souvik Basu
Heritage Institute of Technology
Kolkata, India

Yuanzhu Chen
Memorial University of Newfoundland
St. John's, Canada

Soumyadip Chowdhury
University of Engineering and Management
Kolkata, India

Md. Sharif Hossen
Comilla University
Cumilla, Bangladesh

Meriem Houmer
Moulay Ismail University
Meknes, Morocco

Chih-Lin Hu
National Central University
Taoyuan, Taiwan

Carlos Borrego Iglesias
Universitat de Barcelona
Barcelona, Spain

Abu Zafor Md. Touhidul Islam
University of Rajshahi
Rajshahi, Bangladesh

Md. Khalid Mahbub Khan
University of Rajshahi
Rajshahi, Bangladesh

Vinesh Kumar
Bharati College, University of Delhi
New Delhi, India

Sonam Kumari
Raj Kumar Goel Institute of Technology
Ghaziabad, India

Cheng Li
Memorial University of Newfoundland
Canada

Mohd Yaseen Mir
National Central University
Taoyuan, Taiwan

Rukhsana Naznin
Comilla University
Cumilla, Bangladesh

Mariya Ouaissa
Moulay Ismail University
Meknes, Morocco

Mariyam Ouaissa
Moulay Ismail University
Meknes, Morocco

Muhammad Sajjadur Rahim
University of Rajshahi
Rajshahi, Bangladesh

Siuli Roy
Heritage Institute of Technology
Kolkata, India

Jagdeep Singh
Sant Longowal Institute of Engineering
and Technology, Longowal
Sangrur, India

Mohini Singh
Banaras Hindu University
Varanasi, India

Itu Snigdh
Birla Institute of Technology
Mesra, India

Md. Ibrahim Talukdar
Comilla University
Cumilla, Bangladesh

Anshul Verma
Banaras Hindu University
Varanasi, India

Pradeepika Verma
Indian Institute of Technology (BHU)
Varanasi, India

Zehua Wang
University of British Columbia
Vancouver, Canada

Sanjay Kumar Dhurandher
Netaji Subhas University of Technology
New Delhi, India

Chapter 1

Mobile-Code-Based Opportunistic Networking

Carlos Borrego Iglesias
Universitat de Barcelona, Barcelona, Spain

CONTENTS

1.1 INTRODUCTION

The Internet is the evidence of how flexible and sturdy the TCP/IP network architecture has proved to be to create a global network of networks. Although there are many important issues about this architecture, there are also a good number or reasons to commend it. But there are some networks for which the TCP/IP is not an option because of, among other things, its relatively short timeouts or its routing algorithm. This is the case, for instance, when several devices are able to communicate to each other only once in a while, being unreachable most of the time. The nodes in these types of networks could use their peers to asynchronously forward information for them from node to node to any point of the network, even if this process took a long time waiting for the opportunity to transmit the information. This is not a new idea at all, and many architectures in the literature provide similar solutions, usually under the labels of Challenged Networks, Opportunistic Networking, or Delay and Disruption Tolerant Networking (DTN).

In these networks, a critical issue, if not the most important, to deal with is routing. Because there is not a fixed topography of the network, the decision-making process to choose the next hop among the neighbors cannot rely on it. Instead, routing has to be a dynamic function based on the local information acquired by the router in the very moment of the routing process. Influenced by the success of the Internet, full of general purpose mechanisms, one could think that a holistic "one-fits-all" routing algorithm would be useful here and would establish a standard for opportunistic networks. Unfortunately, the approaches trying to do that are bound to fail. Such a standard, should it be, would lead to deadlocks similar to the ones faced in the current Internet (e.g. multicast routing), because not all the applications have the same requirements and not all opportunistic networks are alike. Different applications may have different goals when it comes to routing the information, such as minimizing the latency time, minimizing the variation in latency time, maximizing the throughput, maximizing the reliability, or minimizing the routing overhead. In fact, most of the successful solutions for doing the routing in these opportunistic networks fit just a particular scenario, normally consisting of a single application [1], and they set the trend. Some of these works claim to provide a generic framework for any type of scenario/application, but at the end of the day they fail to include new routing policies once the network has been deployed. On the other hand, it would be very convenient to let the applications decide the routing instead of trying to design a common algorithm including the associated fundamental properties for all scenarios and applications. This is definitely the way of optimizing the performance of this type of networks: there is no better source for routing information than the applications themselves.

In this chapter, we introduce a general purpose architecture for DTN based upon the idea of letting the application, by means of its messages, decide the routing that will take place in every node. The keystone of this proposal is carrying the routing algorithm code along with every single message. The resulting DTN can be used by different applications, even if they were not foreseen before the deployment of the network. Thus, the DTN is no longer bound to specific applications and becomes a real, open, general-purpose heterogeneous network as it has been on the Internet for connected networks. A network built following this architecture is flexible and open, and application designers can make the most appropriate use of it according to the specific requirements of the application. In short, the evident dependence between an application and its routing protocol is preserved while keeping other applications' interests.

This chapter presents a general architecture for opportunistic networks based on mobile code. A formal definition of all the actors involved in communication networks is presented. A generalization of these communications is described making our proposal a general case of communications information exchange. Several DTN issues such as routing, DTN congestion and error handling are discussed.

1.2 MOTIVATION

We have seen in the previous section the existence of different applications which run on scenarios where there are intermittent connectivity, asymmetric bandwidths, long and variable latency or ambiguous mobility patterns. In these scenarios, the network is created when mobile and/or static nodes holding wireless devices intermittently connect. These network limitations are due to issues such as small wireless radio range, mobility of the nodes, energy constraints, network attacks or noise. Classic network protocols, such as TCP/IP, as explained in previous sections, do not solve these challenging issues. For these purposes different proposals such as the Bundle Protocol [2] and Haggle [3] have been floated. These proposals evolve around the *store-carry-and-forward* paradigm: intermediate mobile or static nodes accept the custody application information until they are able to find new custodians to forward this information.

For every different application which is intended to be run on a DTN scenario, we get from the mentioned proposals, different interesting solutions which include various ways of solving routing problems. These proposals include replication-based protocols, such as epidemic and PRoPHET [1]; forwarding-based protocols, such as MEED [1] and LAROD [1]; and a great many others. We believe that these proposals have been extremely useful to understand the complexity of the problematic issues around DTN networks. However, these different proposals are studied separately, defining a custom DTN network for every different scenario.

We propose a step further in DTN networks. We believe in the necessity of a common network in which different applications may coexist without leaving behind their optimal routing algorithms. Traditional bundle-based proposals [2] allow just one type of routing algorithms. When sharing a unique DTN infrastructure among several applications, a unique routing algorithm must be chosen. We believe this is not efficient and can be improved. Our proposal intends to break this tightly coupled relationship among DTN infrastructures and the applications.

Contrary to well-established protocols like TCP/IP, a unique routing algorithm for many applications has proven to work efficiently, and the Internet is the most compelling proof of it. However, there are different applications in the context of DTN networks, which work optimally just when using different routing algorithms. If we want to let these types of different applications share a unique network infrastructure we need to propose a new paradigm. This paradigm will be described in this chapter.

1.2.1 IP Multicast: A Case Example

Multicasting, in the context of TCP/IP networks and concretely in IPv4, is the sending of information over a TCP/IP internet to multiple destinations with a single datagram. Different applications such as software distribution, multimedia streaming, mobile TV, radio broadcasting, and new media services, with different needs, take advantage of Multicast in order to spread application data all around the Internet.

Datagrams sent to Multicast groups have as their destination IP addresses an Internet address belonging to a special range of addresses called D type. Internet hosts subscribe or unsubscribe dynamically to these Multicast groups to receive or stop receiving application data from the application source.

The network infrastructure is responsible for replicating the Multicast datagrams and intelligently routing these according to the topology of receivers interested in that information, that is, Internet hosts who are subscribed. On one hand, hosts advertise to their local routers about their interest to belong to a given Multicast group, establishing the Multicast group membership. On the other hand, several different protocols have been defined to communicate among local and remote Multicast routers in order to route Multicast traffic from the Multicast source application to the many Multicast clients. Examples on these protocols can be found in [4] and include protocols such as PIM Sparse Mode, PIM Dense Mode, PIM source-specific Multicast, and Bidirectional PIM. These protocols are widely deployed all around the Internet.

Applications willing to use Multicast use the different network infrastructures using the very same protocols to route Multicast datagrams. Unfortunately, no general-purpose Multicast routing algorithm covering every router on the Internet is available. This issue forces sender applications to choose from the different protocols in terms of the better adjustment of their needs and access to the network infrastructure. For example, service providers offer multimedia services to their clients using their network infrastructure which employ optimal Multicast routing protocols chosen in terms of the type of multicasted applications, but do not work well for other types of applications. These networks remain isolated for other applications willing to Multicast since different PIM protocols do not understand each other.

As a result, no global Multicast service is available on the Internet. Unfortunately, Multicast is not the only domain in which the coexistence of different protocols is not entirely solved. In the context of mobile ad hoc networks (MANETs), different routing protocols have been proposed, but when it comes to support heterogeneous MANETs, no common routing paradigm copes with all the desired requirements. Our proposal in this chapter intends to avoid precisely these types of failures in DTN scenarios.

1.3 SCENARIO DESCRIPTION

DTN research tries to provide solutions to problems where network partitions are created due to intermittent connectivity by using the already-described *store-carry-and-forward* switching. We propose in this study an enhanced version of this paradigm based on mobile code. In this section, we first introduce the different actors and functions involved in DTN scenarios. Following this, a six-phase general case of protocol is described.

> *Application list*: We define *applist* as a non-empty set {app1, app2, ..., app$_n$} as the list of n applications present in a network scenario. Applications mainly create messages which are sent over the network from a source to a destination.
>
> *Custodian list*: Let {$c_1, c_2, ... c_c$} be the list of c nodes capable of temporally storing application messages, carrying them, and routing them to some other custodian nodes.
>
> *Application messages*: Let {$mess_{appi,1}, mess_{appi,2}, ..., mess_{appi,m}$} be the list of m messages an application app_i sends over the network. Let $messages_c$ be the list of messages a custodian node c has at a given time.
>
> *Custodian routing information tree (RIT)*. For every custodian node c_c, routing information may be stored locally in the custodian nodes. A RIT may be defined as a root node, RIT_{root} or $t + 1$ number of RIT trees {$RIT_{local}, RIT_{app1}, RIT_{app2}, ..., RIT_{appt}$} where t is the number of the different applications present in the network. These trees contain other RIT trees or values for routing purposes.
>
> *Custodian buffer space*. The amount of space a custodian node c_i has available for application messages, and routing information is defined as $buff(c_i)$.
>
> *Custodian movement model*: Custodian nodes follow movement models m_{cj} which affect important network parameters such as connection window time and encounter statistical distribution.
>
> *Application custodian code*. Let $code_{app,cj}$ be an application code an application *app* intends to execute on a custodian node c_j.
>
> *Neighbor list*: We define *neighbourList*(c_j, t), as the list of p neighbors {$n_1, n_2, ... n_p$} a custodian node c has at a given time t.
>
> *Routing function*: Given an application message $mess_{app,i}$, belonging to application *app*, a custodian node c_j, and its local RIT tree RIT_{cj}, let $f_{routing}(mess_{app,i}, neighbourList(c_j, t), RIT_c)$ be the function which returns a subset of the neighbors of the custodian node c_j which should custody the application message.

In general, network protocols follow three traditional independent phases: Data creation phase, routing and delivery phase. Besides the above-mentioned *store-carry-and-forward* paradigm, we define the following communication paradigm phases (Figure 1.1):

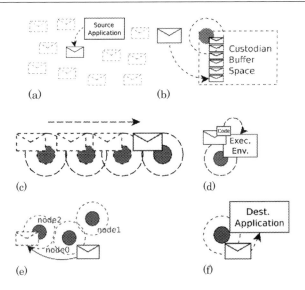

Figure 1.1 (a) *Application message creation phase:* In this phase, application *app* creates $mess_{app,i}$. (b) *Store phase:* Message $mess_{app,i}$ is stored at *buffer* (*custodian c*) in custodian *custodian c*. (c) *Carry phase:* The message follows the custodian *movement*(*custodian c*) movement model. (d) *Process phase:* The custodian node c_j executes local code $code_{appa,cj}$ on behalf of application app_a. (e) *Routing phase:* The message is replicated following its routing algorithm, which in this example returns *node2*. (f) *Delivery phase:* The message $mess_{appa,i}$ is freed from the last custodian buffer (*buff* (c_j)) and delivered to the application layer on custodian c_j.

Data creation phase. We define data creation for application *app* as a process in which application *app* creates *m* application messages $mess_{app,i}$, *i* in [0..*m*]. Application messages may be created as the result of application programs or application custodian codes ($code_{app,cj}$). The format of data messages will be described in Section 1.4.

Store phase: Messages are stored in a custodian node c_j using the $buff(c_j)$. Since this space is limited, the routing problem can be considered as an optimal resource allocation challenge.

Carry phase: Messages are carried from one place to another. Messages perform *active carries*, that is, carrying application *app* information from custodian node to custodian node in terms of the function $f_{routing}(mess_{app,i}, neighbourList(c_j,t), RIT_{cj})$. Instead, a custodian node if in motion, performs a *passive carry* when moving. In this case, there is no software entity performing the transport itself [5]. Instead, there is a physical movement of the custodian node $m(c_c)$ who performs the *carry* action.

Process phase: A custodian node c_j may execute $code_{appi}$, c_j on behalf of the application *app*. For these purposes, a custodian node needs to have execution environments. This phase is equivalent to the *process* phase in the *store-carry-process-and-forward* paradigm that will be explained in Chapter 2.

Forwarding phase: This phase occurs given custodian node c_j and application information message $mess_{app,i}$ belonging to application *app*. This phase is equivalent to the forward phase in the already defined *store-carry-and-forward* paradigm. Message $mess_{app,i}$ is propagated in the network following the following expression. Note that if *forwardToList* = Ø, it means that the message is discarded.

$$forwardToList = f_{routing}\left(mess_{app_i}, neighbourList\left(c_j,t\right), RIT_{c_j}\right), where$$
$$forwardToList \subseteq neighbourList\left(c_j,t\right) \cup c_j$$

Delivery phase: Once a message $mess_{appa,i}$ has arrived to its destination, it is passed to the destination application and removed from the last custodian buffer ($buff(c_{destination})$). Other copies of message $mess_{appa,i}$ may be present in the network in other custodian nodes.

We propose the idea of creating a network to allow different applications coexist without leaving behind their optimal routing algorithms. Our proposal is leaving this routing decision to the very same application. This will let the different applications determine which routing algorithm to apply in every DTN hop. In order to achieve this flexibility, two options must be studied.

The first option involves a deployment of all of the possible routing protocols $f_{routing}(mess_{appa,i}, neighbourList(c_j,t), RIT_{cj})$ in the different DTN custodian nodes $\{c_1, c_2, ..., c_m\}$. This option presents some immediate disadvantages. Routing algorithm deployment is expensive in challenged networks and it is not guaranteed that this deployment of all the necessary $f_{routing}$ functions from the different applications will arrive at every custodian node. Furthermore, newcomer applications will need to make additional routing deployments.

Second, we could let the very same application message $mess$ carry along with itself the routing algorithm $f_{routing}(mess_{appa,i}, neighbourList(c_j,t), RIT_{cj})$ and/or the application custodian code, $code_{appa,cj}$. Theses algorithms must be able to be run in any custodian node. Mobile code is a good candidate to implement a data-driven approach to DTN networks.

In some scenarios with limited storage space, having replicated the same routing algorithm code on every message belonging to the same application could be inefficient. Every time the active message is forwarded, besides message data, the routing code must be sent to the destination platform. These static codes can be cached on the different node execution environments, using as described in [6], a global cache service to efficiently deal with the distribution of agent codes. Caching these codes improves and speeds up message forwarding.

1.4 ACTIVE MESSAGES

Messages containing application data, routing algorithms $f_{routing}(mess_{appa,i}, neighbourList(c_j,t), RIT_{cj})$ and application codes $code_{appa,cj}$ are called active messages. In Figure 1.2 the four fields that define an active message are depicted. These fields are:

Delivery information: In this field the source and destination addresses, active message creation timestamp, and static lifetime is defined.

Application code: The application code $code_{app,cj}$ creates and manipulates application data. This data may be created beyond the source custodian node. Data belonging to an active message is reviewed every time the application code is executed on a custodian node.

Routing code: Routing algorithm code $frouting(app_a)$ chooses from the available neighbors on the list of the potential information custodian nodes. The routing code will be responsible for defining the path the application data will take.

Data payload: Application data created by the source application.

Delivery Information	Application Code	Routing Code	Data

Figure 1.2 Active message fields: control information, routing algorithm, and data payload.

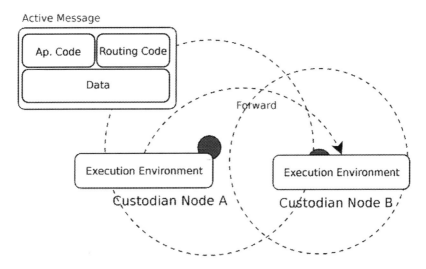

Figure 1.3 Active message forwarding. Custodian Node A forwards an active message containing some data, its routing code, and an application code to Custodian Node B.

We propose a simple paradigm, depicted in Figure 1.3, in which the application information is carried by the active messages. Active messages can be seen therefore as a transmission medium, i.e., each active message is functionally analogous, conceptually speaking, to a carrier pigeon, as in [7]. DTN custodian nodes must include an execution environment which implements the DTN layer that handles delays and disruptions. The application code creates or modifies application data which is encapsulated inside the active messages. Active messages stay in the custodian nodes, which accept their custody, until they are able to be forwarded to another one or they arrive at their destination node. Custodian nodes may employ different convergence layers, the interfaces between the common Bundle Protocol and the specific inter-network protocol suites.

1.5 GENERAL PROTOCOL ARCHITECTURE

Layering protocols is an excellent way to simplify networking designs by the means of separating the latter into functional layers, and defining protocols to delegate each layer's task. In the context of network architectures based on layers, every layer receives from the previous layer its *protocol data unit*. From the moment of this reception, this *protocol data unit becomes* the *service data unit* of the new layer, which upon adding a header from the new layer, a protocol data unit of the new layer is obtained. Each layer interacts, mainly with its neighboring layers, across the interfaces between them. This mechanism is called encapsulation and guarantees the layer principle in which different layers are independent from one another. Traditional de facto standard protocols like TCP/IP break this layer principle for practical reasons, letting some information from one layer be employed by its following layer.

On the other hand, most of the network architectures propose uniqueness when it comes to choosing their different protocols implemented by the different layers. Information traveling on networks defined by these architectures permits different protocols on the different layers, but on every hop, these protocols are limited to the ones chosen by the layer's code. For example, in the context of IP communications, we understand that a datagram is routed on the basis of the information gathered from a certain internal router protocol, such as RIP

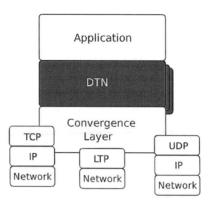

Figure 1.4 General network DTN architecture. The different levels in the DTN layer represent the possibility of employing different routing protocols and different DTN behaviors. The different convergence layers, such as Licklider Transmission Protocol (LTP) [9], provide communication among intermediate nodes when they meet in an opportunistic manner.

or OSPF, but we would not see it as feasible that the datagram could choose among these two mechanisms in order to be routed.

Layered protocols do not only imply just a separation of responsibilities. Layers are placed one after the other and define in which order things should be done. In TCP/IP for example, routing algorithms are executed before the translation of logical addresses into physical addresses while sending a data-gram.

We propose simple DTN architecture which is based on layers, but in a very particular way. This architecture is fully compatible with the DTN proposals defined by the DTN architecture [8] as will be described in the following sections. A layer-based model, according to the definition explained in this section, implies delegation of tasks, encapsulation, and uniqueness and should define a sequence of actions.

In Figure 1.4, our DTN architecture is depicted. An application layer represents the different application which may employ the network. The DTN layer handles problems such as addressing, routing, reliability and custody transfer, congestion, and security very similarly to the DTN architecture with a subtle difference: the DTN layer may be different for every application. Issues such as routing, reliability and custody transfer, congestion and flow control, and security may be treated differently depending on the application.

1.6 SYSTEM ARCHITECTURE ELEMENTS

In this section, the different system architecture elements are described.

RIT manager. Routing information is classified into ontologies, which represent the different domains from the different applications which may coexist in a network. The RIT manager is responsible for the access and storage of this information, which can be created when nodes meet and exchange application information, or with the output from active messages execution codes. The very same active messages may classify, modify, access, or remove information from the *routing information tree* contacting the RIT manager. Local nodes limit, remove, order, index, or purge the content of the RIT depending on their resource constraints.

Local routing information generator. Local routing information may be created by the local nodes for future use of routing algorithms. Algorithms like PRoPHET [10] need local routing information which should be updated at certain times and every time a node is contacted. The information created by the *local routing information generator*, is stored in the *RIT* following a certain and well-defined order in order to allow active messages to find the necessary information for its routing decisions.

Network abstraction: A module called *network abstraction* is defined to permit local nodes interact with different heterogeneous networks. Basic primitives, *send*, *receive*, and *neighbor discovery* are available to the nodes. The primitives *send* and *receive* allow transmitting and receiving bundles of information among nodes which can be of three different types:

- *Active messages*: As introduced in Section 1.4, these are messages containing application data. The content of these messages include application data, application code, and routing code.

- *Beacon messages*: Beacons are sent at certain intervals of time to allow neighbors to be discovered. These messages contain a node identification using Universal Resource Identifiers (URIs) as defined in [11]. Along with the node identifier, basic information that describes the node may also be included in the beacon messages. Which information from the different fields in the RIT should be announced is not a trivial issue. In order to be flexible enough, we allow the very same application messages to flag the desired RIT fields among the RIT application branch to be announced.

Scheduling and forwarding manager: Two queues for incoming and outgoing active messages are managed by the *scheduling and forwarding manager*. A prioritized scheduler manages which active messages should be forwarded and accepted first and for discarding purposes. For ordering or discarding active messages, we will also consider information from other layers, like the application layer. Other proposals for ordering DTN take into account only information from the DTN layer. We believe information from the DTN layer is important, but information from the application layer is crucial to distinguish which are important messages and which are not (Figure 1.5).

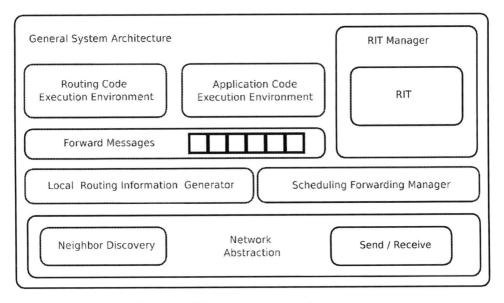

Figure 1.5 The elements that form the DTN proposed system architecture.

Pool of forwarding messages: Active messages are accepted by the local nodes in the *pool of forwarding messages* following the delay and disruption paradigm *store-carry-and-forward*. Storage reserved for these purposes is a key issue in opportunistic networks. Routing algorithms must prevent from making epidemic decisions which may flood buffers containing the pool of messages.

Routing and application code execution environment: As already explained, active messages contain code for routing and application behavior purposes. In order to execute this code, an execution environment must be included on every DTN node of the network. In the following sections, the code execution environment is explained in detail. Two examples of execution environments to be used in DTN scenarios are presented.

In Figure 1.6, the architecture of our proposal is depicted. On one hand, we use mobile nodes which define a delay tolerant network (DTN) by means of exploiting mobile node opportunistic or scheduled contacts. Simultaneously, on the other hand, this network is employed by active messages which are responsible for sensoring tasks. Every mobile node carries a smart device which runs an execution environment for running active messages.

Data obtained from sensing tasks may travel along with the active message as part of its data until the active message reaches a sink node. However, as shown in Figure 1.6, once data are created, they can also become independent from the active message that created it. Consequently, data delivery can be considered from the traditional point of view already studied in the bibliography. It has one particularity, however: the communication paradigm in this case follows a *many-to-one* or *many-to-few* model, where a small group of sink nodes represent the destination nodes. In our scenario, the deployed mobile nodes act as DTN routers, using algorithms which can be found locally in the mobile node and follow mobility models, as explained in Section 1.12. Section 1.10 describes how these routing algorithms can be updated by the very same active messages.

Figure 1.7 shows how the different actors involved in the proposal interact with each other. These actors are:

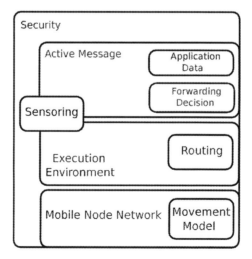

Figure 1.6 A system architecture for DTN mobile code-based WSN.

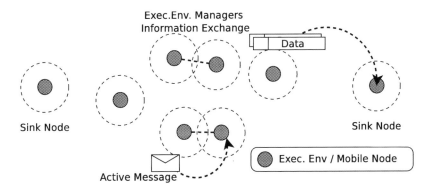

Figure 1.7 General scenario. Active messages travel from DTN node to DTN node creating data messages which are sent to the sink nodes.

Mobile nodes: Robots in our proposal, these carry smart devices which control sensor devices. They are capable of staying in the sensed zone for a given time and are able to come back to the *charging stations* when batteries start to run out of power.

Active messages: These are software entities responsible for sensing tasks. They are capable of migrating from mobile node to mobile node in order to fulfill tasks requested by users.

Execution environments: Active messages need execution environments for running, which themselves run on smart devices. They are responsible for negotiating movement model changes, as described in Section 1.12.

Code execution environment managers: These software entities remain on the code execution environments. They behave as intermediary actors between hardware sensors and active messages. They are also responsible for negotiating movement model changes.

Data messages: Once sensing actions are performed, data are acquired. These data have as a destination the sink nodes, responsible for the data processing. Data messages do not travel inside active messages. They are independent pieces of information that travel from mobile node to mobile node having the sink nodes as destination.

Sink nodes: These are destination nodes for WSN. These nodes are normally stationary and are plugged, connected and located outside the sensed zone.

1.7 SECURITY CONSIDERATIONS

We are aware that security in opportunistic networks is a very important issue [12,13]. While designing our architecture security has been fully taken into account, even if in this chapter it is not directly covered, in this section we want to enumerate the list of security considerations which should be studied. These mechanisms must guarantee the following properties:

Data confidentiality: The confidentiality of the data provided by the application sources and sent to the destination nodes must be warrantied. Otherwise, an attacker could perform some malevolent actions in order to eavesdrop on this information [14].

Data authenticity: Authenticity is vital to avoid any sending or injection of false information destined for a DTN node. It is also important to remark that this authenticity should not be linked to the identity of the node that generated the information [15].

Data integrity: The integrity is crucial to permit the detection of a state where a compromised node has modified the data which come from the application sources, or the data which it forwards to another node in a routing process [16].

Since the active messages are one of the pillars of our proposal, and since their code can influence the movement and the behavior of a node, it is important to protect it to avoid undesirable circumstances. In this case, we consider the following security properties as the most appropriate.

Code confidentiality: The confidentiality of the code is a desirable property to avoid outside attacks, which can include eavesdropping on the radio channel, the goal of which is to profile the kind of information that an application is gathering, and the procedure used to obtain it. Although this requirement can be desirable in some cases, this is not a mandatory property, and it depends on every particular application.

Code integrity: The authenticity of the code of an active message must be warrantied in order to avoid an illegitimate active message code being executed in a node. Otherwise, an insider or outside attack could inject a malicious active message with the purpose of compromising the security of the infrastructure.

Node identification: Since the custody of a message is delegated to DTN nodes, the correct identification of a node is a very important issue. A security mechanism should be defined to avoid an illegitimate DTN node to impersonate a legitimate one.

Key management: In the case symmetric or asymmetric keys are employed in the security mechanisms proposed in this section, a key management protocol should be defined. Issues like revocations lists deployment are complicate tasks in intermittently connected scenarios.

One of the most important aspects of our proposal is the use of the active messages as a way of routing the code that must be used to take the measurements. This can be done by using active messages and by sharing contextual routing information between the nodes. Thus, it seems evident that security must be present in the routing procedures. We impose the following minimum requirements to guarantee the security of the routing:

Routing data authenticity: The authenticity of the contextual routing data shared by nodes must be preserved in order to avoid that any outside attacker injecting incorrect information. This property can also be seen as a way to deter any potential outside attacker, since they cannot provide this property. It is clear that this requirement does not allow us to detect any malevolent data injected by an inside attacker.

Routing data integrity: Integrity must be preserved with the aim of avoiding an attacker modifying the routing data, as this change is not detected by its receiver. Again, this property cannot be preserved in the case of an inside attack, but it is also one way to dissuade any potential outside attackers.

Active messages forwarding non-repudiability: The non-repudiability property in the process of an active message forwarding must be warrantied from the source and the destination perspective. This is usually known as strong fairness and allows us to detect when a compromised node has completed a malicious action from the perspective of the routing of the active messages. For this purpose, all the non-repudiability evidence must be correlated to detect when an inside attack has performed a malicious behavior. This correlation must be done in places considered points of trust.

1.8 REIFICATION

Using a network of mobile nodes as a generalization of robots for sensing purposes is not a new idea [17]. Many proposals have been described or envisaged, hitherto making use of a troop of mobile devices equipped with a bunch of sensors. Some of them even use the very same mobile node network to transmit the acquired samples to a data sink. Yet, most of these systems are oriented to just one application, and all critical mechanisms, such as message routing or the physical movement pattern, are fitted with the specific actions and main goals of that particular application.

Being practical, and considering other network evolutions such as the Internet, it would be much more useful to take advantage of such a physical infrastructure, which would allow several applications to simultaneously use the mobile nodes and their sensors in their own way. The benefits are clear: a general-purpose sensor mobile node network is easier when reusable, and it allows new applications to be implemented after the mobile node network has been deployed. Unfortunately, turning such systems into general-purpose, multiple applications is not very easy. If several different applications coexist, which is the best message routing algorithm for them all? If mobile nodes can alter their pathways to adapt to some applications' requirements, what if there are a handful of applications striving against each other to make the node move under their command? These and other issues have to be considered when designing, and can be very hard to solve.

In this chapter, we propose a general-purpose sensing mobile node network, addressing the previously stated questions and many other issues surrounding the coexistence of applications and the routing of information. The rationale behind this proposal is to have mobile code at two different levels. First is as the user application itself, which can move from mobile device to mobile device as it suits the application best in order to accomplish its goals. Second is at the message level, allowing the very same message to make the routing decision by itself, and independently from other messages' routing policies. The resulting system happens to be a particular case of the wireless sensor network (WSN), i.e., a network of autonomous sensors aimed at monitoring physical or environmental conditions that pass their data through the network to certain locations or data sinks [18], working as a DTN (delay and disruption tolerant networking) [19]. In a DTN, data can be sent between any two nodes of the network, even when middle nodes cannot have concomitant communications. This is possible by following an asynchronous store-carry-and-forward paradigm. A wide variety of data can be acquired by means of sensors, ranging from temperature, humidity, or noise levels, for example, to other higher-level data such as object or human recognition, movement pattern detection, or plague detection; the DTN approach only broadens the scenarios of this particular WSN, allowing a myriad of applications.

One could wonder why a DTN-based network would be better in this case than a more traditional ad hoc or, while we are at it, MANET, network. The main reason for this is the required density of nodes. A DTN approach, for example, does not need a figure of nodes directly proportional to the sensing area to guarantee the operation of the system. Likewise, the sensing area can be extended or shrunken depending solely on the running applications. On the contrary, communication range and sensing area determine the minimum number of nodes to be used in ad hoc networks. Once an ad hoc network has been deployed, it is very difficult to extend the sensing area without adding new nodes.

The architecture and operation of this network is presented within the chapter, which also provides details on how some issues have been resolved. Examples include how population growth is controlled when cloning is considered, how dynamic multi-routing is achieved, or how mobile messages can influence node movement respectfully and fairly to other messages. Security has been considered as well, analyzing the threats and requirements, and giving mechanisms to protect mobile nodes.

The results are significant, and the overall performance of the system has been found to be very reasonable. A number of simulations have been done, from which some interesting outcomes have been drawn, such as a significant improvement in terms of message delivery ratio and latency. Furthermore, a successful proof of concept has been undertaken using real physical mobile nodes to check the feasibility of the proposal. The network has good potential for use in a variety of applications such as underwater environment sensing, unmanned aerial vehicle networks, environment applications, disaster field on emergency scenarios such as earthquakes recovery or terrorist attacks, mines seeking, urban search missions, community development, machine surveillance, accurate agriculture, biological attack reconnaissance, and many others.

The original contributions of this study are: a general-purpose multi-application mobile node sensor network based on mobile code, a dynamic multi-routing schema for allowing different routing algorithms for different applications and an application influenced movement model for minimizing node stagnation and maximizing segregation.

There are interesting publications like [20], which use active messages to sense information in mobile wireless sensor networks in the context of surveillance systems. Basic primitives for active messages missions are restricted to combing instructions in specific areas, while other useful ones, such as search missions, are not considered. Routing decisions for active message forwarding are evaluated in terms of geographical information solely, while other useful information such as congestion, aggregation, movement models, or code persuasion are not taken into account. Data messages remain with the active message until it reaches a sink point, so there is no distinction between how the data messages travel and how the active messages do, which may be inefficient in some scenarios. As a result, active message missions are tightly coupled to mobile nodes, and regions with high concentrations of nodes are created. In order to avoid this, complex mechanisms must be implemented.

Even though there has been considerable ongoing work on WSN and DTN, as it has just been shown in this section, there is not yet a mobile sensor network flexible enough to fit any scenario, including those lacking continuous connectivity, and allowing sharing of the network with other applications. The next sections describe, discuss, and reify precisely a system overcoming these limitations.

As seen in the previous section, there are a wide range of different proposals to implement wireless sensor networks (WSN). Several issues must be taken into account to provide a general-purpose, multi-application, mobile node-based WSN. These issues include sensor placement, sensor code deployment, sensing actions, sensor updating, task scheduling, routing, routing algorithms deployment, defining movement models, speed control, obstacle avoidance algorithms, energy-saving strategies, data fusion/merging/overlap, data division, scheduling and dropping, and finally, data processing.

In our proposal to implement a WSN, active messages play an important role in developing these tasks. Active messages are messages which, besides data, also carry code. We describe in this section the benefits of using active messages, which are executed on mobile nodes to obtain a general-purpose wireless sensor network. In addition to an architectural description, we explain interesting issues to be studied, as well as the limitations of our proposal.

1.9 PRIMITIVE SERVICES TYPES AND TASK DELEGATION

From the point of view of the user, we have identified three different potential uses for employing the network. On one hand, the user would like to search for a given pattern in a given zone in the studied area. On the other hand, the user would also like to perform a sensor comb of a chosen area. Finally, the user would like to arrive at a given point in the studied area and perform some type of unique concrete sensing. We distinguish these three types of behaviors because it will affect the way active messages will be created, how they will migrate, how they will be reproduced, and their communication.

Searching active messages are source-cloned when created. This is to say that the number of copies of the message is decided before entering the affected area. Source cloning minimizes communication among mobile nodes and active messages. Other proposals noted include active message migration at any given point of the studied area. However, population control is a complex issue for solving, due to its high communication requirements. Therefore, since mobile nodes suffer from limited communication and energy constraints, we see in source cloning a good solution for the population control problem.

A second service includes the types of tasks, the aim of which is to comb a given area for different sensing purposes. In this case, the active message must first approach the selected area. Once it has arrived at the selected area, the active message can then clone itself into other active messages. Control of the population is guaranteed, since clones cannot live outside the given area. This is what we call floating cloning, following an idea similar to the context of network-based social applications found in [21] and [10]. One delicate issue is that of the size of the area, as it should not be large, so as to prevent the appearance of highly concentrated regions. This problem will be further discussed in Section 1.12.

Finally, we consider the situation in which an application wants to go to a specific place and for concrete sensing. In this case, there is no need to do any source cloning or floating cloning.

Applications, however, may want to define composite tasks containing different service types. For example, an application may want to define a search behavior in which once the active message finds its target, it can comb the surroundings for additional information. We distinguish then, among behaviors. That is, which of the three different service types the active message is following and also the tasks, the high-level instructions defined by the users.

1.10 DYNAMIC MULTI-ROUTING

In our proposal we have to differentiate among two types of routing. On one hand, active messages jump from node to node, in order to physically arrive at different *points of interest* to perform their tasks. This action can be considered a routing decision, in the sense that the active message makes a decision among the potential neighbors, considering as well the possibility of staying in the current node.

We see this routing as part of the active message model, that is, a complex composite movement model made up of the list of all the n different mobile nodes the active message migrates to, as described in:

$$\text{movement}_{agent} = \left[m_{c_0}, m_{c_1}, \ldots, m_{c_n} \right]$$

On the other hand, sensor data retrieved by the active messages travels heading towards the sink nodes by jumping from mobile to mobile node. In scenarios in which different applications coexist, a single routing algorithm may not be enough to handle all of the applications' needs. This is why routing algorithms must take into account information from the application layer, as well.

This leads us to the necessity of having different routing algorithms on the mobile nodes for data routing purposes. Stored locally on the mobile nodes, the routing algorithms are chosen depending on the application. Given an application message *mess*, and application *app*, a local custodian node c_j and its routing information tree (RIT_{c_j}). This other routing can be expressed as:

$$forwardToList = f_{localrouting}\left(mess_{app_i}, neighbourList\left(c_j, t\right), RIT_{c_j}\right), \text{where}$$
$$forwardToList \subset neighbourList\left(c_j, t\right) \cup c_j$$

That is, routing decisions are made, taking into account the list of the current neighbors and the context information.

There are many proposals which define very efficient solutions for the routing problem in DTN networks for very different DTN scenarios. However, not all applications employing the same WSN need their data to be routed in the same way. A single routing algorithm may not work for various applications. The aim of our approach is not just to improve transmission time, but also to provide a flexible and generic DTN network capable of handling different routing behaviors. The routing algorithms are easily deployed on the different mobile nodes by using the very same active message. Coexistence of different applications willing to use different routing algorithms cannot be easily deployed with other classic DTN proposals such as bundle protocol-based or Haggle-based ones.

An illustrative example of the need for different routing algorithms in a single DTN scenario may be found in emergency rescue coordination applications. In disaster areas, different users such as police officers, firefighters, doctors, nurses, engineers, or rescue teams, among others, along with portable devices such as mobile phones or tablets, may share and use the interconnected network with a mobile robot wireless sensor network. Different applications such as victim location, notification applications, fire coordination and pollution measurement, or radiation location may need different ways of information routing, scheduling, dropping, or aggregating. For example, information that contains a notification to a given user or mobile node may be optimally routed using a probabilistic routing algorithm such as PRoPHET [10]. Alternately, routing decisions for information resulting from sensor tasks may depend on the level of importance of the information seen from the point of view of the application. If the information is important, an epidemic routing algorithm will be used. Otherwise, it can be discarded if the information is obsoleted by some other present in the same node.

1.11 AGGREGATION, SCHEDULING, AND DROPPING

Previously in this chapter, we commented on the advantages of dynamic multi-routing for both levels of routing presented: active message migration and data routing. Traditional routing protocols follow address-centric protocols, in which decisions are based on the destination address. However, other proposals such as [22] also consider the fact that data can be opportunistically aggregated and consolidated at the routing nodes, reducing the impact on the network. Data-centric routing protocols, that is, protocols which make decisions in terms of possible future data aggregation or data consolidation, are being considered in our proposal. In the same way, data routing could be influenced by future nodes storage congestion and more concretely by potential scheduling or dropping policies.

When facing a multi-application schema for a general-purpose wireless sensor network, data aggregation is highly correlated with the different applications. It is quite complex to provide a general solution for the aggregation problem for the different routing algorithms of the various applications. The objective is to make routing algorithms aware of potential future aggregation, dropping, or scheduling actions. Therefore, a combined approach of application-based and context-aware routing algorithms is crucial for these purposes. We are employing the dynamic multi routing application explained in the previous section to include in the different routing protocols ways of obtaining information about aggregation, scheduling actions, and dropping policies.

1.12 APPLICATION INFLUENCED MOVEMENT MODEL

Mobile nodes on a WSN may follow different mobility models. There is a *one-to-many* relationship among a mobile node and an active message; that is, active messages are allowed to congregate in a single mobile node. Other proposals like [20] suggest that both actors should be tightly coupled in a *one-to-one* way. Applications which work in concrete areas could create highly concentrated regions, which are difficult to recover. Our proposal contends that such movement model should be as independent as possible from the rest of the mobile nodes, in order to minimize energy consumption. From a global perspective, we intend to make mobile nodes remain as widely spread out on the affected area as possible but favoring the mobile node contacts. These constraints are affected by the active messages and even by the application data which can subtly modify the mobile node movement model for task accomplishment purposes.

This movement negotiation is handled by the code execution environment managers. Active messages make movement requests to the code execution environment managers, which in turn evaluate these requests in terms of the positions postulated. A queue in which these requests are stored is defined locally in the node. To accept the movement change and therefore introduce a new point in the mobile node queue, the point requested must be inside a circle radius of kd, where d is the distance from the current position of the mobile node to the previously scheduled waypoint. The bigger the factor k is, the more tolerant to mobility the mobile node will be. The mobile node will retrieve the closest position value from this queue until this queue is empty and then proceed with its movement model.

As an example, in Figure 1.8, a code execution environment manager receives a request from three active messages to temporarily modify its movement model. On the lower part of the figure, we see how the movement model is modified by the active messages.

In order to prevent a possible mobile node stagnation caused by queued points which remain far from the mobile node, requests from the active messages can be rejected. In Figure 1.9, an

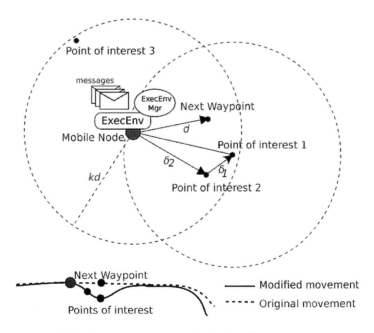

Figure 1.8 Movement model. Active messages request the local node to accept a movement change by suggesting one or more points of interest.

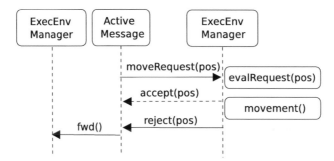

Figure 1.9 Movement request protocol. An active message requests a point of interest, while the local DTN node may or not accept this movement change.

active message performs a movement request, for example, *Point of Interest 3* in Figure 1.8. The code execution environment managers decide to accept it because it is inside the radius *kd*. Once the mobile node has visited the other two points of interest, the queued *Point of Interest 3* remains outside the new circle radius *kd*. The code execution environment manager informs the active message that its request is rejected and the latter migrates to another mobile node, as seen in Figure 1.9.

1.13 ACTIVE MESSAGES: A DISTRIBUTED SENSING INFRASTRUCTURE

Our goal is to provide a general-purpose WSN, multi-application system. Different applications are represented by distributed streams of active messages performing different tasks in the studied zone. In this section, we enumerate these different tasks and analyze the benefits of using the active messages paradigm in the context of WSNs.

Active messages carry the code responsible for managing the sensor retrieval following users' requests. Tasks may be divided into sub-tasks, which can be delegated to other active messages (clones) outside or inside the code execution environment, as seen in Section 1.10.

However, applications may at some point evolve in their needs. In a traditional WSN, this evolution means sophisticated deployment actions in order to modify the way the information is gathered, which are not always easy to achieve. These adjustments may be caused by the applications need to retrieve information from additional sensors, discontinuing some others not needed anymore by the application or even modifying sensor retrieval a posteriori actions.

Active messages are an excellent option to solve these problems. Since the sensor code is traveling with the very active message itself, it is guaranteed that the way the information is gathered, plus a posteriori actions if needed, will be the ones desired by the user. Updates on these codes are made automatically if requested by the application, while visiting the sink nodes or while opportunistically contacting other active messages in the studied area.

The application itself knows best how its data should be routed. As seen in Section 1.11, information retrieved from the mobile sensors travels from mobile node to mobile node in data messages, following a behavior pattern set by the different routing algorithms installed locally on every mobile node. When a new application is willing to use the sensor network, it is possible to include an ad hoc routing protocol for the application. Active messages carry new routing algorithms to the different mobile nodes they visit so that future bundles of data may use them. This deployment of routing algorithms may be seen as additional tasks which the active message must accomplish.

Nodes when finding other mobile nodes exchange context information which will be used by the different routing algorithms, for both active messages and data messages. Willing-to-migrate active messages communicate with a code execution environment manager to query for foreign context information in order to make routing decisions.

REFERENCES

[1] Aruna Balasubramanian, Brian Levine, and Arun Venkataramani. DTN routing as a resource allocation problem. In *Proceedings of the 2007 conference on Applications, technologies, architectures, and protocols for computer communications*, pp. 373–384, 2007.

[2] K Scott, S Burleigh, et al. RFC 5050: bundle protocol specification. *IRTF DTN Research Group*, 2007.

[3] James Scott, Jon Crowcroft, Pan Hui, and Christophe Diot. Haggle: A networking architecture designed around mobile users. In *WONS 2006: Third Annual Conference on Wireless On-demand Network Systems and Services* (pp. 78–86). ARES Team. 2006.

[4] Dana Simonson, John Livdahl, and Rory A Smith. System and method for multicast communications using real time transport protocol (RTP), May 22, 2007. US Patent 7,221,660.

[5] Carlos Borrego, Gerard Garcia, and Sergi Robles. Softwarecast: A code-based delivery manycast scheme in heterogeneous and opportunistic ad hoc networks. *Ad Hoc Networks*, 55:72–86, 2017.

[6] Jiannong Cao, Xinyu Feng, Jian Lu, and Sajal K. Das. Mailbox-based scheme for mobile agent communications. *Computer*, 35(9):54–60, 2002.

[7] D Waitzman. Rfc2549: Ip over avian carriers with quality of service, 1999.

[8] Vinton Cerf, Scott Burleigh, Adrian Hooke, Leigh Torgerson, Robert Durst, Keith Scott, Kevin Fall, and Howard Weiss. RFC 4838, delay-tolerant networking architecture. *IRTF DTN Research Group*, 2(4):6, 2007.

[9] M Ramadas, S Burleigh, and S Farrell. RFC 5326, licklider transmission protocol specification. *IRTF DTN Research Group*, 2008.

[10] Carlos Borrego, Joan Borrell, and Sergi Robles. Hey, influencer! message delivery to social central nodes in social opportunistic networks. *Computer Communications*, 137:81–91, 2019.

[11] Tim Berners-Lee, Roy Fielding, and Larry Masinter. RFC 3986, uniform resource identifier (URI): Generic syntax, 2005. http://www.faqs.org/rfcs/rfc3986.html, 2005.

[12] Adrián Sánchez-Carmona, Sergi Robles, and Carlos Borrego. Improving podcast distribution on Gwanda using Privhab: A multiagent secure georouting protocol. ADCAIJ: Advances in distributed computing and artificial intelligence journal (ISSN: 2255-2863). *Salamanca*, 4(1), 2015.

[13] Adrián Sánchez-Carmona, Sergi Robles, and Carlos Borrego. Privhab+: A secure geographic routing protocol for DTN. *Computer Communications*, 78:56–73, 2016.

[14] Carlos Borrego, Marica Amadeo, Antonella Molinaro, and Rutvij H Jhaveri. Privacy-preserving forwarding using homomorphic encryption for information-centric wireless ad hoc networks. *IEEE Communications Letters*, 23(10):1708–1711, 2019.

[15] Adrián Sánchez-Carmona, Sergi Robles, and Carlos Borrego. Privhab: A privacy preserving georouting protocol based on a multiagent system for podcast distribution on disconnected areas. *Ad Hoc Networks*, 53:110–122, 2016.

[16] Naercio Magaia, Carlos Borrego, Paulo Pereira, and Miguel Correia. *Privo: A privacy-preserving opportunistic routing protocol for delay tolerant networks.* In *2017 IFIP Networking Conference (IFIP Networking) and Workshops*, pp. 1–9. IEEE, 2017.

[17] Carlos Borrego, Sergio Castillo, and Sergi Robles. Striving for sensing: Taming your mobile code to share a robot sensor network. *Information Sciences*, 277:338–357, 2014.

[18] Winston KG Seah, Zhi Ang Eu, and Hwee-Pink Tan. Wireless sensor networks powered by ambient energy harvesting (WSN-HEAP)-survey and challenges. In *2009 1st International Conference on Wireless Communication, Vehicular Technology, Information Theory and Aerospace & Electronic Systems Technology*, pp. 1–5. IEEE, 2009.

[19] Russell J Clark, Evan Zasoski, Jon Olson, Mostafa Ammar, and Ellen Zegura. *D-book: a mobile social networking application for delay tolerant networks*. In *Proceedings of the third ACM workshop on Challenged networks*, pp. 113–116, 2008.

[20] Edison Pignaton De Freitas, Tales Heimfarth, Ivayr Farah Netto, Carlos Eduardo Lino, Carlos Eduardo Pereira, Armando Morado Ferreira, Flávio Rech Wagner, and Tony Larsson. UAV relay network to support WSN connectivity. In *International Congress on Ultra Modern Telecommunications and Control Systems*, pp. 309–314. IEEE, 2010.

[21] Jörg Ott, Esa Hyytiä, Pasi Lassila, Tobias Vaegs, and Jussi Kangasharju. Floating content: Information sharing in urban areas. In *2011 IEEE International Conference on Pervasive Computing and Communications (PerCom)*, pp. 136–146. IEEE, 2011.

[22] Ruhai Wang, Zhiguo Wei, Qinyu Zhang, and Jia Hou. LTP aggregation of DTN bundles in space communications. *IEEE Transactions on Aerospace and Electronic Systems*, 49(3):1677–1691, 2013.

Chapter 2

Opportunistic Emergency Scenarios
An Opportunistic Distributed Computing Approach

Carlos Borrego Iglesias

Universitat de Barcelona, Barcelona, Spain

CONTENTS

2.1 AN OPPORTUNISTIC APPROACH FOR MOBILE-CODE-BASED DISTRIBUTED COMPUTING

Distributed computing in general and grid computing in particular [1, 2] has consolidated as a technology capable of solving some of the most challenging scientific projects of our century. The needs of these projects usually include complex computation of data obtained from different sources and stored in large storage resources. The main goal of grid computing, precisely, is to share these resources among different institutes and virtual organizations across high-speed networks and to distribute and coordinate its processing [3].

Wireless sensor networks (WSNs), on the other hand, are a technology that can be very useful when it comes to acquiring and transporting data collected in widely spaced areas. These networks consist of different nodes carrying different sensors along with autonomous computational devices which transmit data through the network to some specific locations or data sinks.

This chapter analyzes how both technologies, grid computing and wireless sensor networks, can be combined into an integrated WSN and computer grid infrastructure allowing new functionalities. The cornerstone of this conjugation is using delay and disruption tolerant networking (DTN) concepts [4, 5] along with mobile code to create an intelligent grid network [6, 7] capable of routing and managing processes depending on the context. Some other recent proposals integrate WSNs and grid computing as well. However, our proposal comes from the network perspective. We consider WSNs nodes as intermittent connected nodes, containing asymmetric bandwidths, long and variable latencies, and ambiguous mobility patterns. This new perspective contributes to the creation of a novel concept of intelligent grid computing networks, going beyond the possibilities of the reviewed literature in some current scenarios, and providing promising prospects for supporting future grid services.

The routing decision [8] making and execution policies travel with the messages, instead of being static and exactly the same for all nodes. These policies, in the shape of mobile code [9], can take into account the context of the nodes to choose the behavior that fits best in each situation [10]. All in all, the system acts more like an ant colony, with differentiated autonomous parts acting locally but with a cooperative aim, rather than a traditional and more inflexible system. Thus, using mobile code makes the grid network an intelligent system, pliable enough to adapt to new scenarios of grid computing. However, the proposed system cannot be considered a silver bullet for all grid computing. In highly connected grids, with low latencies and where data do not need to be processed before getting to the execution destination, mobile code would introduce an unnecessary extra overhead and other unwanted side effects.

2.1.1 Integrating DTN WSNs in Grid Computer Infrastructures Using Mobile Code

In this section we propose a way to integrate intermittently connected WSNs in computer grid infrastructures. On one hand, the data acquired by the WSNs can be processed and stored using traditional grid services inside connected regions. On the other hand, and as a more challenging option, sensors can be considered as grid sources of data commanded by the very grid users themselves. We propose the job behavior that evolves beyond the traditional *read-process-and-storage* model – that is, reading from a data source, processing, and storing the result – to a possibility of having a *create-read-process-and-storage* model, in which jobs are able to create their input information.

For both ways of integration, there is a challenging issue which must be solved. Data must arrive from the WSN to nodes connected to the grid, known as sink nodes. However, connectivity among the nodes belonging to a WSN can rarely be available, due to large latencies, high error rates, or even very small bandwidths. Examples of these scenarios include outer space sensors and computer facilities, and isolated environment sensors.

Traditional grid scenarios use different types of data sources which include large hadron colliders like the LHC [11], biomedical equipment or supplies for studies like [12], or even simulated data obtained from the very same grid resources in the context of the same domains. Most of them are constantly connected to the Internet or have a straightforward way to store its data to grid connected resources. Likewise, grid services are implemented based on networks with a constant Internet connection, taking for granted a continuous network connection.

By considering WSNs as a potential grid source of information, we analyze in this study how to include WSNs in grid computing, taking into account that they may be not fully permanently connected to the rest of the grid infrastructure. The result is to obtain what we call a *delay and disruption tolerant computer grid source*.

In Figure 2.1, we can see how grid infrastructures and WSN technologies may coexist. On the left-hand side of the figure, a WSN is depicted connected to a grid infrastructure by a computing element, a door to the sensor network. This sensor network is then seen as a grid source becoming grid-wide available. The WSN is composed by several sensors which, as a global network, are intermittently connected to the Internet and could have a poor

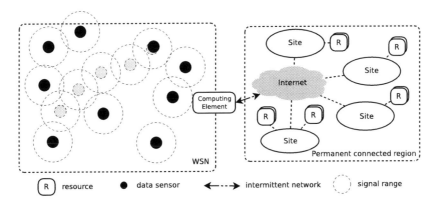

Figure 2.1 Coexistence of WSN and grid infrastructure. On the left-hand side, an intermittently connected sensor grid is connected to a permanent connected grid infrastructure.

connection among them. Some of these sensors may adopt ambiguous mobility patterns and therefore act as DTN data mules [5], sensors in movement which gather information from other sensors which act as opportunistic router nodes.

The traditional DTN scheme follows the well-known *store-carry-and-forward* paradigm [5]. We introduce a new paradigm: the *store-process-carry-and-forward*. We can benefit from the fact that in DTN the information can be blocked waiting in some DTN node which temporarily is disconnected. This waiting time can be extremely long, and we can make good use of it. If the application data contain information to be processed, we could make the local node process the data while it waits for some other node to arrive.

Distributed computing application information – that is, information created on behalf of a user and retrieved from the sensors – is carried inside messages from the sensors back to the computing element. Custodian nodes – that is, nodes which custody application information until it can be forwarded to another node – handle delays and disruptions. The application layer will create application data which will be included in one DTN message. These messages will stay in a custodian node until they are able to migrate to another one. Since delays in DTN networks can be huge, messages containing application data may wait for a long time. Once a potential custodian node becomes available, the message will try to be forwarded to the newcomer custodian node.

As seen in Figure 2.2, traditional grid layers – application layer, collective layer, resource layer, connectivity layer, and fabric layer – can be completed by adding an execution

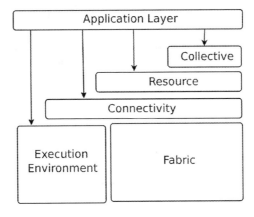

Figure 2.2 Grid layers.

environment layer. Mobile code can be seen as an alternative to the grid classic fabric layer. Its aim is to provide an interface to local resources, the sensors. There is a subtle difference with traditional grid architectures. This alternative to the fabric layer in this case is directly connected to the application layer in order to allow applications direct access to the information retrieved by the local sensors present in the nodes.

2.1.2 Grid Job Management

Information created in grid sources are processed using a computing infrastructure. Jobs input data in experiments like ATLAS [13] can reach to be tens of thousands of times larger than the output. In scenarios in which we include the grid sources, such as delay and disruption WSNs, we believe that computing locally may save loads of data transfer. This is especially interesting, since these transfers are very expensive.

In the case in which WSNs are isolated, for example a sensor network in deep space, bringing data outside the network in order to compute it can be really expensive. Instead, we propose a model in which raw data, information extracted directly from the grid sources, are kept and processed locally.

Traditional grid users define their jobs by specifying their executables, their input data, their output data definition, requirements such as the site the user wants to send the job to, and the virtual organization. The computing element receives the job specification along with the files the user has decided to send to the site. This information is passed to the different job managers. A typical job manager would receive a job request and would submit the job to a local queuing system, such as pbs [14].

In our approach for sensor grid scenarios, the job manager's behavior would be different. We have defined and implemented a computer element job manager which, instead of submitting a job to a queuing system, will create an active message and will launch it to the WSN. Active messages are messages capable of performing processing on their own and consequently behave autonomously. These messages will travel from node to node in order to accomplish their tasks.

Creating this active message is quite a challenging issue. We need to clearly specify its behavior in terms of routing decisions, sensor retrieval, data management, and task scheduling. The user, by means of the job description, can explicitly specify these issues; otherwise, the job manager will be responsible for properly creating it.

Looking more closely at the intelligence of this message, from one hand we have to differentiate between the application code, that is, the behavior the active message will require which defines the actions and tasks it will perform. These tasks include, for example, searching and combing actions in the wireless sensor network area and processing data [15]. As a result, application data are created. From the other hand, the active message will need to make routing decisions whenever finding different nodes on the networks. Traditionally, DTN approaches leave routing decisions for DTN nodes. In our proposal, intelligence comes from allowing routing code to travel with the very active message, allowing different applications employing the same WSN to use different routing algorithms. Finally, both input and output may need to travel with the active message. This mapping is represented in Figure 2.3.

2.1.3 Store-Process-Carry-and-Forward Paradigm

The active message implements the *store-process-carry-and-forward* paradigm:

- Once an active message arrives to a node, it waits *stored* in a queue of messages, until another node is available for forwarding.

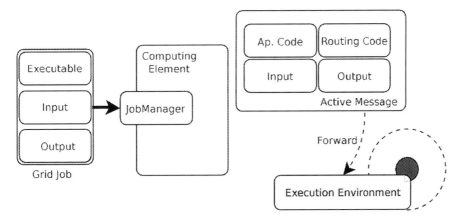

Figure 2.3 Active message creation. The grid job manager creates an active message from the grid job specification.

- When the active message is waiting for a node to be forwarded, if the input files of the grid job are available, *process of data* can be started.
- When the message is forwarded, it behaves like a *data mule* [5], making an *active carry*, i.e. carrying information containing the grid job output. Instead, a custodian node in motion, as it could be a robot, satellite or an animal, performs a *passive carry* when moving. In this case, no software entity is performing the transport itself. Instead, there is a physical movement of the custodian node who produces the *carry* action.
- The very active message contains internally a routing algorithm to decide when and where to be *forwarded*. The routing algorithm is part of the active message. This way we allow different applications using the same infrastructure to behave differently when it comes to routing decisions, contrary to classical networks such as TCP/IP, in which routing static elements implement and perform the routing decisions.

2.1.4 Processing Models

Bringing back the concept of data processing while waiting to migrate, it could happen that when the message is able to be forwarded from the current custodian node, the code execution has not yet finished.

We have three behavior possibilities here to define how to handle job management:

- *Run & Go.* When a new node is available for forwarding, jobs do not abort their execution. Active messages try to be forwarded once the job has finished; however, the new node still might not be available for forwarding. This option is used when node density is high. In Figure 2.4, a new node appearance is represented by an arrow. In Figure 2.4(a) concretely, a job finishes its execution and waits for the next node to be forwarded.
- *Abort & Go.* Jobs could be aborted when a new custodian node appears, losing all the processed information. The active message could eventually discard the computing effort already done, finding it more convenient to be forwarded to a new node. This is represented in Figure 2.4(b). Jobs will start again in the new custodian from scratch. This option is useful when contacts are scarce and processing cost is low.

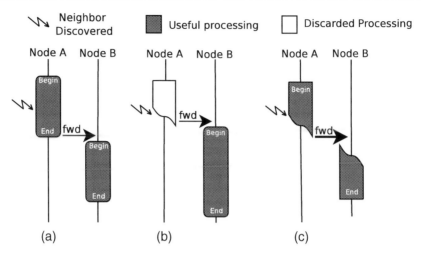

Figure 2.4 Processing models.

- *Stop, Go, & Resume.* Jobs stop their execution when they are aware of a new suitable custodian available, forwarding themselves to the new custodian and resuming their processing in the new node. As seen in Figure 2.4(c), the job is stopped but resumed in the new node.
  - *Weak mobility* is limited to the code and the data state. When arriving in the new execution environments, the code would be restarted.
  - *Strong mobility* allows code to be moved at a very low level, such as stack pointers and instruction pointers to different execution environments.

Before the active message code is invoked, its job behavior is queried. This behavior defines how an active message wants to proceed if the message is ready to be forwarded but the processing is not yet finished. The *EEM* is responsible for finding the application code present in the active message and for executing it. This code represents the way the active message will behave in terms of sensor retrieval and task accomplishment in a custodian node.

In this way, the *EEM* has control of all the computing running on the local execution environment and can stop, suspend, or prioritize any computation from any active message. Jobs prioritization is not covered in this chapter, but it is a very interesting open issue.

2.1.5 Storage

Storage service in grid computing is responsible for managing and holding data generated by grid sources or computing jobs as well as to provide this data grid-wide upon request. In our proposal, we have two options for creating the active message concerning the storage of the information created and/or processed.

- *Store the information in the disconnected region.* This information will be physically stored in one of the DTN nodes from the disconnected region. The very active message will choose which would be the physical location of the data, which could be any of the DTN nodes the active message will visit, a subgroup of them, or a given one.
- *Store the raw data outside the disconnected region.* Instead of leaving the data inside the disconnected region, the data is routed to a DTN gateway, a special node that belongs to both the disconnected region and the connected region. Subsequently, the data is stored in a storage element outside the disconnected region.

By any means, information needs to be forwarded from one node to another, traveling all around the disconnected region. Thus, the storage problem becomes a routing problem, hence, finding the appropriate DTN node to leave the data once created. Many articles such as [16] cover the still-open topic of DTN routing.

In any case, the file, after being stored, needs to be cataloged so it can be located in the future. The active message will need to reach the connected region in order to inform the file catalog about the information placement.

2.1.6 The Routing Issue

In our intelligent system, routing is the decision process in which the best path inside a given network is calculated. In traditional networks, routing decisions aim to minimize latency of message delivery and improve the delivery ratio, without overlooking local and network resource consumption. In the proposed scenario, routing is a very delicate issue. Once an active message arrives at the WSN, latency time to arrive at its destination and the delivery ratio are important.

However, maximizing task performance in our scenario could be more important than improving latency times or delivery ratio. This performance is different from application to application and depends on several factors such as processing time, network congestion, flow control, etc. Letting the very same application perform the routing decision can improve task performance.

As an example, in Figure 2.5 an active message performs a combing task for *zone 1*. This message must choose among two potential custodians, S_2 and S_3, in order to be forwarded. The routing code for this active message chooses custodian S_2 to improve *zone 1* combing task over choosing S_3, even if this latter seems to have a direct contact with its destination. The very active message routing code prioritizes task completion over delivery latency. Without an application perspective, this decision would be very difficult to make by a local routing algorithm running in a custodian node. This allows the applications to intelligently employ the network in a very *personal* way.

2.1.7 Implementation

The execution environments on which active messages run are called platforms. We need to implement on every sensor grid node an execution environment platform to let active messages carry grid-level information to run on. In order to implement active messages described in the previous sections, we need some special platforms: besides being capable of executing

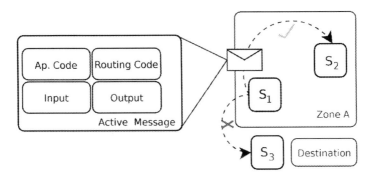

Figure 2.5 Dynamic routing. An active message selects node S_2 as its next custodian node.

code, platforms must allow the code to be forwarded from one sensor node to another, stopping their code and resuming its execution upon arrival, and permit the ability of carrying their data and state with them.

Among the different available execution environment platforms, we decided to use Mobile-C [17], a Foundation for Intelligent Physical Agents (FIPA) [18] complaint platform for mobile C/C++ codes in networked intelligent mechatronic and embedded systems. The benefits in comparison to other execution environments like Jade based on Java are mainly achieved through the reduction in memory consumption. Moreover, in general terms, the impact on system resources is very small.

We have implemented a job manager[1] in the context of gLite grid middleware [19]. This job manager accepts traditional grid job requests and, by parsing jobs requirements, accordingly creates active messages. We are using gLite's grid job description language to allow the user – besides the traditional *InputSandbox* and *OutputSandbox* – to specify the application code and the routing code in the creation of the active message. One example is described as follows:

```
JobType = "ActiveMessageJob";
ApplicationCode = "combing.cc";
RoutingCode = "myprophet.cc";
StdOutput = "output.txt";
StdError = "error.txt";
InputSandbox = {"combing.cc","myprophet.cc"};
OutputSandbox = {"output.txt","error.txt" };
```

Once active messages are created, they travel along with the grid's input data if any, from node to node, performing their tasks and retrieving information data. Execution environments custody active messages until another node becomes available for forwarding. The very same active message chooses its node path by executing the Mobile-C *MC – AddAgent()* method. Afterwards, it waits until a new node becomes available. Considering DTN scenarios, the active message can stay in this state for a long time. We want to let active messages choose for themselves their next hop, that is, the routing decision. This is done by modifying the Mobile-C code to allow for having a well-known method for these purposes.

2.2 OPPORTUNISTIC EMERGENCY SCENARIOS APPLICATIONS

The widespread use of portable devices that are generally equipped with wireless-enabled communications, GPS receivers, and/or touch screens has remarkably improved outdoor applications in a great variety of scenarios. Although the most common network configuration is ad hoc, or mobile ad hoc (MANET), new communication paradigms are emerging to fill the void for some specific settings. This is the case of Delay and Disruption Tolerant Networking (DTN) [3], which is extremely useful when no concomitant network links connect the source and destination at transmission time. Emergency and disaster recovery systems are an example of an application domain where having no network infrastructure makes DTN significantly extend possible applications. Applications based on DTN can coexist with other solutions in order to create network infrastructure and restore network connectivity. Furthermore, the DTN approach provides cheaper, easier, and ready-to-use deployment solutions.

Although DTN has strong foundations such as the Bundle Protocol [20], and many groups have been working on its formalities for some years, there are still a number of issues to solve. Some of the most problematic issues include routing, lifetime control, and security, which need solutions quite different from the ones normally used on the Internet. The rationale for this is that the diversity of applications running on such a limited connected network calls for a number of different mechanisms to solve the specific problems presented by each application. As opposed to what happens on the Internet, no general-purpose mechanisms exist which satisfy the requirements of all applications at once. A possible way of facing this challenge is by adding code to every bundle of information sent over the network to make (autonomously and in a context-aware fashion) all the decisions regarding the transportation of that particular piece of data. Mobile code [21] is a well-known technology designed for precisely this.

Mobile code is a good candidate for implementing a data-driven approach to DTN networks. This technology is able to migrate with their code and data from host to host and to continue their execution upon reaching its destination. Mobile code needs execution environments to run the execution environments. Mobile-C [17], Jade [18], and Agentscape [22] are popular examples of platforms supporting mobile code. However, these platforms cannot be directly used for implementing DTN networks. To begin with, forwarding procedures have to be redesigned for the decision-making on which next intermediate node, also known as the custodian node, to be forwarded to. We propose leaving this decision to the bundle itself.

We present in this chapter an extension to the Bundle Protocol based on mobile code to implement a heterogeneous data-driven DTN. While fully respecting the Delay-Tolerant Networking Architecture [23], the keystone of this integration of technologies is to move the routing algorithm from the host to the bundle. We will provide details on how this approach significantly improves these networks. We propose a scheduling infrastructure which minimizes delays as well as improves reliability by prioritizing bundles carrying important application data. We also describe several mechanisms to solve specific issues with DTN networks concerning routing, congestion, data aggregation, routing code deployment, and DTN lifetime control.

Emergency and disaster recovery scenarios are very convenient to demonstrate the new possibilities of this combination of DTN concepts and mobile code. This study uses the case of the Mobile Agent Electronic Triage Tag (MAETT) [15], to illustrate how applications can profit from this new and improved architecture. However, the results can be used in many other domains.

2.2.1 Scenario Description

Disaster recoveries after emergencies, such as terrorist attacks or meteorological calamities, are difficult to conduct. Connected areas may become precipitously disconnected. Using DTN networks is an excellent way of rapidly deploying communication networks. Other studies, like [24], already use DTN networks to coordinate victims from emergency scenarios. In this section we will define a scenario based on [24] to show the advantages of using code block extensions in DTN networks.

Different users such as police, firefighters, doctors, nurses, engineers, and rescue teams, among others, along with their portable devices such as mobile phones or tablets, create the intermittently connected network. Opportunistic contacts among the different users permit the different applications to utilize the network for very different purposes.

Rescue applications conducted by doctors, nurses, and rescue teams classify the different victims during an initial evaluation of their health condition while in the confines of the emergency area. Statuses attributed to victims are 0 for deceased, 1 for seriously injured, 2 for injured, and 3 for mildly injured. This process is called *triage*. Figure 2.6 depicts this application.

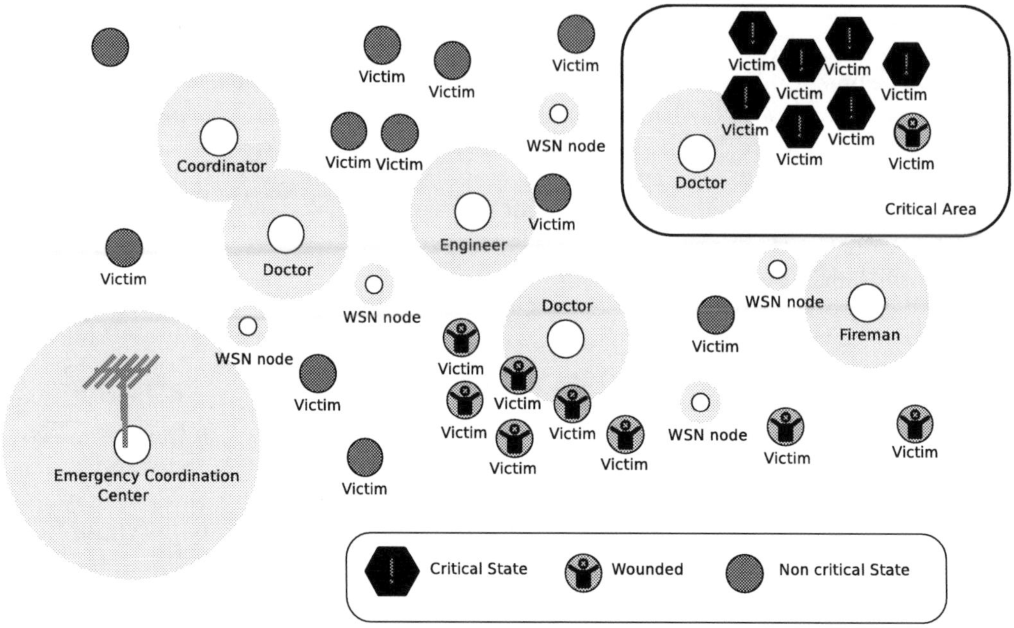

Figure 2.6 Different users in an emergency scenario create a DTN network. A sink node, the Emergency Coordination Center, collects the information from the different victims. A critical area where high density of victims is located on the top-right part of the figure.

To manage and organize injury statuses and locations, we propose sending bundles with the information of every victim that has been classified. The goal of this application is to allow this information to arrive as soon as possible at the *Emergency Coordination Center*, hopping from device to device when a shorter route is detected.

Other applications such as notification applications, wireless sensor applications, applications used for firefighting, pollution measurement, or radiation detection could employ the same network. The coexistence of these applications is depicted in Figure 2.6. Coexistence of applications by allowing different users to share a single network decreases cost and creates a favorable atmosphere for node contacting. However, this coexistence of applications and having simultaneous users introduces some challenges that will be described in this section.

2.2.2 Dynamic Routing and Routing Algorithm Deployment

In emergency scenarios, different applications may coexist. For example, data from victim rescue application operations may share the network with wireless sensor data from firefighters or other emergency response teams. Different applications may need different types of routing algorithms. For example, information that contains a notification to a given user or mobile node may be most favorably routed using a probabilistic routing algorithm such as PRoPHET [16].

Alternately, routing decisions for information resulting from sensor tasks may depend on the level of importance of the information seen from the point of view of the application. If the information is important, an epidemic routing algorithm will be used to improve the delivery performance. Instead, it can be discarded if the information is made obsolete by other information present in the same node. The routing decision may travel with the bundles themselves instead of being static and exactly the same for all nodes in the network. These policies, in the shape of mobile code, may consider the local context to choose the behavior that fits best in each situation. Routing algorithms are carried using the *bundle routing code blocks* described in the previous section.

We may differentiate among three types of routing algorithm deployment in emergency scenarios.

First, the *bundle routing code (BRC)* is the routing algorithm employed by the bundles to choose where the bundle should be forwarded to. These routing algorithms may travel along with the bundles themselves, as explained in this section. However, we propose an alternative way of deploying this code in case this paradigm cannot perform properly for a given application or a given DTN scenario. New applications arriving to the emergency zone may perform a routing algorithm deployment before sending its data. After the deployment, bundles may carry Routing Code Blocks with references to already deployed routing algorithms, setting the Reference/Value bit and expressing the path to the routing algorithm. During the deployment phase a broadcast [25] bundle containing a Routing Code Block with the routing algorithm is sent to the network. Additionally, the "*Local copy*" bit should be set to allow the routing algorithm to be copied locally to the Routing Information Tree (RIT) in the appropriate path indicated in the *RIT path* field.

Second, some routing algorithms need to periodically execute code on every custodian node. For example, a notification application for users inside the emergency area may employ a probabilistic routing algorithm such as PRoPHET [16] to route its information. This kind of routing algorithm needs to execute a code on the custodian nodes to update delivery probabilities for every node contacted as well as to decrease these probabilities if the nodes are not contacted. These codes may be deployed by including a *Custodian Routing block* in a bundle. In this bundle, the periodicity of the code should be indicated by using the *Periodicity field*.

Third, an application may install in the custodian nodes an aggregation function for network traffic reduction and energy consumption saving purposes using the *Custodian Routing block*. Data aggregation is an essential paradigm for DTN emergency scenarios such as DTN wireless sensor networks that sense information from the emergency area. The aim of this procedure is to merge the data coming from different custodian nodes, eliminating possible redundancy, improving DTN congestion by minimizing the number of transmissions and therefore improving energy consumption. For example, different data belonging to an application from a contamination sensor network placed all along the affected area may be aggregated following different mathematical functions such as average, summation, maximum, etc. These strategies are fully application dependent; therefore, a general solution for every application employing the network will probably be inefficient.

Application routing algorithms may recognize these advantages in advance and take them into account as an extra routing criterion. Address-centric routing algorithm approaches – that is, finding the shortest path between the source and the destination can be evolved or combined with an information-centric approach in which data is consolidated by application-dependent redundant functions.

This paradigm shifts the focus from the traditional address-centric approaches of networking (finding short routes between pairs of addressable end nodes) to a more information-centric approach (finding routes from multiple sources to a single destination that allows in-network consolidation of redundant data).

For these purposes, bundles may choose to delay its routing for a period of time in order to maximize application aggregation. Since data aggregation cannot be performed among different applications, this aggregation delay is an application-related issue and may be calculated in the bundle routing code.

2.2.3 Alleviate DTN Congestion

In scenarios like disaster management, network congestion avoidance is a top priority. We consider the routing problem to be an optimal resource allocation challenge. We propose using the *RIT update code block* to allow the bundles to inform local and outer custodian nodes about different application-related issues.

Each time a user leaves the Emergency Coordination Center, they set a timer indicating when they expect to return. This timer, called Time to Return (TTR) by the authors of [15], automatically decreases its value as time goes by. If a user detects another user with a smaller TTR, it will try to forward its information to this user and thus improve the expected time of arrival at the Emergency Coordination Center. For example, in Figure 2.7, we see how the sender application from the first custodian node chooses the lower custodian node (*doctor0*) as its *next-hop* because it has been informed by another bundle (*Bundle2*) about a low *Time to Return* to the Emergency Coordination Center. Forwarding bundles to the upper custodian node (*doctor2*) could cause congestion.

This is carried out by allowing the bundle agent to execute a concrete method present in the code included in the bundle extension, (the RIT update code extension).

This information will be used by other bundles to update the information on outside variables which may cause congestion, which could then be employed by their routing algorithms. Congestion is highly correlated with routing algorithms, and the infrastructure proposed is an excellent way of informing the different routing algorithms from outer platforms.

2.2.4 DTN Lifetime Control

In emergency scenarios, lifetime control may vary from application to application. We propose a mechanism beyond the Bundle *Lifetime* field to indicate when the bundle's payload is no longer valid. This mechanism is code-based and allows the application to control the bundle's lifetime from an application perspective. Using the *lifetime code extension* in conjunction with the *RIT update code extension*, applications are allowed to cancel application flows and inform both the intermediate custodian nodes and the sender application.

In Figure 2.8 we see how an application sends a bundle containing some application data. Once this bundle has arrived at the fourth custodian node, it executes a special code using

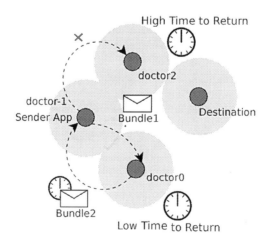

Figure 2.7 Bundle2 forwarded from doctor0 node carries information about doctor0's Time to Return (TTR). After reading this information, Bundle1 message on top chooses doctor0 as its next custodian.

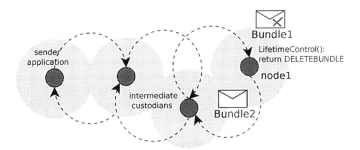

Figure 2.8 Node node1 executes the *lifetime code extension* for *Bundle1*, deletes it, and sends an information bundle with a *RIT update code extension* to cancel the original bundle flow in the intermediate custodian nodes and the sender custodian node.

the *lifetime code extension* to recognize that the application information carried is no longer valid. This could be due, for example, to some context change which could be evaluated only locally and from the point of view of the application. In this case, the bundle generates a control message using the *RIT update code extension* to the sender application, and while visiting every DTN node it will inform all of the bundles containing application information from the same flow (if any). When the bundle arrives at the source DTN node, it will definitely cancel the application flow. Since local context information differs from one custodian node to another, we propose a solution flexible enough to allow the middle DTN custodian nodes to cancel an application flow, improving DTN congestion as bundles are cancelled.

2.2.5 Dynamic Prioritized Scheduling

Since delays in DTN networks can be very big, bundles containing application data may pile up at the DTN layer. Once a potential custodian DTN node becomes available to a given custodian node, the latter will try to forward some of their bundles to the contacted custodian node. It is possible that other bundles are intending to be forwarded to the same destination custodian node at the same time. Policies like FIFO may not be enough for some DTN scenarios. To regulate these situations, a prioritized scheduler inside the custodian node is defined. This scheduler will decide bundle order forwarding and discarding policies.

In the scenarios described, while two users inside the emergency area are next to each other, bundles containing information about victims, for example, may be forwarded to another custodian node. Usually, the time it takes to forward this information is minimal due to the users being in motion. This is why it is useful to first forward bundles containing victims with a given state over others.

In the same way, if the buffer intended for storing bundles gets full, there must be a criterion for choosing which bundles should be discarded. The majority of the proposals are static in the sense that all custodian nodes behave in the same way. Our proposal provides criteria for dropping bundles in the case that the buffer gets full, but with a valuable dissimilarity. We let bundles themselves carry these criteria and update custodian nodes in order to permit application dynamism when it comes to bundle prioritizing or dropping.

Our proposal consists of including bundle agents running on user's devices, a prioritized scheduler that orders bundles while taking into account the victim's state. The prioritized scheduler must then differentiate among the varying victim states.

It is necessary to consider that the criteria used to schedule bundles could change at any given time. Let us consider the possibility of a bigger disaster which could eventually collapse disaster recovery services. In this case, disaster recovery coordinators could decide that the information on the casualties, that is, victims of level 0, are no longer important. Or, in a more pessimistic scenario, the coordinator could be forced to let victims of level 1 die. These criteria could change dynamically within the time of the disaster.

By using the *bundle priority extension block*, described in Section 2.2.5, bundles carry information not only about the victims but also about the criteria to schedule themselves. If the bundle agent finds that the criteria in a platform has been made obsolete by the one carried by the bundle, it will update itself in the RIT.

Bundles belonging to different applications should not compete directly among themselves. A round robin structure containing different queues for every application is proposed. The platform will go around this structure obtaining the first *n* elements of each queue, depending on the weight assigned to every application. This weight is determined by how important the application is and how it is controlled by the local bundle agent.

NOTE

1 Source code can be found at http://ccd.uab.es/_cborrego/active-sensor-grid-src-1.2.tar.gz

REFERENCES

[1]   Carlos Borrego and Sergi Robles. *Relative information in grid information service and grid monitoring using mobile agents*. In *7th International Conference on Practical Applications of Agents and Multi-Agent Systems (PAAMS 2009)*, pp. 150–158. Springer, 2009.

[2]   Carlos Borrego and Sergi Robles. A store-carry-process-and-forward paradigm for intelligent sensor grids. *Information Sciences*, 222:113–125, 2013.

[3]   Carlos Borrego, Sergi Robles, Angela Fabregues, and Adrián Sánchez-Carmona. A mobile code bundle extension for application-defined routing in delay and disruption tolerant networking. *Computer Networks*, 87:59–77, 2015.

[4]   Carlos Borrego, Marica Amadeo, Antonella Molinaro, and Rutvij H Jhaveri. Privacy-preserving forwarding using homomorphic encryption for information-centric wireless ad hoc networks. *IEEE Communications Letters*, 23(10):1708–1711, 2019.

[5]   Russell J Clark, Evan Zasoski, Jon Olson, Mostafa Ammar, and Ellen Zegura. *D-book: A mobile social networking application for delay tolerant networks*. In *Proceedings of the Third ACM Workshop on Challenged Networks*, pp. 113–116, 2008.

[6]   Carlos Borrego, Joan Borrell, and Sergi Robles. Hey, influencer! Message delivery to social central nodes in social opportunistic networks. *Computer Communications*, 137:81–91, 2019.

[7]   Carlos Borrego, Sergio Castillo, and Sergi Robles. Striving for sensing: Taming your mobile code to share a robot sensor network. *Information Sciences*, 277:338–357, 2014.

[8]   Carlos Borrego, Gerard Garcia, and Sergi Robles. Softwarecast: A code-based delivery manycast scheme in heterogeneous and opportunistic ad hoc networks. *Ad Hoc Networks*, 55:72–86, 2017.

[9]   Naercio Magaia, Carlos Borrego, Paulo Pereira, and Miguel Correia. *Privo: A privacy-preserving opportunistic routing protocol for delay tolerant networks*. In *2017 IFIP Networking Conference (IFIP Networking) and Workshops*, pp. 1–9. IEEE, 2017.

[10] Adrián Sánchez-Carmona, Sergi Robles, and Carlos Borrego. Privhab+: A secure geographic routing protocol for DTN. *Computer Communications*, 78:56–73, 2016.

[11] J Schovancova, S Campana, DC Van Der Ster, J Elmsheuser, T Kouba, R Medrano Llamas, G Sciacca, F Legger, E Magradze, G Negri, et al. Atlas distributed computing automation. Technical report, ATL-COM-SOFT-2012-157, 2012.

[12] Erwin Laure, A Edlund, F Pacini, P Buncic, M Barroso, A Di Meglio, F Prelz, A Frohner, O Mulmo, A Krenek, et al. Programming the grid with glite. Technical report, 2006.

[13] Julia Andreeva, C Borrego Iglesias, S Campana, A Di Girolamo, I Dzhunov, X Espinal Curull, S Gayazov, E Magradze, MM Nowotka, L Rinaldi, et al. Automating atlas computing operations using the site status board. *Journal of Physics: Conference Series*, vol. 396:1–6, 2012.

[14] Paulo F Jarschel, Jin H Kim, Louis Biadala, Maxime Berthe, Yannick Lambert, Richard M Osgood, Gilles Patriarche, Bruno Grandidier, and Jimmy Xu. Single-electron tunneling pbs/inp neuromorphic computing building blocks. *arXiv preprint arXiv:1908.08602*, 2019.

[15] Carlos Borrego, Adrián Sánchez-Carmona, Zhiyuan Li, and Sergi Robles. Explore and wait: A composite routing-delivery scheme for relative profile-casting in opportunistic networks. *Computer Networks*, 123:51–63, 2017.

[16] Aruna Balasubramanian, Brian Levine, and Arun Venkataramani. *DTN routing as a resource allocation problem*. In *Proceedings of the 2007 Conference on Applications, Technologies, Architectures, and Protocols for Computer Communications*, pp. 373–384, 2007.

[17] Moazam Ali and Susmit Bagchi. Probabilistic normed load monitoring in large scale distributed systems using mobile agents. *Future Generation Computer Systems*, 96:148–167, 2019.

[18] Tatiana Pereira Filgueiras, Leonardo M Rodrigues, Luciana de Oliveira Rech, Luciana Moreira Sá de Souza, and Hylson Vescovi Netto. Rt-jade: A preemptive real-time scheduling middleware for mobile agents. *Concurrency and Computation: Practice and Experience*, 31(13):e5061, 2019.

[19] Akanksha Ahuja, Sharad Agarwal, and David Lange. Growth and evolution CMS offline computing from run 1 to HL-LHC. 2020.

[20] K Scott, S Burleigh, et al. RFC 5050: bundle protocol specification. *IRTF DTN Research Group*, 2007.

[21] Dario Bruneo, Salvatore Distefano, Maurizio Giacobbe, Antonino Longo Minnolo, Francesco Longo, Giovanni Merlino, Davide Mulfari, Alfonso Panarello, Giuseppe Patanè, Antonio Puliafito, et al. An IoT service ecosystem for smart cities: The# smartme project. *Internet of Things*, 5:12–33, 2019.

[22] Xinxing Zhao, Chandra Sekar Veerappan, Peter Loh, and Caleb Wee. Towards cross-platform detection of cyberattacks using micro-agents. *International Journal of Cyber-Security and Digital Forensics*, 8(2):108–120, 2019.

[23] Vinton Cerf, Scott Burleigh, Adrian Hooke, Leigh Torgerson, Robert Durst, Keith Scott, Kevin Fall, and Howard Weiss. Rfc 4838, delay-tolerant networking architecture. *IRTF DTN Research Group*, 2(4):6, 2007.

[24] Abraham Martín-Campillo, Jon Crowcroft, Eiko Yoneki, and Ramon Martí. Evaluating opportunistic networks in disaster scenarios. *Journal of Network and Computer Applications*, 36(2):870–880, 2013.

[25] Carlos Borrego, Joan Borrell, and Sergi Robles. Efficient broadcast in opportunistic networks using optimal stopping theory. *Ad Hoc Networks*, 88:5–17, 2019.

Chapter 3

Reactive and Proactive Routing Strategies in Mobile Ad Hoc Network

Md. Ibrahim Talukdar and Md. Sharif Hossen

Comilla University, Cumilla, Bangladesh

CONTENTS

3.1 INTRODUCTION AND RELATED WORKS

Wireless networks are classified as infrastructure-based networks with a central access point, and ad hoc networks with no access point. Mobile ad hoc networks (MANETs) [1, 2] are wireless in nature with no infrastructure, where each node/device may act as a source or a router or a destination, etc., and the bridge of information forwards packets for nodes that are not in the transmission range. These nodes or devices can have different speeds, transmission ranges, data rates, and packet sizes. Some of the characteristics [3] of the ad-hoc network are autonomous, dynamic topology, multi-hop, etc. These networks are also constrained to the transmission ranges, packet losses, security, Quality of Service (QoS), hidden terminal problems, etc. MANET protocols can be categorized into three schemes, such as proactive, reactive, and hybrid (a combination of both). Regardless of various protocols, only Ad Hoc On-Demand Distance Vector (AODV), Destination-Sequenced Distance-Vector Routing (DSDV), and Dynamic Source Routing (DSR) are examined here.

A myriad of works is performed on an ad hoc network. Authors [4] examine the reactive protocols in [5]; the proactive and reactive protocols are analyzed using NS-3. The authors [6] evaluate MANET protocols using OPNET. Moreover, the energy efficiency of the AODV protocol is analyzed in [7]. Special kinds of ad hoc networks, e.g., Vehicular Ad hoc Networks (VANETs), are more realistic in intelligent transport systems, which are more popular today. In [8, 9], the routing protocols are evaluated for the VANET environments.

Ad hoc networks have a variety of application areas [10] but are not limited to vehicle communication, distributed computing, earthquake/tsunami-hit areas, sensor networks, military applications, multimedia applications, space exploration, tracking of rare animals, biological detection, undersea operations, Personal Area Network (PAN), Bluetooth, etc. Hence, MANETs have emerged as a prime research area. Diverse applications of these networks in many real-time scenarios stimulate us to judge the protocols of MANETs. There are many papers on the performance measurement of ad hoc with the basis of different metrics and variations. The key motivation behind this chapter is the better understanding of protocols where they are applicable or not with challenged and crucial variants like mobile nodes, velocity, data rates, and packet sizes. In this chapter, MANET protocols like AODV [11, 12], DSDV [13], and DSR [14] are examined to select the best method for establishing an emergency communication when needed. Here, the significance of ad hoc routing protocols are examined in terms of the QoS parameters like packet delivery, delay, routing overhead, and throughput with varying some parameters such as the number of nodes, packet sizes, data rates, and the velocity of nodes for random simulation area with random waypoint mobility [15]. In the remaining sections, Section 3.2 describes the ad hoc routing protocols with the mobility models; Section 3.3 briefly discusses the methodology and the simulation environment; Section 3.4 depicts the QoS parameters and the graphical results with decent analysis; Finally, Section 3.5 recapitulates the chapter.

3.2 STATE OF THE ROUTING STRATEGIES WITH MOBILITY MODELS (MM)

Routing is a very crucial part for MANETs. Basically, the mobile ad hoc networks are composed of two routing protocols, namely, proactive and reactive. Moreover, the amalgamation of both introduces the hybrid routing protocol (Figure 3.1).

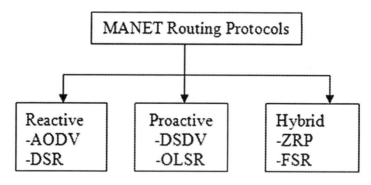

Figure 3.1 Routing protocols in MANET.

(a) Proactive: The proactive [16] protocols follow a table-driven approach wherein all the nodes endlessly search for routing information within a network. In this protocol, each node retains one or more routing table representing the entire network topology. Hence, the routing table maintenance is difficult for a large topology. These protocols require more overhead, which consumes more bandwidths. Some of these protocols are derived from the link-state [17] routings. Optimized Link State Routing (OLSR), DSDV, etc., are proactive routing protocols in ad hoc networks.

(b) Reactive: The reactive protocols, also termed as on-demand routing protocols, do not keep track of the routing information until any communication takes place. In the reactive [18] approach, the routing information is accumulated as per need, and the route determination relays on forwarding the route queries all over the network. The routes are uncovered by flooding the route request. AODV, Area Border Routing (ABR) [19], DSR, Temporally Ordered Routing (TORA) [20], etc., are reactive protocols.

3.2.1 Protocols under Investigation

This subsection precisely describes the considered protocols under investigation in this chapter.

3.2.1.1 Ad Hoc On-Demand Distance Vector (AODV)

AODV discovers the route on user demand fashion, that is, it starts searching for a route during the transmission of any packet to the destination. It discovers a route, generates the messages, and maintains the established route. In AODV, no overhead packet is required if the path exists from one node to another. It disallows the routing loops by using the sequence numbers. Besides, it uses a broadcast technique for searching a route and the unicast mechanism for responding from the route. It relays on the DSDV and the DSR wherein the information on the sequence number and the periodic beaconing of the DSDV and the route finding technique of the DSR are used. However, the DSR and the AODV have the following dissimilarities:

(i) The DSR conveys the complete routing information of packets, whereas the AODV keeps only the destination address which implies that the AODV requires less routing overheads than the DSR.

(ii) The route response methods of the DSR convey the address of every node along the path, whereas only the sequence number and the destination IP addresses are carried by the AODV route reply policy.

AODV protocol follows two phases as depicted in the following.

(a) *Route Discovery of AODV*: To understand this process, suppose the source node 1 tries to communicate with the destination node 7 where the origin launches a route discovery through broadcasting the route request with the inclusion of source ID (Sid), the destination ID (Did), the source sequence (SSeq), the destination sequence (DSeq), time to live (TTL), etc. to all its neighbors (Figure 3.2).

The neighboring nodes compare the DSeq of Route Request (RREQ) with their corresponding arrival in route cache. For greater DSeq of RREQ, the neighbor node responds to the source with an RREP carrying the path to the destination. Generally, a node (say, node 3) having a route to the destination (node 7) in its cache with higher

DSeq as compared to RREQ sends a Route Reply (RREP) back to the source. Thus, the route 1-3-6-7 is kept in node 1. Moreover, the destination node also onwards an RREP back to the origin.

(b) *Route Maintenance of AODV*: In the route maintenance mechanism of AODV, when a broken link is detected, the origin and the endpoint are notified by disseminating an RERR like as the DSR scenario. During the broken link between two nodes (say, node 3 and 5) as depicted in Figure 3.3 on the route 1-3-5-7, both nodes will send an RERR to the source and the endpoint. The major point of AODV over DSR and AODV is that it reduces routing overhead through avoiding source routing.

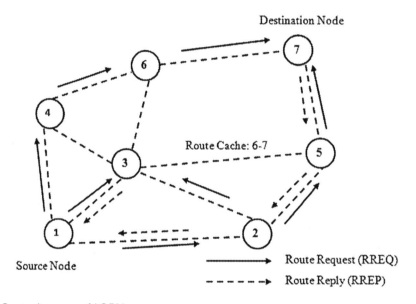

Figure 3.2 Route discovery of AODV.

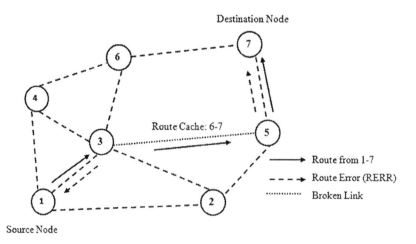

Figure 3.3 Route Maintenance of AODV.

3.2.1.2 Dynamic Source Routing (DSR)

DSR routing uses the idea of source routing wherein the sender has the entire node-to-node path to the destination. In DSR, the routing information is not needed to be maintained at every host. It has two phases, path finding and route maintenance, which work jointly to discover and support source routes to random destinations in the MANET. When the destination route is unknown to a node, it uses a flooding-based route discovery strategy with RREQ where each one accepts RREQ and resends it to the neighboring nodes until the destination, D, is found. Whenever RREQ reaches to D, D responds to the RREQ with an RREP packet to the origin by traversing this path backwardly. However, it is not suited for larger networks, since it consumes most of the bandwidth of the networks [21]. The DSR also follows two phases like the AODV.

(a) *Route Discovery of DSR*: In the DSR route discovery phase, the source node, say, node 1, tries to communicate with the destination node (say node 7), wherein the source starts a route discovery through broadcasting an RREQ to its adjacent nodes, i.e., 2, 3, and 4. Similarly, the intermediate nodes follow the same process. After reaching at node 7, it adds its own address and reverses the route to the destination using unicast fashion. The destination node accepts the first received route from a node as the best path and then caches the other nodes (Figure 3.4).

(b) *Route Maintenance of DSR*: The route maintenance is required during the broken link between nodes 6 and 7. The damaged link can be detected by either passive or active monitoring of the link to a node. In this case, an intermediate node sends a route error packet to the origin, and then the source reinitiates the path discovery (Figure 3.5).

3.2.1.3 Destination-Sequenced Distance-Vector (DSDV)

The Bellman-Ford approach is utilized in the DSDV with some amelioration. Also, the DSDV, a proactive protocol, is derived from the routing information protocol (RIP) and placed into an ad hoc routing. It selects the best path to the destination using the distance vector shortest path. It adds new attributes and sequence numbers to the routing table, and using this

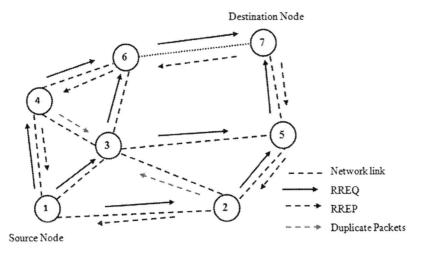

Figure 3.4 Route Discovery of DSR.

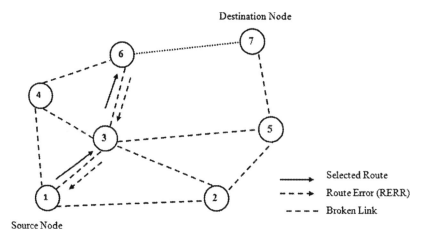

Figure 3.5 Route Maintenance of DSR.

sequence, it handles the infinite loop problems. Due to the large overhead, it consumes more network resources and battery power. This protocol is suitable for a small network and does not support multicasting [22].

3.2.2 Mobility Models (MM)

There are lots of MMs, e.g., Random Walk, Random Direction, Random Waypoint (RWP), etc., to move nodes independently within the simulation area. Among these models, the RWP is chosen to capture the performance of the networks. It adds the concept of pause times in a node before changing its speed and direction. Without pause time, it acts like Random Walk [23].

3.2.2.1 Random Walk (RW)

In RW, the nodes move in an unpredictable fashion. Here, each node conveys random speed and direction from the predefined ranges until it reaches the destination. Whenever any node reaches the destination, it receives new speed and direction from that range. In RW, the nodes move freely without any restriction, and no tracking records of the nodes are kept.

3.2.2.2 Random Waypoint (RWP)

The working principle of the RWP is very similar to the RW mobility, except it adds the notion of pause time while moving nodes. With zero pause time, the RWP and the RW are the same. Very often the RWP faces non-uniform node distribution problems. However, by using a smooth RWP (SRWP) technique with the concept of temporal dependency, it improves the performance of the RWP.

3.2.2.3 Random Direction (RD)

RD overcomes the problem that arises in the RWP. In the RD mobility, a mobile node proceeds in a specific direction until it touches the destination border of the simulation area with a fixed speed. It suggests an even distribution of the nodes in the specified simulation region.

3.3 METHODOLOGY AND ENVIRONMENT SETTING

The simulation methodology and environment settings are described briefly in this section.

3.3.1 Simulation Tool

The network simulator 2, called NS-2, along with Linux, is used as a simulation tool. It uses an object tool command language shell (OTCL) as the user interface which allows the scripts written in TCL format to be executed. It is a discrete event simulator along with the object-oriented concepts and used by the ad hoc networking community and researchers to determine the performance of routings. It uses two languages: (1) C++, which runs in the back end, and (2) tool command language (TCL), which runs in the front end [24]. Moreover, the components of NS-2 include NS, NAM (network animator), pre-processing, and post-analysis for the simulator, visualization of NS output, handwritten TCL, and the trace analysis using AWK scripts.

3.3.2 Random Traffic and Mobility Generation

The CBRGEN tool under "~ns/indep-utils/cmu-scen-gen" is found to generate the random traffic among the nodes using TCP or CBR connection. Moreover, the SETDEST tool under "~ns/indep-utils/cmu-scen-gen/setdest/" is used to produce the traces of the nodes by the random movement with the velocity of the nodes to any location (not fixed) within the considered wireless region. The node's mobility is distributed by a random waypoint fashion [24]. However, the traffic connections and nodes' mobility can be made manually for a small network.

3.3.3 Running the NAM File and Data Sending

After running the TCL script for a specified network environment, the NAM and the trace files are generated. Then, using AWK scripts, PDR, delay, t-put (throughput), etc., are examined. The NAM is used for tracking real-world traces. Here is the scenario of running the NAM file (Figure 3.6).

3.3.4 Simulation Environment Setup

In this research, the wireless network environment is constructed for moving the nodes' CBR traffic patterns, and the RWP mobility models with specified simulation area, time, etc. The wireless channel with the random waypoint is considered as the mobility model. Table 3.1 shows the general simulation parameters in detail.

Table 3.2 shows the variations of the mobile nodes, packets, data rates, and velocity of nodes that are used in environment simulation.

3.4 RESULT AND DISCUSSION

3.4.1 Evaluation Metrics

The following metrics are examined to analyze the performance of protocols.

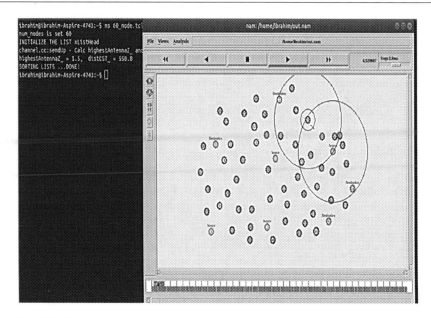

Figure 3.6 The NAM file and data sending.

Table 3.1 General parameters for simulation.

Parameters	Values
Routing Protocols	AODV, DSDV, and DSR
Mobility Mode	Two Ray Ground
Antenna	Omni Antenna
Simulation Time	160 sec
Channel	Wireless Channel
Mobility Model	Random Way Point
Simulation Area	1100 × 800
Traffic	CBR/UDP
Packet Size	1500 Bytes
MAC	MAC/802-11
Mobile Nodes	30
Mobility Speed	6 m/s
Data Rates	0.1 Mbps
Performance Metrics	Packet Delivery Ratio, End-to-end Delay, Normalized Routing Overhead, Throughput
Simulator	NS 2.35

(i) Packet delivery ratio (PDR): PDR, a vital issue of a network, can be expressed as the number of packets received at the destination over the packets sent from the source. Higher packet delivery ensures better performance [25]. It is determined as

$$PDR = \frac{\text{Total packet received}}{\text{Total packet sent}} \times 100\%$$

Table 3.2 Variation of parameters.

Parameters	Values
Mobile Nodes	20, 30, 40, 50, and 60
Packet Sizes	512, 1000, 1500, 1700, and 200 Bytes
Data Rates	0.1, 2, 5, 10, and 15 Mbps
Speeds	6, 10, 20, 25, and 35 m/s

(ii) End-to-end delay: It is the time between the packet generated by the source node and the packet received by the destination node. It can be expressed as follows at millisecond (ms) [26].

$$\text{End-to-end Delay} = \sum_{i=1}^{n} \frac{\text{Received time} - \text{Sent time}}{\text{Numbers of packets received}}$$

(iii) Overhead: The scheme defines how many redundant packets are relayed to convey one packet. In other words, it can be expressed as follows.

$$\text{Normalized Routing Overhead} = \frac{\text{Total data packets sent}}{\text{Total data packets received}}$$

(iv) Throughput (T-put): The average throughput, in Mbps or Kbps, refers to how fast a packet can be sent from one to another. High t-put is desirable for every network. It is directly proportional to PDR.

3.4.2 Performance Analysis through Varying the Number of Nodes

For varying the number of nodes, it is clear from Figure 3.7 that the AODV protocol achieves slightly higher delivery as compared to the DSR. It is also noticeable that the delivery probability is likely in increasing order with the increase of the mobile nodes. Among the protocols, the DSDV provides the least performance in case of the delivery.

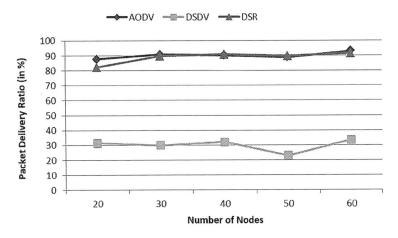

Figure 3.7 PDR with varying the number of nodes.

From Figure 3.8, it is obvious that the DSDV outperforms the AODV and the DSR with varying the number of nodes as it services few messages to the destination. Moreover, the AODV approximately exhibits lower delay than the DSR.

Figure 3.9 depicts that with the increase of the number of mobile nodes, the normalized routing overhead is increasing for the AODV, and the DSDV where the DSDV shows the highest overhead value and the DSR exhibits the least overhead (approximately zero).

Again, for varying the number of nodes, it is seen that the AODV ensures the best throughput and DSDV performs the least among the protocols shown in Figure 3.10. Moreover, it is blatant that the throughput is proportional to the number of mobile nodes.

3.4.3 Performance Analysis through Varying the Packet Sizes

Figure 3.11 depicts that the AODV presents a higher delivery ratio with respect to the packet size as it does for the nodes' variation. Notice also that the DSDV delivers lower packets than the AODV and the DSR.

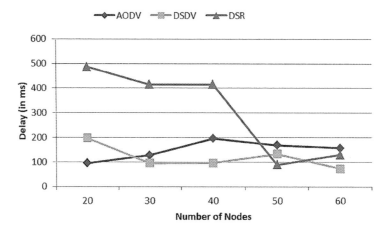

Figure 3.8 Delay with varying the number of nodes.

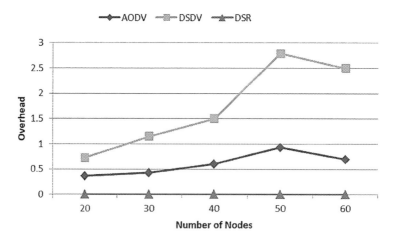

Figure 3.9 Overhead with varying the number of nodes.

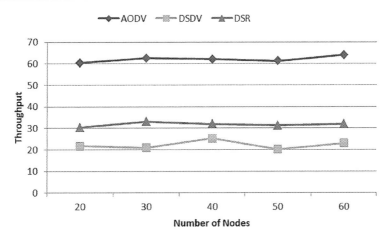

Figure 3.10 T-put with varying the number of nodes.

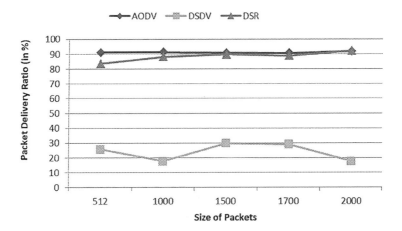

Figure 3.11 PDR with varying the packet sizes.

Figure 3.12 depicts that the average delay is fluctuating for varying the size of packets. The average delay is extremely high for the DSR with respect to the packets' variation because the DSR collects and maintains all the possible routes to a destination node.

The DSDV consumes a very large number of extra packets for data delivery as illustrated in Figure 3.13. It shows the highest overhead value among all the protocols. As with the node's variation, the variation of packet size also shows that the DSR has the least overhead value (approximately zero).

The AODV routing achieves a slightly higher throughput value than the DSR. Here, the DSDV shows the lowest throughput compared to the AODV and the DSR. With a packet size of 512 bytes, all the protocols reach their maximum gain. However, the throughput is decreasing with the increase of packet sizes for all the protocols, as shown in Figure 3.14.

3.4.4 Performance Analysis through Varying the Data Rates

From Figure 3.15, it seems that the values overlap with each other. Actually, the PDR is at maximum for the AODV, lowest for the DSDV, and medium for the DSR. Additionally, it is blatant that the packet delivery is decreasing with the increase of data rates for all the

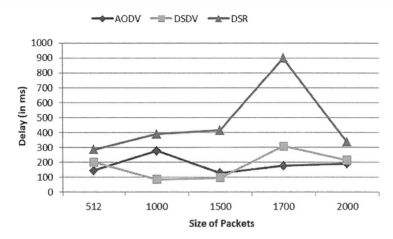

Figure 3.12 Delay with varying the packet sizes.

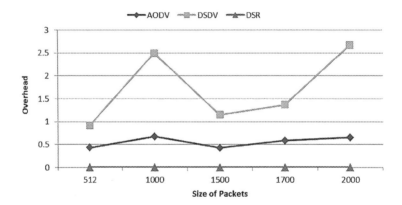

Figure 3.13 Overhead with varying the packet sizes.

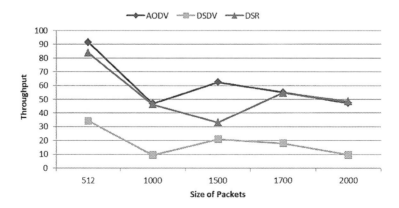

Figure 3.14 T-put with varying the packet sizes.

protocols. However, the AODV outperforms the DSDV and the DSR in terms of the packet delivery with the alteration of the data rates, as shown in Figure 3.15.

Figure 3.16 indicates that the average delay is increasing overall with the increase of the data rates. DSR badly exhibits high latency due to complete the route maintenance from the source to the destination. Proactive protocol (i.e. DSDV) shows the lowest delay.

The AODV routing exhibits lower overhead than the DSDV when the data rates are within 0.1 and 2 MB. With the increase of data rates, the overhead is improving for the DSDV than the AODV. The DSR is not affected by the data rates and exhibits constant, and approximately zero overhead (Figure 3.17).

Figure 3.18 depicts that the AODV achieves maximum throughput with the variety of data rates while the DSDV shows lower t-put. In contrast to the message delivery, the throughput is increasing for varying the data rates.

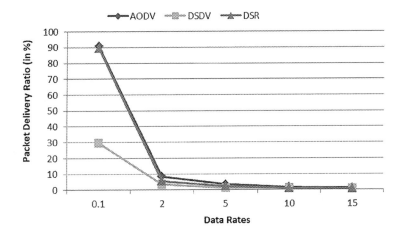

Figure 3.15 PDR with varying the data rates.

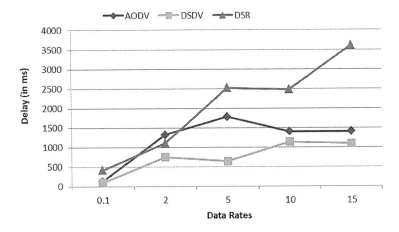

Figure 3.16 Delay with varying the data rates.

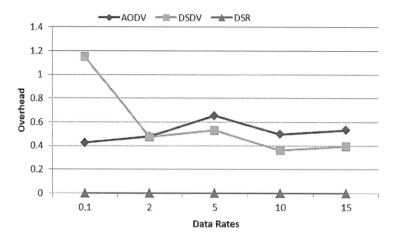

Figure 3.17 Overhead with varying the data rates.

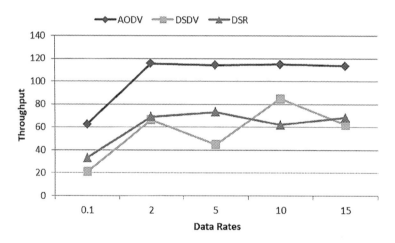

Figure 3.18 T-put with varying the data rates.

3.4.5 Performance Analysis through Varying the Speed of Nodes

Figure 3.19 illustrates that the changes of the nodes' speed significantly affect the packet delivery. The AODV shows the maximum packet delivery than others wherein the DSDV exhibits the least delivery. But PDR is decreasing with the increase in the number of nodes for reactive protocols.

Figure 3.20 depicts that the delay is fluctuating for the reactive protocols. The average delay is minimum for DSDV protocol, as are all the variants under consideration.

In all cases (Figures 3.9, 3.13, 3.17, and 3.21), it is obvious that the DSR protocol exhibits the least (approximately zero) overhead regardless of all the variants. The AODV routing shows lower overhead compared to the DSDV, as shown in Figure 3.21.

Figure 3.22 depicts that the throughput of the AODV routing is significantly higher than the DSR and the DSDV, as discussed previously, for varying the number of nodes, the packet sizes, and the data rates. The DSDV shows the least throughput, like other cases.

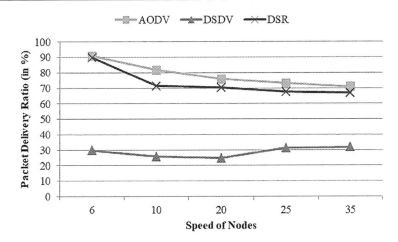

Figure 3.19 PDR with varying the speed of nodes.

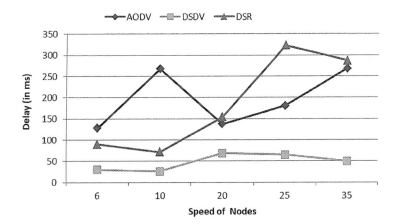

Figure 3.20 Delay with varying the speed of nodes.

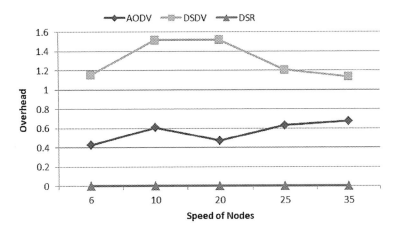

Figure 3.21 Overhead with varying the speed of nodes.

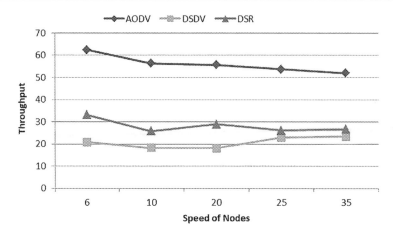

Figure 3.22 T-put with varying the speed of nodes.

3.5 CONCLUSION

To ensure which protocol is best for constructing mobile networks in an emergency situation, this research is performed by comparing the performance of proactive and reactive techniques using NS-2. This chapter reveals the performance of AODV, DSDV, and DSR for varying the number of nodes, packet sizes, data rates, and node's speeds within a specified area on the basis of packet delivery, delay, overhead, and throughput. The pictorial analysis shows that the AODV protocol outperforms the DSDV and the DSR in terms of delivery with all the variants. Also, it ensures better throughput among all the protocols considered here which blatantly indicates that the delivery is often changing proportionally with the throughput. Again, the graphical results depict that the DSDV protocol achieves desirable delay among the protocols since its delivery is very low compared to the reactive protocols, whereas the DSR shows higher delay compared to the AODV since its header is larger than the AODV. Here, the simulation results also show that the DSR experiences no overhead, and the AODV shows lower overhead compared to the DSDV. In addition, each routing protocol has its pros and cons and should be used as the application requires. It is rarely possible to name one protocol as being better than the others in any network under all criteria. Considering the overall performances, the AODV performs well as compared to other routing approaches for considered areas. However, the security issues of the routing techniques are not considered here. The investigation in this chapter has been done under no attack, so there is a possibility of changing the performance of the routing approaches during the security issues.

ACKNOWLEDGEMENTS

The authors would like to thank the ICT Division of the Bangladesh Government for awarding a research fellowship to Md. Ibrahim Talukdar for this research.

REFERENCES

[1] Talukdar, M. I. (2021). Performance Improvements of AODV by Black Hole Attack Detection Using IDS and Digital Signature. *Wireless Communications and Mobile Computing*, 2021(1), pp. 1–13.

[2] Salam, T. (2020, January 21). Performance Analysis on Homogeneous LEACH and EAMMH Protocols in Wireless Sensor Network. *Wireless Personal Communications.* doi:10.1007/s11277-020-07185-6.

[3] Venkatesan, T. P. (2014, April). *Overview of Proactive Routing Protocols in MANET. Fourth International Conference on Communication Systems and Network Technologies*, pp. 173–177. doi:10.1109/CSNT.2014.42.

[4] Mohapatra, R. K. (2018, October). *Performance Analysis of Reactive Routing Protocols in MANET under CBR Traffic using NS2. International Conference on Advances in Computing, Communication Control and Networking (ICACCCN)*, pp. 352–356. doi:10.1109/ICACCCN.2018.8748851.

[5] Raman Chandan, R. (2018). Performance Evaluation of AODV, DSDV, OLSR Routing Protocols using NS-3 Simulator. *International Journal of Computer Network and Information Security*, 10(7), pp. 59–65. doi:10.5815/ijcnis.2018.07.07.

[6] Aneiba, A., Melad, M. (2016). Performance Evaluation of AODV, DSR, OLSR, and GRP MANET Routing Protocols Using OPNET. *International Journal of Future Computer and Communication*, 5(1), pp. 57–60.

[7] Anand, M. (2019). Efficient Energy Optimization in Mobile Ad Hoc Network (MANET) Using Better Quality AODV Protocol. *Cluster Computing*, 22(S5), pp. 12681–12687. doi:10.1007/s10586-018-1721-2.

[8] Ur Rehman, M. (2018). Performance and Execution Evaluation of VANETs Routing Protocols in Different Scenarios. *EAI Endorsed Transactions on Energy Web*, 5(17), pp. 154458–154455. doi:10.4108/eai.10-4-2018.154458.

[9] Poonia, R. C. (2018). A Performance Evaluation of Routing Protocols for Vehicular Ad Hoc Networks with Swarm Intelligence. *International Journal of System Assurance Engineering and Management*, 9(4), pp. 830–835. doi:10.1007/s13198-017-0661-1.

[10] Hossen, M. S., Rahim, M. S. (2019). Analysis of Delay-Tolerant Routing Protocols Using the Impact of Mobility Models. *Journal of Scalable Computing: Practice and Experience*, 20(1), pp. 39–48.

[11] Hossen, M. S., Rahim, M. S. (2015). On the Performance of Delay-Tolerant Routing Protocols in Intermittently Connected Mobile Networks. *Rajshahi University Journal of Science & Engineering*, 43, pp. 29–38.

[12] Swidan, A. (2016). Mobility and Direction Aware Ad-Hoc on Demand Distance Vector Routing Protocol. *Procedia Computer Science*, 94, pp. 49–56. doi:10.1016/j.procs.2016.08.011.

[13] Wang, J. (2010). A Secure Destination-Sequenced Distance-Vector Routing Protocol for Ad Hoc Networks. *Journal of Networks*, 5(8), p. 942.

[14] Suriya, C. (2014). A Study on Dynamic Source Routing In Ad Hoc Wireless Networks. *International Journal of Advanced Networking and Applications*, 5(5), p. 2046.

[15] Zhong, Y. (2003). *Dynamic Source Routing Protocol for Wireless Ad Hoc Networks in Special Scenario Using Location Information. International Conference on Communication Technology Proceedings*, Vol. 2, pp. 1287–1290. doi:10.1109/ICCT.2003.1209766.

[16] Mohseni, S. (2010, April). *Comparative Review Study of Reactive and Proactive Routing Protocols in MANETs. 4th IEEE International Conference on Digital Ecosystems and Technologies*, pp. 304–309. doi:10.1109/DEST.2010.5610631.

[17] Hong, X. (2002). Scalable Routing Protocols for Mobile Ad Hoc Networks. *IEEE Network*, 16(4), pp. 11–21. doi:10.1109/MNET.2002.1020231.

[18] Araghi, T. K. (2013, September). *Performance Analysis in Reactive Routing Protocols in Wireless Mobile Ad Hoc Networks Using DSR, AODV and AOMDV. International Conference on Informatics and Creative Multimedia*, pp. 81–84. doi:10.1109/ICICM.2013.62.

[19] Gupta, S. (2015). Performance Comparison of Proactive Routing Protocols: OLSR, DSDV, WRP. *International Journal of Advanced Research in Computer Science*, 6(8) pp. 73–77.

[20] Mafirabadza, C. (2017). Efficient Power Aware AODV Routing Protocol for MANET. *Wireless Personal Communications*, 97(4), pp. 5707–5717. doi:10.1007/s11277-017-4804-0.

[21] Das, S. (2000). *Performance Comparison of Two On-Demand Routing Protocols for Ad Hoc Networks. Proceedings IEEE INFOCOM 2000. Conference on Computer Communications.*

Nineteenth Annual Joint Conference of the IEEE Computer and Communications Societies (Cat. No.00CH37064), Vol. 1, pp. 3–12. doi:10.1109/INFCOM.2000.832168.

[22]   Madhusrhee, B. (2016). Performance Analysis of AODV, DSDV and AOMDV Using WiMAX in NS-2. *Computational Methods in Social Sciences*, 4(1), pp. 22–28.

[23]   Hossen, S. (2016, January). *Impact of Mobile Nodes for Few Mobility Models on Delay-Tolerant Network Routing Protocols. International Conference on Networking Systems and Security (NSysS)*, pp. 1–6. doi:10.1109/NSysS.2016.7400704.

[24]   Issariyakul, T. (2011). *An Introduction to Network Simulator NS2*. New York, NY: Springer.

[25]   Hossen, M. S. (2019). DTN Routing Protocols on Two Distinct Geographical Regions in an Opportunistic Network: An Analysis. *Wireless Personal Communications*, 108, pp. 839–851. doi:10.1007/s11277-019-06431-w.8.

[26]   Talukdar, M. I., Hossen, M. S. (2019). Selecting Mobility Model and Routing Protocol for Establishing Emergency Communication in a Congested City for Delay-Tolerant Network. *International Journal of Sensor Networks and Data Communications*, 8(1), pp. 1–9. doi:10.4172/2090-4886.1000163.

Chapter 4

Secure Hierarchical Infrastructure-Based Privacy Preservation Authentication Scheme in Vehicular Ad Hoc Networks

Mariya Ouaissa, Mariyam Ouaissa and Meriem Houmer
Moulay Ismail University, Meknes, Morocco

CONTENTS

4.1 INTRODUCTION

Wireless networks have experienced remarkable progress in recent years; they are undeniable today. Their appearance and the advancement of communication and information technologies are giving rise to the so-called Intelligent Transport Systems (ITS). The principal aim of these systems is to make the road more efficient. Moreover, one of the main strengths of ITS is to enable a level of cooperation between participants in the road network by equipping vehicles with wireless communication equipment. This type of wireless network refers to vehicular networks [1].

Vehicular Ad hoc Networks (VANETs) [2] are a new form of Mobile Ad hoc Networks (MANETs) that aim to provide communications between vehicles or with infrastructure located at roadside. These networks are described as a dynamic topology according to the addition or departure of a vehicle from the network. In these networks, vehicles are equipped with wireless short and medium-range communication. Actually, the vehicles can communicate with each other with two methods, either Vehicle to Vehicle (V2V) or Vehicle to Infrastructure (V2I) where the vehicle communicates with the equipment next to the road named Road Side Unit (RSU) [3]. VANETs are used to meet the communication needs applied to transport networks to improve driving and road safety for road users.

In VANET, vehicles exchange messages and communicate with each other in a wireless environment. This situation can give rise to internal or external security attacks which can have the objective of rendering the network non-functional, of causing an accident. For that reason, the preservation of the security of information exchanged between vehicles is a crucial necessity. Communication must go through the analysis of the potential of security threats, and the design of a robust architecture capable of dealing with these threats. In this context, the implementation of vehicular networks requires an effective security mechanism in order to satisfy security requirements such as authentication, integrity, and the privacy preservation of a user [4].

This chapter presents a hierarchical authentication protocol with the aim of ensuring basic security requirements such as integrity, confidentiality, non-repudiation, and availability, as well as the preservation privacy of users by using an identifier or a real identifier. First, our solution allows vehicles and RSUs to authenticate with the certification authority so that they are legal entities in the network. Then a mutual authentication between vehicles and RSUs makes it possible to minimize access to malicious entities in the network.

The proposed protocol combines the symmetrical and asymmetrical approaches, where the symmetrical approach used the Elliptic Curve Diffie Hellman (ECDH) algorithm for reliable key exchange that is implemented in order to create and share the secret key, whose objective is to ensure the security of the exchange of the parameters for the asymmetric system (Private Key/Public Key), authentication packets, and vehicle certificate using the Edwards-curve Digital Signature Algorithm (EdDSA). In addition, the creation of sub-lists of revoked certificates based on vehicle type makes it possible to minimize the response time by looking for a certificate whether it is revoked or not.

The remainder of this chapter is organized as follows. The next section details the related works. Section 4.3 presents the background, including communication architecture and security requirements in VANETs. Section 4.4 describes several preliminaries used in our solution. In Section 4.5, we propose a secure hierarchical infrastructure-based privacy preservation authentication scheme in VANET. The security of the designed protocol was checked by the verification tool Automated Validation of Internet Security Protocols and Applications (AVISPA), in Section 4.6. Section 4.7 evaluates and analyzes the performances of the existing protocols as well as our proposition. We draw our conclusion in Section 4.8.

4.2 RELATED WORKS

Authentication represents an essential cryptographic mechanism that provides confidence between vehicles and infrastructures in vehicle ad hoc networks. However, an improved authentication process will effectively detect malicious nodes and then maintain VANET security. Therefore, various authentication mechanisms that ensure protected communication in the network were suggested. In this section, we discuss them briefly.

The basic idea of the technique based on anonymous certificates is given by Raya and Hubaux [5]. The authors use anonymous certificates (e.g. pseudonyms) to hide the real identity of users. The anonymous certificate does not include any information about the real identities of the users, but privacy may be violated because the messages contain an exchanged key which makes it possible to track the vehicle's true identity.

The group signature is an alternative to achieve security and preserve privacy in VANETs. In this technique, a group manager is responsible for managing the group. The members may enter or exit the group dynamically. Upon registering and joining a group, a member can sign anonymously on the behalf of the group, and the recipient uses the public key to validate the signature but will never know who sent the packet. However, there are exceptional cases where the group manager may reveal the identity of a sender of any group signature. The group signing approach has emerged to overcome the disadvantage of the anonymous certificate technique. The first protocol of this technique to be implemented within vehicle networks is the Group Signature ID-based Signature protocol [6].

The authors of [7] aim to guarantee the identification, authentication, non-repudiation, and integrity of the roadside unit when transmitting messages from RSU to vehicles (I2V). An identification aggregation will be carried out by several RSUs and without the intervention of a trusted third party. Their algorithm first ensures the identification of RSUs by the Elliptic Curve Diffie-Hellman's algorithm, which verifies both that the nearest RSU possesses the same secret key and that the vehicle authenticates the signed message utilizing the Digital Signature Elliptic Curve. The ECDH-ECDSA provides a greater degree of protection even if its aggregation takes roughly 40 ms further than the basic ECDSA method.

Vijayabharathi and Malarchelvi [8] present an Expedite Message Authentication Protocol (EMAP) based Hash Message Authentication Code (HMAC) for vehicular networks. The authors propose a new fast process of revocation checking to minimize the computation process and avoid overhead problems.

A new conception of Public Key Infrastructure (PKI) for the authentication process is proposed in [9]. The authors design an infrastructure PKI-based symmetric encryption in order to reduce the treatment time and eliminate overhead for authentication.

Das et al. [10] offer a hierarchical protocol to reach the objectives of scalability and certification in VANET. This protocol built a structure of tree with a hierarchy of Certification Authority (CA) to operate the VANET. Authors suggest two kinds of nodes; the powerful nodes are the certifying authorities, and the leaf nodes are vehicles.

The Lightweight Identity Authentication Protocol (LIAP) was proposed in [11] in order to supply a fast mutual authentication between the roadside unit and the vehicle, and also to guarantee the vehicle's conditional privacy. This protocol achieves the transfer authentication process by employing a secret dynamic session mechanism and avoids the utilization of the encryption/decryption operations in the roadside unit and the vehicle.

The scheme proposed in [12] constitutes a group of vehicles and RSU through the use of self-authentication without the need of a certification authority. It uses a Group Key (GK) to improve the efficiency of certification, and the protocol selects deniable group key agreement method to avoid attacks into legal vehicles.

The authors in [13] propose a new protocol based on the complexity of two popular mathematical problems to deal with the problems existing in previous Mobile Wireless Networks' (MWNs) handover authentication protocols. Security analysis illustrates that the proposed protocol is protected from various threats and can satisfy a number of security requirements. A hierarchical revocable authentication protocol based random oracle model within the

Diffie-Hellman (DH) hypothesis is presented in [14]. The evaluation of the protocol shows that it saves highly computation overheads and meets the security requirements.

Our proposal also aims to ensure privacy and authentication in a vehicular ad hoc network. The privacy is preserved by ensuring anonymity. In this study, we use an ECDH algorithm for reliable key exchange and the EdDSA algorithm to speed up the execution of the authentication process, especially at the key management level, message signing, and verification of this signature.

4.3 BACKGROUND

This section introduces the VANET communication system architecture and discusses the VANET's safety requirements.

4.3.1 System Architecture of VANET

The vehicular ad hoc network is a highly mobile ad hoc network integrated by ITS in order to improve traffic efficiency, minimize traffic congestion, avoid accidents, and make easy access to news, information, and entertainment while driving. However, recent research in vehicular networks is exploring all aspects of communication. We distinguish three modes of communication in a VANET, which are vehicle-to-vehicle communication (V2V), vehicle-to-infrastructure communication (V2I) and infrastructure-to-infrastructure communication (I2I) [15].

- Vehicle-to-vehicle communication (V2V) is based on a simple inter-vehicle communication that utilizes the OBUs (On-Board Units) installed on each vehicle to communicate with each other. This communication can be established without fixed infrastructure relays.
- Vehicle-to-infrastructure communication (V2I) allows better use of shared resources and increases the services provided (for example, data exchange, Internet access, remote diagnostics to repair a vehicle, etc.) thanks to access points deployed at the roadside. These access points are named RSUs (road side units) and are located in certain critical sections of the road, such as traffic lights or stop signs, in order to enhance road safety and traffic efficiency, and also to enjoy driving.
- Infrastructure to infrastructure communication (I2I) provides communication between RSUs or between RSU and base station. It increases the communication range and connects all vehicles in the network.

As the range of infrastructure is limited, vehicles can be used as relays to extend this distance and avoid the multiplication of base stations at each corner of the road. Therefore, the combination of the communication modes (V2V, V2I, and I2I) may achieve a very interesting and economical hybrid communication. In order to establish all these communications, a VANET consists of three main components (OBU, RSU, and CA) [16] required to provide communications in the network (Figure 4.1).

- The *On-Board Unit* (OBU) is a sensor mounted in vehicles. This device provides important information to vehicle control units for automatic driving assistance. This on-board unit is used to exchange information with other OBUs or with road side units (RSUs). It includes a set of high-tech hardware and software components such as GPS,

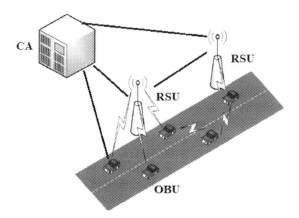

Figure 4.1 Network model.

radar, cameras, various sensors, and others. Its role is to ensure the location, reception, calculation, storage, and sending data over the network.

- The *Road Side Unit* (RSU) is an infrastructure placed along roads. Its main role is to inform nearby vehicles by broadcasting traffic conditions, weather, or specific road conditions (maximum speed, overtaking, etc.). They can also play the role of a base station by relaying information sent by vehicles.
- The *Certification Authority* (CA) represents the trusted authority in the VANET. It plays the role of a server which ensures the security of the various services such as the issuing of certificates, keys distribution, and the storage of certain data. Moreover, the CA maintains secure management of the VANET by verifying the vehicle authentication, the user ID, and the OBU ID in order to prevent any damage to any vehicle.

4.3.2 Security Requirements in VANET

All the constraints weighing on a system need to be understood to satisfy its security requirements. This step then makes it possible to quantify the security criteria. Regarding vehicular networks, the implementation of a security policy requires a good understanding and apprehension of the characteristics and constraints of these networks.

Since vehicular networks are regarded as a subclass of mobile ad hoc networks, they have the same characteristics, with a few exceptions. In contrast to traditional ad-hoc networks, VANETs have positive features such as the better-organized order of movement, in addition to the energy constraint which poses no problem in the vehicular network, because the nodes are connected to the vehicle's electrical system, thus reducing a certain number of known attacks in ad hoc networks aiming at consuming the energy sources of the nodes. In addition, the time and position in VANET are known and are used for the transmission of messages and also for their security. In addition, physical access to the network is restricted by an authority to drivers and the service center, which can reduce access to malicious entities [17].

Vehicular ad hoc networks have to deploy security requirements, including authentication, confidentiality, integrity, availability, non-repudiation, and privacy to cope with security threats. These requirements use one or more security mechanisms designed to detect or prevent a security attack [18–20].

4.3.2.1 Authentication

This requirement is one of the most critical system requirements. In VANET, it is crucial to consider more information about the transmission node, as well as its identity, properties, and geographical location. Therefore, the source node and the message broadcasting in the network must be authenticated. The main purpose of authentication is to control the authorization levels of the vehicle in the network. Two forms of authentication exist:

Entity authentication that helps to determine the type of equipment that is in communication. It may be another vehicle, an RSU, or other equipment.

ID authentication that enables the node to define the transmitters of a given packet by specifying a unique identifier for each vehicle. Using this technique prevents a vehicle from claiming to have multiple identifiers and pretending to be more than one vehicle, and thereby perpetrating an attack on the network.

Several attacks have been designed on authentication in VANETs, such as:

GPS spoofing: This consists of forcing the antenna, or module receiving the signal, to emit a false GPS signal in order to change the location of the vehicle. That has an effect on VANET routing, and the attacker can actually trick vehicles into believing they are at another location.

The Sybil attack: This represents the case in which the malicious node is capable of claiming multiple entities called Sybil nodes or false nodes. Thus, by posing as these different identities, the malicious node can more easily compromise the general functioning of the vehicle network.

Node impersonation: The attacker exploits the identity of another vehicle or RSU, phone, or laptop in other to impersonate a legitimate node and take advantage of their privileges.

4.3.2.2 Confidentiality

The objective of confidentiality is to make sure that the information is available only to authorized entities. It is, therefore, the protection of data transmitted against passive attacks. In order to provide this service, the mechanism used is cryptography, in particular encryption algorithms. Encryption can be achieved by symmetric or asymmetric cryptography. In the case of symmetric cryptography, the stakeholders must share a secret key. This key is used to encrypt and decrypt the message. One of the major problems with symmetric cryptography is the creation of a secure channel for distributing the private key. In the instance of asymmetric cryptography, each vehicle has a pair of keys: public and private keys. The private key is known only by the source vehicle, while the public key is shared with all entities in the network.

Confidentiality can be vulnerable to many attacks, like:

Monitoring Attack: In this attack, the malicious tracks the entire network, listening to all communication that occurs in vehicle-to-vehicle (V2V) mode or vehicle-to-infrastructure (V2I) mode. Once important data or information is received, the attacker forwards it to the interested person.

Traffic Analysis Attack: This is the method by which network traffic is analyzed to deduce data from communication patterns. The traffic information capturing from this attack are vehicle ID, location of the user, the traveling route of the user, and many others.

4.3.2.3 Integrity

This security requirement ensures that the transmitted messages are not willingly or inadvertently altered. The message receiver checks the received message. It ensures that the sender's identifier remains the same throughout the transaction and that the message received is the one that was sent. Integrity protects against destruction and alteration of the message during transmission. If a corrupted message is accepted, it is considered that there has been an integrity violation. To implement integrity, the system should prevent attacks against message corruption, since the content of the message must always be reliable. Certain security protocols use the electronic signature to make sure that the packet has not modified during the transaction. Thus, upon arrival of the message, the signature is verified to judge the integrity of the message.

Several attacks on integrity are distinct:

Bogus Information attack: The main goal of the Bogus Information attack is to forward inaccurate or bogus information in the network for the attacker's personal gain. The attacker can be an intruder (outsider) or a legitimate user (insider). For instance, the malicious node may send warning messages to the other nodes indicating that traffic conditions are heavy to make its travel easier on the road.

Timing attack: This attack depends on estimating and analyzing the time required to conduct some process in safety applications to detect secret information. In VANETs, times are a critical constraint. Thus, all data broadcasted between vehicles require arriving at a specific time. However, when the intruder (malicious vehicles) gets a packet, he adds an interval of time before sending it to the final destination, in order to create a delay.

4.3.2.4 Availability

The goal of availability is to guarantee permanent access to a service or resources. The network and applications must remain available even if the network crashes. This requirement not only protects the network, but it also allows it to withstand fault. Consequently, services must remain available until the failure has been solved. To attain all receivers of messages, an efficient routing protocol is required. Certain messages must remain contained at a definite time, so as not to mislead the vehicles if the information is no longer relevant.

Several applications require a rapid response from the sensors or the ad hoc network because the delay will make the message obsolete, which can cause dire situations. Accordingly, it is essential that resources are available at the right time and place. Availability may face different attacks, among them:

Denial of Service (DoS): It is the most common assault in traditional networks and in VANETs. The main goal of the DoS attack is to render the services or resources of the network unavailable for an indefinite time. This attack can be made through injecting several useless data into the network (flooding attack). The scope of such a kind of attack is generally broad, implying that this attack will involve several nodes. Therefore, a DoS attack can spread on a huge geographical zone through several nodes using multi-hop routing.

Black hole: The attacker in the black hole attack claims to possess an optimal path to the destination and suggests routing the message through this route, and then cheats the protocol for routing. This attack has the effect of destroying or misusing the packages captured without transmitting them to the destination.

4.3.2.5 Non-repudiation

This requirement prevents an entity from denying having participated in communication. It protects the system against the denial of a node, which indicates that it did not participate in communication when it did. Non-repudiation allows the receiver to prove that he received the message from a third party. Thus, for each message received, the sender can be clearly identified. The non-repudiation guarantees that the evidence of an incident or action is obtained, preserved, and made available to settle disagreements over whether or not an action took place. Consequently, non-repudiation relies on authentication. This helps the system to recognize the sender of the malicious packet. There are several threats in terms of non-repudiation, such as:

   Message Tampering: Attackers may alter messages transmitted in the V2V or V2I communications during this attack. Message tampering attacks are induced by authentication vulnerabilities.

   Message Suppression/Fabrication: Attackers are able to deactivate themselves from responding to any messages in two ways, either by disabling the connection between vehicles or by modifying the application.

4.3.2.6 Privacy Preserving

In the vehicular network, the driver's privacy is an essential aspect. Thus, the personal and private information of a user should not be accessible by other users. However, the driver's identity must not be inferred from the vehicle information (such as speed, geographical location, the destination, etc.) propagated in the network. Through this information, the position of the driver, vehicle movements, and tracking are more sensitive and must be taken into account with care. Like authentication, privacy also faces the GPS spoofing attack and many others such as:

   Location tracking: The aim of the malicious node is to violate the privacy of a user by trying to find confidential information about it. The attacker tracks the trajectory of the target vehicle in order to get its position or route followed by that driver.

   Tunneling: A GPS signal fails temporarily in tunnels, which leads to a temporary failure to location information. The malicious node exploits this weak point and injects false position information into the vehicle's database before receiving the original position update.

4.4 PRELIMINARIES

We present in this section the cryptographic algorithms that we use in our proposal. They are the Elliptic Curve Diffie-Hellman (ECDH) Protocol and the Edwards-curve Digital Signature (EdDSA) Algorithm.

4.4.1 Elliptic Curve Diffie-Hellman (ECDH)

Elliptic Curve Diffie-Hellman is a key agreement protocol that allows two entities to create a secret key that will be used for private key operations. The public key created by each entity is shared with the other. The methods on the elliptical curves are used to generate the secret key. We assume that Alice and Bob agree to a standard key cryptographic protocol for data exchange. It is considered that they had no contact and that the mode of communication available to them through the channel is only public. Both exchange public data or public cryptographic keys. Each of them has a private key used to generate the shared key called a public key. The two individuals agree on the same domain parameters. The steps of this protocol are as follows [21]:

1. Alice and Bob choose a common elliptical curve E on a prime field Fp. They also have a base point G $\in$ E (Fp) so that the subgroup generated by G has a greater group cardinality. This determines the strength of the method involved.
2. Alice chooses an integer a. It is a secret key that is not shared with anyone. This is Alice's private key. It then uses point multiplication and calculates the public key Ta = a.G and sends Ta to Bob.
3. Bob also selects an integer b which becomes his private key, calculates Tb = b.G by multiplying points and sends Tb to Alice.
4. Alice calculates a.Tb = a.b.G. This is done by multiplying points from Alice's secret key with Bob's shared key.
5. Bob multiplies points between his private key and Alice's public key and calculates b.Ta = a.b.G. The only data that a spy can obtain concern the elliptical curve E, the finite field Fp, and the points G, a.G, and b.G.

4.4.2 Edwards-curve Digital Signature (EdDSA)

The Edwards-curve Digital Signature Algorithm is a digital signature method based on the Twisted Edwards curves using a Schnorr signature variant. A signature protocol published in 2011 by Bernstein et al. [22], EdDSA was originally intended to be used with the Curve25519 elliptical curve, but it may very well be used with any other elliptical curve. A signature constructed from this protocol is a couple (R, s) and is generated from:

E – the parameters defining the elliptical curve used.
P – a point on the high-order curve.
n – the order of point P.

The *EdDSA key-pair* consists of the following:

- The *private key* is generated from a *random integer*, named *seed* (which has to use a similar bit length as a curve order). The seed is first hashed, then the last few bits, corresponding to the curve *cofactor*, are cleared; then the highest bit is cleared, and the second-highest bit is set.
- The public key *PubKey* is a point on the elliptic curve, calculated by the EC point multiplication: *PubKey* = *PrivKey* * P (the private key, multiplied by the generator point P for the curve).

Algorithms 4.1 and 4.2 respectively present the signature and the verification of signatures performed by this protocol.

ALGORITHM 4.1   EDDSA SIGNATURE

Inputs: the private key, a; the associated public key, Q; the message, m
Outputs: (R, s) the signature associated with m
    $H(a, m) \leftarrow hash(a, m)$
    $R \leftarrow H(a, m)P$
    $H(R, Q, m) \leftarrow hash(R, Q, m)$
    $s \leftarrow (H(a, m) + H(R, Q, m)a)[n]$
    return (R, s)

ALGORITHM 4.2   EDDSA VERIFICATION

Inputs: a message, m; the associated signature, (R, s); the public key associated with
      the private key which served to sign the message, Q.
Outputs: returns true if the signature is correct, false otherwise
    $H(R, Q, m) \leftarrow hash(R, Q, m)$
    $U \leftarrow 8sP$
    $V \leftarrow 8R + 8H(R, Q, m)Q$
    if $U \neq V$ then
        return False
    end if
    return True

We can show that the signature produced by Algorithm 4.1 will be validated by Algorithm 4.2.

$$
\begin{aligned}
R + H(R,Q,m)Q &= H(a,m)P + H(R,Q,m)aP \\
&= \big(H(a,m) + H(R,Q,m)a\big)P \\
&= sP
\end{aligned}
$$

4.5 PROPOSED SCHEME

This section presents the description of the phases of our secure hierarchical infrastructure scheme in the VANET.

4.5.1 Protocol Description

Our solution takes place in four phases:

- First is the main authentication phase between the different network entities which allows VANET entities to authenticate with the CA in order to have a public key certificate for use in communication. This phase is executed by using a symmetrical approach in order to secure the authentication packets exchanged. In addition, the EdDSA algorithm for the generation of the public/private key pair, the generation of the signature of a message, and its verification.
- The second phase is the authentication and communication phase between OBUs and RSUs, where OBUs authenticate for a second time with RSUs for the purpose of having two different keys. The first is a shared secret key for I2V or V2I communication, while the other key is used for V2V communication.

- The phase of message exchange between OBUs or V2V communication, using the K_{Gi} group key issued by RSUs to encrypt the messages transmitted, and the use of the private key to sign the hash of the message in order to ensure integrity and non-repudiation. In addition, the use of a certificate makes it possible to guarantee the authentication and the identity verification of the sender.
- The last phase is the revocation phase, of which the CA sends a list of revoked certificates to the OBUs. A certificate is revoked in the event that a vehicle declares the theft or loss of its private key or in the event that an RSU suspects the behavior of a vehicle. The vehicle receiving the list of revoked certificates can check the validity of a sender's certificate by searching for it in its list. In our solution, we proposed to share the main list of revoked certificates to sub-lists according to the type (Professional, Private, Personal, etc.) of the vehicle, which is a field in the certificate, in order to reduce the response time. When a vehicle searches for a certificate whether it is revoked or not, instead of searching the entire main list, it searches only in the list corresponding to the type of vehicle.

The notations used in our proposal are illustrated in Table 4.1.

4.5.2 Registration Phase

In this work, we consider the same registration and authentication process to be used for mutual authentication between CA→RSUs and CA→OBU. The vehicle to authenticate and to have a public key certificate must execute a series of steps:

Table 4.1 The different notations used in the proposed solution.

Notation	Description
$C_{RSUi}, C_{OBUi}, C_{CA}$	RSU_i cookie, OBU_i cookie, CA cookie
KP_{OBUi}	The OBU_i private key
$KPUB_{OBUi}$	The OBU_i public key
Enk et Dek	Encryption and decryption algorithms, respectively
Hash	The hash calculated after using the hash function defined in SA
K_{OBUi}	The OBU_i's secret key
K_{CA}	TA's secret key
IDr_{OBUi}	The real OBU_i identifier
ID_{OBUi}	OBU_i pseudo identifier
Sig_{OBUi}	The digital signature of the OBU_i
H	The clock associated with the message to determine its freshness; it is the moment when the message msg is sent
$Cert_{OBUi}$	Vehicle certificate issued by CA
$Type_{OBUi}$	The type of vehicle (Professional, Public, Personal, etc.)
K_{RSUi}	The RSU_i public key
Texp	The lifetime of the certificate.
ID_{CA}	The identifier of the certification authority
Sig_{CA}	The certificate signature produced by the CA private key
K_{Gi}	The group key of vehicles belonging to the same RSU_i, used for V2V communications
K_s	The session key issued by the RSU

- *Step 1*: $OBU_i \rightarrow CA$: C_{OBUi}, SA_{OBUi}.

  The OBU_i sends a message containing a cookie C_{OBUi} used to confirm that the OBUi is communicating with the CA and SA_{OBUi} presented in the SA block that represents the list of algorithms cryptographic supported by OBUi.

- *Step 2*: $CA \rightarrow OBU_i$: C_{CA}, SA_{TA}, ID_{OBUi}.

  After the reception of the first message, the CA responds with a message similar to the first message, which contains a C_{CA} cookie with the same purpose as C_{OBUi}, SA_{CA} as the type of cryptographic algorithm chosen and used in encrypted exchanges, and ID_{OBUi} as a session identifier.

- *Step 3*: $OBU_i \rightarrow CA$: C_{CA}, Nonce, $\{I_{OBUi}.P\}$, ID_{OBUi}.

  The OBU_i receiving the message and checks the validity of the C_{OBUi} cookie. The OBU_i and CA agree together and publicly on an elliptical curve E (a, b, K), they choose a finite field K in (Z/pZ) and an elliptical curve, and they choose together, and always publicly, a point P located on the curve. Then the OBU_i chooses an integer I_{OBUi} and sends a message that contains a header C_{CA}, which is the same in the previous message, the point of the elliptical curve $I_{OBUi}P$ and a nonce as a random for the design of the keys, and ID_{OBUi} as its session identifier.

- *Step 4*: $CA \rightarrow OBU_i$: C_{OBUi}, Nonce, $\{I_{CA}.P\}$.

  After reception of parameters and verification of the validity of C_{CA}, the CA generates an integer K_{CA} and returns a message which contains $I_{CA}P$. At this point, the OBU_i and CA can calculate their secret keys K_{OBUi} and K_{CA}, respectively.

$$K_{OBUi} = I_{OBUi}\left(I_{CA}P\right) = I_{CA}\left(I_{OBUi}P\right) = K_{CA} = \left(I_{OBUi}I_{CA}\right)P$$

- *Step 5*: $OBU_i \rightarrow CA$: Enk_{KOBUi} (C_{OBUi}, IDr_{OBUi}, ID_{OBUi}, Hash{S1, S2, S3, S4}).

  After establishing the secret key, the OBU_i sends a message containing a header C_{OBUi}, a real identifier IDr_{OBUi}, and the session identifier ID_{OBUi}, plus a hash of the four previous steps using the hash function defined in SA (SHA-256) and all encrypted with the secret key (AES-128).

- *Step 6*: $CA \rightarrow OBU_i$: Enk_{KCA} (ID_{OBUi}, Hash{S1, S2, S3, S4}, $KPUB_{CA}$).

  Upon receiving this message and after decrypting it with the CA's secret key, the CA retrieves the IDr_{OBUi} in order to generate a certificate and returns the message with a hash of the four previous messages.

  After checking the fingerprints (the hash) sent by the two entities, these fingerprints must be the same in order to validate and confirm the agreement on the keys and the algorithm to be negotiated, by carrying out an integrity check and authenticating the messages traded. The CA public key is used to verify the signature of the latter in the certificate issued to the OBU_i.

- *Step 7*: The OBU_i requests a certificate from CA in order to communicate with the other entities in the network.

- *Step 8*: $CA \rightarrow OBU_i$: Enk_{KCA}($KPUB_{OBUi}$, KP_{OBUi}, SigOBUi, t, $Cert_{OBUi}$).

  Upon receiving the message of the request, the CA checks the ID_{OBUi}, executes the EdDSA algorithm to generate the public/private key pair and the signature for the OBU_i, and sends a message encrypted with its secret key which contains the pair of keys associated with the vehicle, the signature of the vehicle, and the certificate.

- The OBU_i decrypts the message, extracts the key pair, the signature, and the certificate, and sends a Finished message to the CA indicating the end of the authentication process.

- Upon receipt the previous message, CA sends back a Finished message to confirm the end of the process.

Figure 4.2 shows the different messages exchanged during the authentication phase.

The CA records in its database the information relating to the OBUi <ID_{OBUi}, IDr_{OBUi}, KP_{OBUi}, $KPUB_{OBUi}$> to use in the event that a tracking has been launched against a vehicle in bad behavior, for example, in order to obtain the true vehicle identity.

The format of a certificate:

$$Cert_{OBUi} : \{ID_{OBUi}, KPUB_{OBUi}, Type_{OBUi}, T_{exp}, ID_{CA}, Sig_{CA}\}$$

4.5.3 Authentication and Communication between OBU$_i$ and RSU$_i$

A vehicle with a certificate issued by the CA can communicate with the other network RSUs, and this can be done after mutual authentication between the OBU$_i$ and the RSU$_i$ (Figure 4.3).

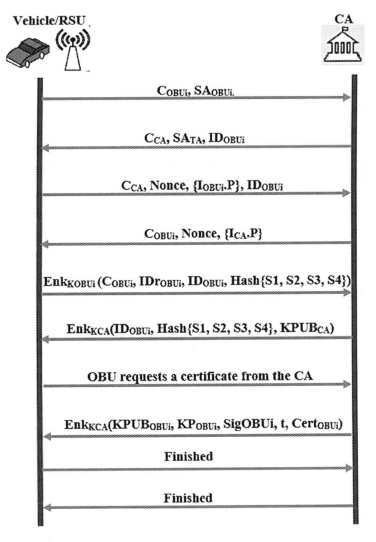

Figure 4.2 Registration phase.

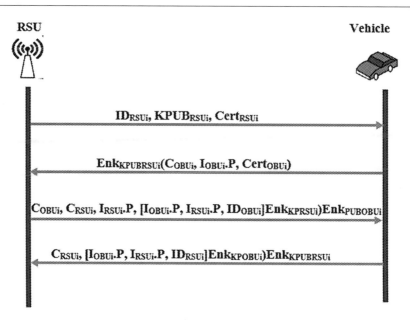

Figure 4.3 Authentication and communication between OBU$_i$ and RSU$_i$.

- *Step 1*: RSU$_i$ → OBU$_i$: (ID$_{RSUi}$, KPUB$_{RSUi}$, Cert$_{RSUi}$)

  The RSU$_i$ periodically broadcasts a message that includes its ID$_{RSUi}$ identity, its K$_{RSUi}$ public key, and its CertRSUi certificate issued by the CA. This certificate allows the OBU$_i$ to verify the validity of the RSU$_i$ public key. By intercepting this message, a OBU$_i$ retrieves the public key from the RSU$_i$ and sends a series of authentication and secret key generation messages following the ECDH algorithm.

- *Step 2*: OBU$_i$ → RSU$_i$: Enk$_{KPUBRSUi}$(C$_{OBUi}$, I$_{OBUi}$.P, Cert$_{OBUi}$).

  The OBU$_i$ sends a first message containing a C$_{OBUi}$ cookie to fight against DoS attacks and I$_{OBUi}$P (point of the elliptical curve generated using ECDH algorithm); thus, its Cert$_{OBUi}$ certificate issued by the CA.

- *Step 3*: RSU$_i$ → OBU$_i$: (C$_{OBUi}$, C$_{RSUi}$, I$_{RSUi}$.P, [I$_{OBUi}$.P, I$_{RSUi}$.P, ID$_{OBUi}$]Enk$_{KPRSUi}$)Enk$_{PUBOBUi}$.

  The RSU$_i$ when receiving the first message from OBU$_i$, decrypts (C$_{OBUi}$, I$_{OBUi}$.P, Cert$_{OBUi}$)KPUB$_{RSUi}$ using its private key KP$_{RSUi}$ and checks the validity of the certificate of OBU$_i$ Cert$_{OBUi}$. The RSU$_i$ then sends a message which contains its signature, the C$_{OBUi}$ vehicle cookie, and its C$_{RSUi}$ cookie, where the ID$_{OBUi}$ is a pseudo identifier used to protect user privacy.

- *Step 4*: OBU$_i$ → RSU$_i$: (C$_{RSUi}$, [I$_{OBUi}$.P, I$_{RSUi}$.P, ID$_{RSUi}$]Enk$_{KPOBUi}$)Enk$_{KPUBRSUi}$.

  The OBU$_i$ checks the signature [I$_{OBUi}$.P, I$_{RSUi}$.P, ID$_{OBUi}$]KP$_{RSUi}$ using the public key of the RSU$_i$ and checks the C$_{OBUi}$. If the cookies are different, it can directly interrupt the communication because it perceives that it is exchanged with a malicious entity. Otherwise, it continues the exchange by sending his signature [I$_{OBUi}$.P, I$_{RSUi}$.P, ID$_{RSUi}$] KP$_{Vi}$ and the C$_{RSUi}$. Similarly for the RSU$_i$, when it receives the message, it first checks the C$_{RSUi}$ cookie if it is valid, and it decrypts the signature.

At this time, the OBU$_i$ and the RSU$_i$ can calculate the symmetric and shared secret key K$_S$, used for communications between an OBU$_i$ and an RSU$_i$. In addition to this key, the RSU$_i$ generates another K$_{Gi}$ key for vehicles after successful authentication. This key will allow vehicles to communicate with each other in the same coverage area of the RSU$_i$.

The RSU_i communicates with neighboring RSUs to obtain their group keys (V2V communication key). At the same time as the delivery of the K_{Gi} key to the vehicles, it sends the keys of the neighboring RSUs obtained as well as their identifiers (a set of keys). Consequently, the OBU_i will have a list of group keys from neighboring RSUs in addition to that of its RSU_i. When receiving a message from an OBU_j, the OBU_i retrieves the identifier of the RSU_j in order to determine the corresponding key for decryption.

The use of two different keys for both I2V and V2V communication ensures confidentiality especially at the level of I2V communication, because the use of a group key causes certain drawbacks. For example, an OBU_i requesting service with an RSU_i encrypts its request with the group key, and in the case that an OBU_j receives this information, it can decrypt the message and know its content, which violates the confidentiality of the message. However, with the use of the secret key, the OBU_i can send a request, and only the RSU can decrypt it using its key shared with the OBU_i, thus preserving the concept of confidentiality.

4.5.4 Communication between Vehicles

To send messages to other vehicles, the OBU_i encrypts the message with the K_{Gi} key delivered by its RSU_i, signs the message content using the EdDSA algorithm with its private key, and sends its certificate with the adding of a parking meter (Clock) for the message freshness (Figure 4.4).

$$OBU_i \rightarrow OBU_j : (ID_{OBUi}, Cert_{OBUi}, Data(Msg, h), Sig_{OBUi}(Hash(Data))Enk_{KGi}, ID_{RSUi}$$

When receiving a message, the OBU_j checks the RSU_i identifier, retrieves the corresponding key and decrypts the message with the K_{Gi} key corresponding to the RSU_i identifier. It verifies the sender's signature using the EdDSA algorithm and the time interval between the current moment and the moment when the OBU_i sent the message. This interval must not exceed a predefined threshold (e.g. 300 ms). In addition, it checks the validity of the certificate for the current date. If the certificate is not valid, it ignores the message; otherwise, the message is accepted. The sender's current position can be added to the sent message in order to avoid replay attacks, where message receivers including the RSU can compare between the sender's current position and the position to which he sent the message in order to deduce whether this entity is malicious or not.

4.5.5 Revocation

A vehicle whose behavior is detected to be abnormal – for example, misconduct or the sending of false safety messages, etc. – is marked malicious. The RSUs are responsible for this task and inform the CA to die whether or not their certificate is revoked. In addition to its behavior, a vehicle may lose its private key, or its key may be stolen. So in these cases, the vehicle certificate must be revoked and registered in the Certificate Revocation List (CRL) revoked certificate list in order to inform the RSUs and vehicles belonging to the VANET. This is illustrated in Figure 4.5.

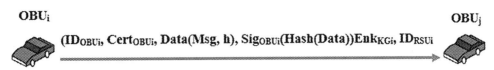

OBU$_i$ **OBU$_j$**

$(ID_{OBUi}, Cert_{OBUi}, Data(Msg, h), Sig_{OBUi}(Hash(Data))Enk_{KGi}, ID_{RSUi}$

Figure 4.4 Communication between vehicles.

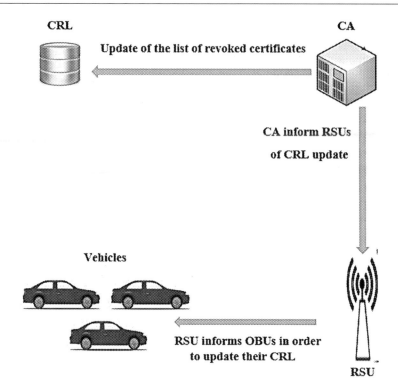

Figure 4.5 The different steps of the revocation phase.

1. The CA sends the revoked certificate for all RSUs.
2. The RSUs send the revoked certificate to the OBU$_i$.
3. The OBU$_i$ updates its CRL following the previous request.

In this phase, we are proposing a new strategy for managing CRLs, in order to improve the response time to search for the validity of a certificate. We will proceed by dividing the main CRL into several sub-CRLs based on the type cited in the public key certificate; for example, one under CRL for professional vehicles, one under CRL for private vehicles, etc.

When updating these sub-CRLs, the revoked certificate will register in a sub-list according to its type, and when an OBU$_i$ wants to verify the certificate, it will not need to browse the entire main list, it searches in the sub-list corresponding to the type of OBU$_i$ certificate.

4.6 SECURITY ANALYSIS

In this section, we analyze the security analysis and the formal verification to illustrate that our model can reach the security objectives and requirements.

4.6.1 Analysis of Security Requirements

- *Mutual authentication between the CA and the OBUs and between the CA and the RSUs*: It is ensured by the authentication process where the OBU authenticates with the CA (and the CA authenticates with the OBU) and ensures that the latter is the entity it claims by using cookies, in order to issue a public key certificate for the OBU.

- *Mutual authentication between RSUs and OBUs*: It is carried out in the second phase of the solution, where OBUs and RSUs exchange a series of messages in order to authenticate each other and obtain a secret key. This secret sharing is carried out using the ECDH algorithm.

  In addition to this key, the RSU generates a group key K_{Gi} for all the OBUs in its zone, which is used for V2V communication.

- *Negotiating and choosing a cryptographic policy*: It allows an algorithm to be easily excluded in the event of vulnerability, in order to replace the cryptographic policy used by a new policy supported by the two entities. To remedy the vulnerability of an algorithm, we can define a duration during which a user can authenticate, and in case the defined threshold is exceeded, we will change the cryptographic policy used.

- *Confidentiality of authentication packets*: Our solution ensures the confidentiality of authentication packets, by encrypting them using the secret key shared between the OBUs and CA.

- *Protection against replay attacks*: The use of nonce as random events in order to ward off replay attacks, where the malicious replaying entity cannot calculate shared secrets.

- *Protection against denial of service (DoS) attacks*: The use of cookies allows our solutions to remedy DoS attacks. Malicious OBUs generally start by spoofing the addresses of other OBUs, then massively sending authentication requests in order to exhaust CA resources, or using the latter as an amplifier or attack relay to OBUs from which they initially usurped their address. Therefore, the exchange of cookies during the authentication phase is essential, so that the CA server expends its resources only if the OBU returns the CA cookie, to confirm that it is in use. This limits exchange with the alleged entity, which limits this type of attack.

- *Confidentiality of the exchanged packets*: This is ensured by encrypting the exchanges with the secret key K_S in V2I or I2V communications and with the group key K_{Gi} in V2V communications.

- *Non-repudiation*: The signature makes it possible to verify the identity of the sender of this fact. Each message exchanged in the VANET must be signed with the private key of the sender. This key is generated by the CA and OBUs after successful authentication, and the signature is verified through the sender's public key and the use of the EdDSA algorithm. Signing allows the receiver of a message to authenticate the identity of the sender and to ensure non-repudiation.

- *Privacy preservation*: The use of pseudo identifiers or real vehicle identifiers allows users to preserve their private lives by exchanging messages anonymously. These pseudo identifiers are issued by the CA during the authentication process.

- *Availability*: DoS attacks aim to exhaust the resources of a server and make the server inaccessible. The use of cookies makes it possible to counter these types of attacks and therefore to ensure server availability.

- *Integrity*: The hash functions ensure data integrity. In our solution, the hash of the four messages sent in Steps 5 and 6 makes it possible to verify the integrity of these exchanges.

4.6.2 Formal Verification

We used the AVISPA (Automatic Validation of Internet Security Protocols and Applications) [23] tool to provide a formal modular and expressive language to specify the protocols and their security properties. It integrates a different back end which uses a variety of automatic machine analysis techniques. The AVISPA tool uses the Dolev-Yao intrusion model, where the attacker can spy on all transmitted messages, usurp the identity of a legitimate entity (attack by identity theft), and modify or inject messages, but he considers that the cryptography is

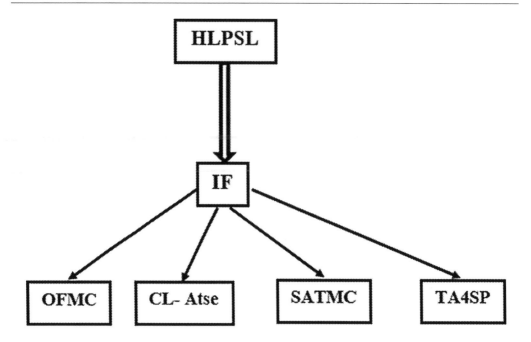

Figure 4.6 AVISPA architecture.

perfect, i.e. the attacker cannot break the cryptography. The AVISPA framework is shown in Figure 4.6. The first step in using the tool is to present the analyzed protocol in a special language called HLPSL (High Level Protocol Specification Language). The HLPSL specification of the protocol is translated into the lower-level language called IF (Intermediate Format). This translation is done by a translator called HLPSL2IF. This step is completely transparent to the user.

The IF specification of the protocol is used as an input to the four different back ends: On-the-fly Model-Checker (OFMC), CL-based Attack Searcher (CL-AtSe), SAT-based Model-Checker (SATMC) and Tree-Automata-based Protocol Analyzer (TA4SP). These back ends perform the analysis and formatting of the output that contains the results. These results indicate whether there are problems in the protocol or not.

The primary goal of our proposed scheme is to verify that it can provide a reliable key exchange between the entities of the VANET in order to secure the registration, authentication, and data transfer phases by using back-end servers. In Figures 4.7 through 4.10, we present respectively the roles of vehicle, CA, RSU, and the authentication goals.

After running this specification with OFMC and CLAtSe back ends, we can conclude that the proposed scheme can accomplish our goal and can resist those malicious attacks, such as replay attacks, secrecy attacks, and DoS attacks, under the test of AVISPA. The outputs of the model checking results are shown in Figures 4.11 and 4.12.

4.7 PERFORMANCE EVALUATION

This section evaluates the performances of our authentication protocol with other existing protocols according to security and computation overhead.

```
role vi (Vi, CA, RSU : agent, K: symmetric_key,
H : hash_func,
SND, RCV: channel (dy))
played_by Vi
def=
local
State: nat,
Cv, SAv, Cca, SAca, IDv, IDrv, Nonce, Certv, Sigv, KPv, KPca, KPrsu,
Crsu, IDrsu, Certrsu: text,
Finish : message,
KPUBv, KPUBca, KPUBrsu: public_key
const
      r1, r2, r3 : protocol_id

init State := 1
transition
1. State = 1
/\SND (start) =|>
State' := 2
/\Cv' := new()
.
.
.
2.  State = 2
/\SND (Cca', Nonce', IDv') =|>
State' := 3
/\Cv' := new()
/\RCV (Cv', Nonce')
3. State = 3
/\SND ((Cv'. IDv'. IDrv'. H(Cv', SAv',Cca', SAca', IDv',Cca',
Nonce', IDv', Cv', Nonce')). K')
 =|>
.
.
.
State' := 5
/\SND(Finish)
/\RCV(Finish)
5. State = 5
/\RCV (IDrsu', KPUBrsu', Certrsu') =|>
.
.
/\SND ((Crsu'. IDrsu').KPv')
end role
```

Figure 4.7 Role of vehicle.

4.7.1 Comparison of Security Performance

We have compared the security performance of existing authentication protocols with our protocol. As shown in Table 4.2, the proposal can provide the most comprehensive security performance and check the security level.

4.7.2 Computation Overhead

This part evaluates the computation cost required by related protocols and our scheme. In this context we used the rational arithmetic C/C++ library (MIRACL) [24] installed on a computer with a 3.2G HZ CPU and 8G of memory in order to calculate the execution time of such single operations used in existing and proposed protocols [25]. Table 4.3 illustrates the operations and their computation overheads.

The overheads are the sum of the time consumed on both the vehicle side and the RSU side. Table 4.4 indicates the operation numbers of the computation overheads for the existing

```
role ca (CA, Vi, RSU: agent, K: symmetric_key,
H : hash_func,
SND, RCV: channel (dy))
played_by CA
def=
local
State: nat,
Cv, SAv, Cca, SAca, IDv, IDrv, Nonce, Certv, Sigv, KPv, KPca: text,
Finish : message,
KPUBv, KPUBca: public_key
const
     r1, r2, r3 : protocol_id

init State := 0
transition
1. State = 0
/\RCV (start) =|>
State' := 1
.
.
2.  State = 1
/\RCV (Cca', Nonce', IDv') =|>
State' := 2
/\Cv' := new()
/\SND (Cv', Nonce')
3. State = 2
.
.
/\Certv' := new()
/\SND ((IDv'. KPUBca'. H(Cv', SAv',Cca', SAca', IDv',Cca', Nonce',
IDv', Cv', Nonce')). K')
4.  State = 3
.
.
/\SND(Finish)
end role
```

Figure 4.8 Role of CA.

```
role rsu (Vi, CA, RSU : agent, K: symmetric_key,
H : hash_func,
SND, RCV: channel (dy))
played_by Vi
def=
local
State: nat,
Cv, IDv, Certv, KPv, KPrsu, Crsu, IDrsu, Certrsu: text,
KPUBv, KPUBrsu: public_key
const
     r1, r2, r3 : protocol_id

init State := 5
transition
1. State = 1
.
.
/\RCV ((Crsu'. IDrsu').KPv')
end role
```

Figure 4.9 Role of RSU.

```
goal

authentication_on r1
authentication_on r2
authentication_on r3

end goal
```

Figure 4.10 Analysis goals of our scheme.

```
% OFMC
% Version of 2006/02/13
SUMMARY
  SAFE
DETAILS
  BOUNDED_NUMBER_OF_SESSIONS
PROTOCOL
  /home/span/span/testsuite/results/proposed-final.if
GOAL
  as_specified
BACKEND
  OFMC
COMMENTS
STATISTICS
  parseTime: 0.00s
  searchTime: 0.47s
  visitedNodes: 131 nodes
  depth: 10 plies
```

Figure 4.11 Results reported by the OFMC back end.

protocols [12–14] and the proposed protocol. We demonstrate the computation cost for OBU and RSU in Figure 4.13. It is observed that the proposed protocol is faster than the protocols adopted in [12–14]; this is due to the choice of lightweight operations to make the mutual authentication.

4.8 CONCLUSION

The security of vehicular networks is a prerequisite for their deployment. Given that the fact that these networks are a category of wireless networks and the importance of the information exchanged which can endanger the lives of users, the security of these networks does not stop to receive particular interest from research communities in academic and industrial circles. Among the problems that have arisen in these networks is the problem of privacy preservation of users. Our solution allows vehicles (OBUs) and RSUs to authenticate with the certification authority so that they are legal entities in the network. This authentication faces attacks such as denial of service attacks and replay attacks using cookies and nonce, respectively. In addition, double authentication between OBUs and RSUs makes it possible to minimize access to malicious entities in the network. Also, the use of a symmetric key during V2I and V2V communication makes it possible to reduce the calculation time and the verification time compared to other protocols using the asymmetric approach. The use

```
SUMMARY
  SAFE

DETAILS
  BOUNDED_NUMBER_OF_SESSIONS
  TYPED_MODEL

PROTOCOL
  /home/span/span/testsuite/results/proposed-final.if

GOAL
  As Specified

BACKEND
  CL-AtSe

STATISTICS

  Analysed   : 1 states
  Reachable  : 1 states
  Translation: 0.09 seconds
  Computation: 0.00 seconds
```

Figure 4.12 Results reported by the CL-AtSe back end.

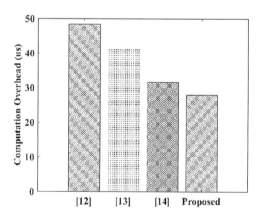

Figure 4.13 Computation cost of different schemes.

of the EdDSA algorithm makes it possible to speed up the execution of the authentication process, especially at the level of key management, message signature, and verification of this signature. On the other hand, the creation of sub-lists of revoked certificates based on vehicle type makes it possible to minimize the response time by looking for a certificate whether it is revoked or not. In summary, our solution ensures basic security requirements

Table 4.2 Security performance comparison.

Security features	[10]	[11]	[12]	[13]	[14]	Proposed Scheme
Mutual authentication	No	Yes	Yes	Yes	Yes	Yes
Provides message confidentiality	No	No	Yes	Yes	Yes	Yes
Provides message integrity	No	No	Yes	Yes	Yes	Yes
Providing non-repudiation	No	No	Yes	No	No	Yes
Privacy preservation	No	Yes	No	Yes	Yes	Yes
Resistance against the DoS attack	No	No	No	No	No	Yes
Resistance against the replay attack	No	Yes	No	Yes	Yes	Yes

Table 4.3 Computation overhead of a single operation.

Operations	Description	Time (milliseconds)
PM	Point Multiplication	2.258
BP	Bilinear Pairing	6.443
H	Hash (SHA-256)	0.021
EXP	Exponentiation in Bilinear Group	3.212
ENC	AES-128 Encryption	0.902
DEC	AES-128 Decryption	7.357
MM	Modular Multiplication	1.657
MP	Modular Square Root	2.942
MTP	Map-to-Point Hash Function	2.258
SING/VER	Signature/Verification EdDSA	3.21

Table 4.4 Comparison of computation overheads.

Protocols	Operation Numbers		Total (ms)
	Vehicle's side	RSU's side	
[12]	PM + 2MTP + 5BP	2BP + 4MTP	48.352
[13]	4PM + 5H + 2EXP	3BP + 5H + 2EXP	41.419
[14]	7PM	7PM	30.95
Our	2ENC+DEC+SING/VER+2PM	ENC+2DEC+SING/VER +PM	27.971

such as integrity, confidentiality, non-repudiation, and availability, as well as the preservation of users' privacy by using the real identifier.

REFERENCES

[1] Chavhan, S., Gupta, D., Chandana, B. N., Khanna, A., and Rodrigues, J. J. (2019). IoT-based Context Aware Intelligent Public Transport System in a metropolitan area. *IEEE Internet of Things Journal*, 7(7), 6023–6034.

[2] Zeadally, S., Hunt, R., Chen, Y. S., Irwin, A., and Hassan, A. (2012). Vehicular ad hoc networks (VANETS): Status, results, and challenges. *Telecommun. Syst.*, 50(4), 217–241.

[3] Eze, J., Zhang, S., Liu, E., and Eze, E. (2017, September). *Cognitive radio technology assisted vehicular ad-hoc networks (VANETs): Current status, challenges, and research trends*. In *2017 23rd International Conference on Automation and Computing (ICAC)* (pp. 1–6). IEEE.

[4] Sheikh, M. S., Liang, J., and Wang, W. (2019). A survey of security services, attacks, and applications for vehicular ad hoc networks (VANETs). *Sensors*, 19(16), 3589.

[5] Raya, M., Aziz, A., and Hubaux, J. P. (2006, September). *Efficient secure aggregation in VANETs*. In *Proceedings of the 3rd International Workshop on Vehicular Ad hoc Networks* (pp. 67–75).

[6] Lin, X., Sun, X., Ho, P. H., and Shen, X. (2007). GSIS: A secure and privacy-preserving protocol for vehicular communications. *IEEE Transactions on Vehicular Technology*, 56(6), 3442–3456.

[7] Bendouma, A. and Bensaber, B. A. (2017, May). *RSU authentication by aggregation in VANET using an interaction zone*. In *Proceedings of IEEE International Conference on Communications (ICC)* (pp. 1–6). IEEE.

[8] Vijayabharathi, V. and Malarchelvi, P. S. K. (2014, February). *Implementing HMAC in expedite message authentication protocol for VANET*. In *Proceedings of International Conference of Information Communication and Embedded Systems (ICICES)* (pp. 1–5). IEEE.

[9] Sakhreliya, S. C. and Pandya, N. H. (2014, December). *PKI-SC: Public key infrastructure using symmetric key cryptography for authentication in VANETs*. In *Proceedings of IEEE International Conference Computational Intelligence and Computing Research (ICCIC)* (pp. 1–6). IEEE.

[10] Das, A., Chowdary, D. R., and Rai, A. (2011). An efficient cross authentication protocol in VANET hierarchical model. *Int. J. Mob. Adhoc Netw.*, 1(1), 128–136.

[11] Li, J. S. and Liu, K. H. (2013). A lightweight identity authentication protocol for vehicular networks. *Telecommunication systems*, 53(4), 425–438.

[12] Han, M., Hua, L., and Ma, S. (2017). A self-authentication and deniable efficient group key agreement protocol for VANET. *KSII Transactions on Internet and Information Systems*, 11(7), 3678–3698.

[13] He, D., Wang, D., Xie, Q., and Chen, K. (2017). Anonymous handover authentication protocol for mobile wireless networks with conditional privacy preservation. *Science China Information Sciences*, 60(5), 052104.

[14] Li, X., Han, Y., Gao, J., and Niu, J. (2019). Secure hierarchical authentication protocol in VANET. *IET Information Security*, 14(1), 99–110.

[15] Goyal, A. K., Agarwal, G., and Tripathi, A. K. (2019). Network architectures, challenges, security attacks, research domains and research methodologies in VANET: A survey. *International Journal of Computer Network and Information Security*, 10(10), 37–44.

[16] Kaibalina, N. and Rizvi, A. E. M. (2018, October). *Security and privacy in VANETs*. In *Proceedings of IEEE 12th International Conference on Application of Information and Communication Technologies (AICT)*. IEEE.

[17] Elsadig, M. A. and Fadlalla, Y. A. (2016). VANETs security issues and challenges: A survey. *Indian Journal of Science and Technology*, 9(28), 1–8.

[18] Zhu, L., Zhang, Z., and Xu, C. (2017). *Security and privacy preservation in VANET*. In: *Secure and Privacy-Preserving Data Communication in Internet of Things*. *Springer Briefs in Electrical and Computer Engineering*, Springer, Singapore (pp. 53–76).

[19] Al Junaid, M. A. H., Syed, A. A., Warip, M. N. M., Azir, K. N. F. K., and Romli, N. H. (2018). *Classification of security attacks in VANET: A review of requirements and perspectives.* In *MATEC Web of Conferences*, (Vol. 150, 06038). EDP Sciences.

[20] Hasrouny, H., Samhat, A. E., Bassil, C., and Laouiti, A. (2017). VANet security challenges and solutions: A survey. *Vehicular Communications*, 7, 7–20.

[21] Fournaris, A. P., Zafeirakis, I., Koulamas, C., Sklavos, N., and Koufopavlou, O. (2015, May). *Designing efficient elliptic curve Diffie-Hellman accelerators for embedded systems.* In *IEEE International Symposium on Circuits and Systems (ISCAS)* (pp. 2025–2028), IEEE.

[22] Bernstein, D. J., Josefsson, S., Lange, T., Schwabe, P., and Yang, B. Y. (2015). EdDSA for more curves. *Cryptology ePrint Archive*, 2015.

[23] AVISPA Project, http://www.avispa-project.org/

[24] MIRACL Library, http://www.shmus.ie/index.php.

[25] Wang, P., Liu, Y., and Lv, S. (2019). An improved lightweight identity authentication protocol for VANET. *Journal of Internet Technology*, 20(5), 1491–1504.

Chapter 5

Simulation Tools for Opportunistic Networks

How to Set Up and Simulate the ONE Simulator

Md. Sharif Hossen

Comilla University, Cumilla, Bangladesh

CONTENTS

5.1 INTRODUCTION

Delay/Disruption-Tolerant Networks (DTNs) [1], also called Intermittently Connected Mobile Networks (ICMNs) [2], are mobile ad hoc networks where there is no consistent path from one node to another, which is essentially required for successful communication. These networks can be seen in the areas of satellite communication [3], tracking wildlife [4], military applications, vehicular networks, etc. [5, 6].

Since the connection from one node to another node is not known in mobile networks, the route is selected by choosing intermediate links, and hence it is a critical problem of DTN communication. These intermediate links follow the "store-carry-and-forward" approach in which nodes keep the information in their buffers and then carry them until finding a suitable node to forward [7]. Figure 5.1 illustrates the approach of the store-carry-and-forward approach, where source A sends a message copy to the neighboring node B which stores the copy into its buffer. Then, it carries the copy and forwards to the neighboring node C. Repeating this process, node C reaches the destination D and then finally sends it to D.

Routing for DTNs is classified into two ways: forwarding- and replication-based. This chapter represents an algorithmic and pictorial discussion and analysis of forwarding and

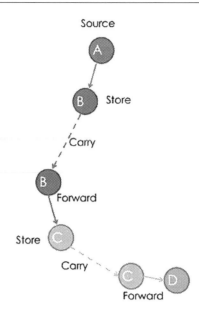

Figure 5.1 Store and forward strategy.

replication-based routing approaches, for example, first-contact, direct-delivery, Epidemic, and PRoPHET.

The chapter has been outlined as follows. Section 5.2 illustrates the routing algorithms with pictorial representation. Section 5.3 includes the discussion of simulation tools. Section 5.4 gives details of how to install and run the ONE simulator. Section 5.5 illustrates the default settings in the ONE simulator. The simulation configuration of ONE is mentioned in Section 5.6. Section 5.7 elaborates on the simulated results with performance evaluation. Section 5.8 provides the concluding remarks.

5.2 ROUTING PROTOCOLS AND MOBILITY MODELS

This section illustrates the algorithmic and pictorial discussion of DTN routing protocols which can be [8] either forwarding-based or replication-based. Forwarding routings do not occupy many network resources since only a copy remains in the network at the network lifetime [9,10]. Existing forwarding-based routings are direct-delivery [11] and first-contact [9,12]. In the first-contact routing, a single message is forwarded to the first available contact to meet the destination. As shown in Figure 5.2(a), at first say, source node S encounters node C, then C forwards the message directly to D. In direct-delivery, S itself delivers message copies only to the final recipient D as shown in Figure 5.2(b).

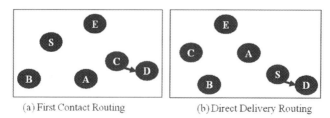

(a) First Contact Routing   (b) Direct Delivery Routing

Figure 5.2 Forwarding-based routing.

ALGORITHM (FIRST-CONTACT)

A node can forward only a single message copy.

Step 1: if first available contacts are found,
then a node forwards a single message copy to that contact and deletes its own copy form buffer,

else

a node keeps the copy in its own buffer and finds next available contact.

Step 2: Continue Step 1 until the node meets the destination.

ALGORITHM (DIRECT-DELIVERY)

Step 1: if Source node directly finds the destination node,
then Source forwards the copy to it and stops the communication.

else

source node moves and waits for the destination.

Step 2: Continue Step 1 until the node meets the destination.

ALGORITHM (EPIDEMIC)

Step 1: if there are no message copies in common,
Each node can forward message copies to its neighboring nodes.

else

no copies can be forwarded.

Step 2: Continue Step 1 until the node meets the destination.

ALGORITHM (PROPHET)

Each node can forward a message copy to another.

Step 1: if a connection's state changes and a new connection is created to a node, then the node finds its peer and collects all essages from this peer, and finds new copies of message that this routing does not hold.

Step 2: if a node has not already traversed another then,
the node can forward a message copy to another.

else

no message copies can be sent.

Step 3: Continue Step 1 and 2 until the node meets the destination.

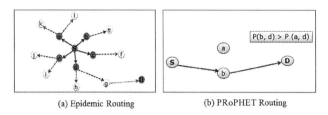

(a) Epidemic Routing   (b) PRoPHET Routing

Figure 5.3 Replication-based routing.

In replication-based routing, redundant copies of the same message remain in the network. Examples of these routings are Epidemic [12], PRoPHET [13], MaxProp [14], RAPID [15], Spray and Wait (SNW) [16], and Spray and Focus (SNF) [17]. In this chapter, only Epidemic and PRoPHET have been used for investigation.

Epidemic routing consumes many network resources because of its flooding nature, where a node can send more than one copy to its neighbors which do not have the same message. In Figure 5.3(a), S sends copies to its neighbor nodes; these nodes then forward copies to the next nodes assuming no copy in common to the nodes.

In PRoPHET routing, depending on the higher value of the probability of a node to contact the destination, the source can forward a message to a suitable node as shown in Figure 5.3(b).

In this investigation, the Shortest Path Map-Based Movement (SPMBM) mobility model has been used that uses a map-described environment to restrict node movement. This model uses the Dijkstra algorithm to find the shortest route to reach the destination. Although all the points on a map have equal probability, depending on users' movements or preferences, map data may contain some points of interest (POI) groups gathering people in different places, e.g., tourist attractions and supermarkets. This model shows good performance compared to Random Walk and Random Direction [18].

5.3 SIMULATION TOOLS

In this section, three types of simulation tools are briefly discussed.

5.3.1 ONE Simulator

Opportunistic Network Environment (ONE) is an event-driven Java-based simulator which is particularly used for analyzing DTN routings [11]. Real-time nodes' traces can be seen using a graphical user interface. Reports are generated in a text file. All the settings can be set using different file extensions internally and externally. Hence, the version of 1.5.1 RC2 of the simulator has been used [19]. Figure 5.4 shows the simulation scenario of the ONE simulator.

5.3.2 NS-2

A network simulator called NS-2 along with Linux is used as a simulation tool. It has an object tool command language shell (OTCL) as the user interface, which allows scripts written in TCL format to be executed. It is a discrete event simulator along with the object-oriented concepts and is used mostly by the ad hoc networking community and researchers to determine the performance of routings. It uses two languages: (1) C++, which runs in the back end, and (2) tool command language (TCL), which runs in the front end. Moreover, the components of NS-2 include NS, NAM (network animator), pre-processing, and post-analysis for the simulator, visualization of NS output, handwritten TCL, and trace analysis using AWK scripts, respectively. Figure 5.5 shows the scenario of black hole simulation [20, 21].

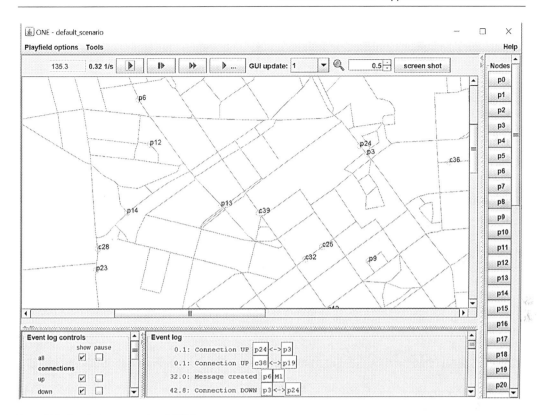

Figure 5.4 Simulation scenario of ONE.

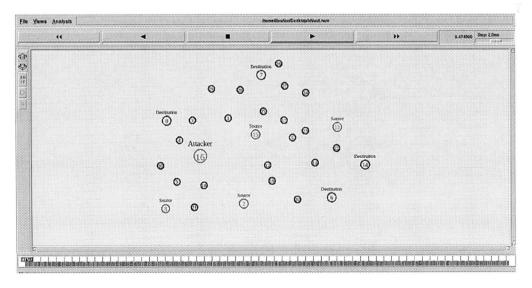

Figure 5.5 The NAM scenario for black hole simulation.

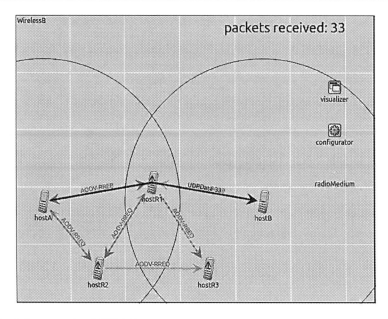

Figure 5.6 Simulation scenario of OMNeT++.

5.3.3 OMNeT++

OMNeT++ is an event-driven framework to build and implement the emulated scenario of mobile ad hoc networks. It is developed using the C++ language. Figure 5.6 shows the simulation scenario of OMNeT++.

5.4 HOW TO INSTALL AND RUN THE ONE SIMULATOR

This section will discuss how to install and run the ONE simulator. The simplest way to install the ONE simulator is to use Eclipse.

The following steps are needed to install the ONE simulator.

i. Download and install Java JDK
ii. Download ONE simulator
iii. Download and extract Eclipse and
iv. Integrate the ONE simulator in Eclipse

These four steps are discussed briefly with screenshots.

i. *Download and install Java JDK*
First, go to the following link or search on Google by typing "Java JDK":
https://www.oracle.com/java/technologies/javase-downloads.html
Then, download and install the JDK version as your computer's configuration.
ii. *Download ONE simulator*
Download the ONE simulator from any of the following two links.
http://akeranen.github.io/the-one/
https://www.netlab.tkk.fi/tutkimus/dtn/theone/

The file that you will download is a zip file. Consider naming the file "ONE.zip". As of this writing, the latest version release of ONE is 1.6.0. After extracting the file, a folder named "ONE" will be created that will be used later during the integration of the ONE simulator in Eclipse.

iii. *Download and extract Eclipse*

Go to the following link or search on Google by typing "eclipse for java developers". https://www.eclipse.org/downloads/packages/release/kepler/sr1/eclipse-ide-java-developers

You can download the eclipse according to your PC's operating systems.

iv. *Integrate the ONE simulator in Eclipse*

The step-by-step integration of the ONE simulator in Eclipse is illustrated as follows: First, create a Java project named "ONESim" in Eclipse as follows.

File → New → Java Project

This means that you click on the "File" tab, then click "New", and then "Java Project".

After clicking "Java Project", enter the project name as "ONESim". Click "Next" and then the "Finish" button. A Java project named "ONESim" will be created.

Now, ONE simulation will be integrated into Eclipse. Open the ONE folder that you have extracted. Copy the "src" folder in the extracted file. Paste and replace the "src" folder in the "ONESim" Java project.

It needs to add some jar files. Jar files are included in the "lib" subfolder of the "src" folder. To integrate the jar files, right-click on the "ONESim" project, then go to the following director (Figure 5.7).

Build Path → Configure Build Path

Then, a window will appear, and there, click on "ModulePath". After that, click on the button "Add JARs" on the right side and then a "JAR Selection" window will appear. Here, it is necessary to select the jar files under the "lib" subfolder of the "src" folder. Then, click the "OK" button. Finally, click the "Apply and Close" button. It is now integrated the ONE simulator completely. It is shown in Figure 5.8. It is necessary to download a "junit" jar file that will not be found into the "lib" folder. The link is as follows:

https://repo1.maven.org/maven2/junit/junit/4.12-beta-3/junit-4.12-beta-3.jar

To run the ONE simulator, right-click on the java project and choose "Run As" and then "Run Configurations".

Hence, a window will appear as shown in Figure 5.9. Here, select the main class as "DTNSim" in the Main tab. Then, go to the Arguments tab. Select the "Other" button under the "Working director" and click on the "Workspace" button. Finally, select the "src" folder,

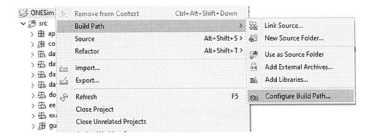

Figure 5.7 Directory of "Configure Build Path".

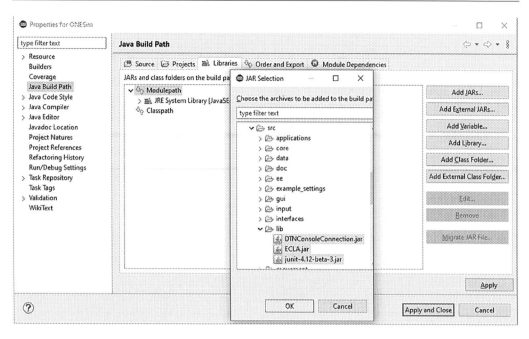

Figure 5.8 Adding jar files.

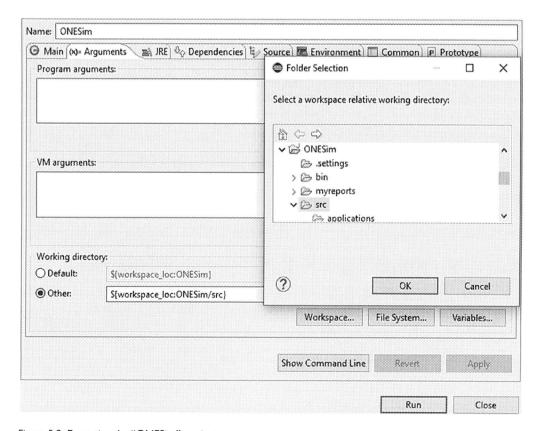

Figure 5.9 Executing the "ONESim" project.

click the "OK" button, and then the "Run" button. Now, the simulation scenario of ONE simulator, shown in Figure 5.4, will appear. This process of installing and running ONE simulator is shown at https://www.youtube.com/watch?v=pH8MKr0pjeE.

5.5 DEFAULT SETTINGS IN THE ONE SIMULATOR

There is a default setting file ("defaultsettings.txt") for the simulation where user can set up the configuration of ONE simulation settings. Figure 5.10 shows scenario settings where scenario name, update interval, and simulation time are given.

Nodes in the network can form a group which requires an interface to communicate within a range with different transmit speeds. Two types of interfaces, Bluetooth and High speed, can be used as a simple broadcast interface. The number of host groups can be described as "Scenario.nrofHostGroups=6".

If the user does not provide any information, a common setting for all groups will be applied. Here, a common group does not contain any group name; just have only "Group", as shown in Figure 5.11.

For example, the first three groups shown in Figure 5.12 are not assigned by the number of nodes, buffer space, interface, TTL, so they use the aforementioned common settings shown in Figure 5.11.

```
Scenario.name = default_scenario
Scenario.simulateConnections = true
Scenario.updateInterval = 0.1
Scenario.endTime = 43200
```

Figure 5.10 Scenario settings.

```
Group.movementModel = ShortestPathMapBasedMovement
Group.router = SprayAndFocusRouter
Group.bufferSize = 5M
Group.waitTime = 0, 120
Group.nrofInterfaces = 1
Group.interface1 = btInterface
Group.speed = 0.5, 1.5
Group.msgTtl = 300 minutes
Group.nrofHosts = 40
```

Figure 5.11 Common setting for all groups.

```
# group1 (pedestrians) specific settings
Group1.groupID = p

# group2 specific settings
Group2.groupID = c
# cars can drive only on roads
Group2.okMaps = 1
# 10-50 km/h
Group2.speed = 2.7, 13.9

# another group of pedestrians
Group3.groupID = w
```

Figure 5.12 First three groups in "defaultsettings.txt" file.

The remaining three groups consist of two nodes. Hence, the total number of nodes is $(40 \times 3) + (2 \times 3) = 126$. In the message creation parameters as given below, this information is needed in "Events1.host=0,125", as shown in Figure 5.13. Message generation rates are determined by updating the events' interval.

The number of reports and reports' names with the directory of the generated reports can be given as shown in Figure 5.14.

5.6 SIMULATION CONFIGURATION FOR THE ONE SIMULATOR

Simulation settings are specified in Table 5.1. Four parameters have been considered fixed for any change.

Message generation rate = 2 (msg/min)
Number of nodes = 80
Buffer size = 5 MB
Message TTL = 300 minutes

For example, for varying message generation rates 2, 3, 4, 5, and 6, three parameters are considered constant as mentioned earlier (number of nodes = 80, buffer size = 5 MB, message

```
# How many event generators
Events.nrof = 1
# Class of the first event generator
Events1.class = MessageEventGenerator
# Creation interval in seconds (one new message every 25 to 35 seconds)
Events1.interval = 25,35
# Message sizes (500kB - 1MB)
Events1.size = 500k,1M
# range of message source/destination addresses
Events1.hosts = 0,125
# Message ID prefix
Events1.prefix = M
```

Figure 5.13 Event generations in "defaultsettings.txt" file.

```
# how many reports to load
Report.nrofReports = 2
# length of the warm up period (simulated seconds)
Report.warmup = 0
# default directory of reports
Report.reportDir = reports/
# Report classes to load
Report.report1 = ContactTimesReport
Report.report2 = ConnectivityONEReport
```

Figure 5.14 Report settings.

Table 5.1 Simulation settings.

Parameters	Value
Simulation duration	12 hr
Message copies	2, 3, 4, 5, 6 (msg/min)
Nodes	20, 50, 80, 100, 120
Buffer size	5, 10, 15, 20, 30 (MB)
Message TTL	60, 120, 180, 240, 300 (min)
Movement model	SPMBM
Message size	500 KB to 1 MB
Forwarding-based routing	First-contact, direct-delivery
Replication-based routing	Epidemic, PRoPHET
Area	4500 × 3400 m

TTL = 300 minutes). Two types of groups have been considered, pedestrians (P) with speed of 0.5–1.5 m/s, and cars (C) with speed of 2.7–13.9 m/s. Bluetooth with speed 250 kbps and range 10 m has been used as an interface.

5.7 SIMULATED RESULTS AND DISCUSSION WITH ANALYSIS

In this section, the performance analysis of the aforementioned routings is discussed. Hence, the analysis is done on five performance metrics, namely delivery probability, average latency, overhead ratio, average hop count, and average buffer time. These performance metrics are obtained with varying four parameters: message generation rates, number of nodes, buffer size, and time-to-live.

After completing the simulation, the calculated delivery probability, average latency, overhead ratio, average hop count, and average buffer time are stored in the file "MessageStatsReport.txt". This file can be found in the following directory of the project "ONESim" of your computer's drive.

ONESim → src → reports → MessageStatsReport.txt

Figure 5.15 shows the generated message stats reports.

5.7.1 Delivery Probability

Delivery probability is calculated by dividing the number of messages received by the destination and the messages sent from the source. With increasing message copies, mobile nodes, and time-to-live, the direct-delivery routing shows better performance, Epidemic indicates lower results, as shown in Figures 5.16, 5.17, and 5.19, respectively. However, for varying the buffer sizes, first-contact exhibits good delivery, while Epidemic shows lower delivery as shown in Figure 5.18. In all cases, Epidemic represents lower delivery.

5.7.2 Average Latency

Average latency is calculated from the difference of time between messages produced by the source and the messages got to the destination. With the increase of message copies, direct-delivery routing shows lower latency when the number of message copies is 2, while Epidemic routing shows lower latency when the message copies are greater than 2, as shown

sim_time: 432000
created: 1461
started: 58268
relayed: 31006
aborted: 27261
dropped: 30948
removed: 0
delivered: 343
delivery_prob: 0.2348
response_prob: 0.0000
overhead_ratio: 89.3965
latency_avg: 4775.2942
latency_med: 3352.6000
hopcount_avg: 4.5190
hopcount_med: 4
buffertime_avg: 1407.0311

Figure 5.15 Message stats reports.

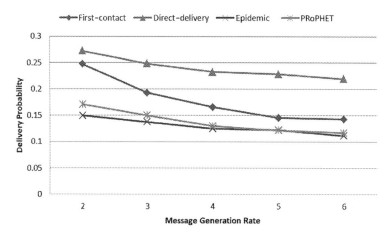

Figure 5.16 Delivery probability vs. message generation rate.

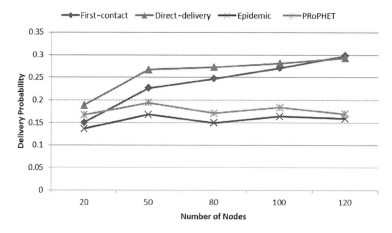

Figure 5.17 Delivery probability vs. mobile nodes.

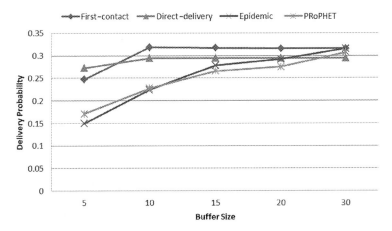

Figure 5.18 Delivery probability vs. buffer size.

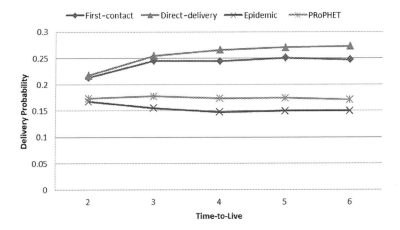

Figure 5.19 Delivery probability vs. TTL.

in Figure 5.20. Direct-delivery shows lower latency compared to the other considered routings when the mobile nodes are more than 50, whereas Epidemic indicates lower latency when mobile nodes are no more than 50, as shown in Figure 5.21. With the increase of buffer sizes, i.e., 5, 10, 15, 20, and 30 MB, direct-delivery routing shows lower average latency while first-contact shows higher latency, as shown in Figure 5.22. Direct-delivery shows the lowest latency when TTL = 6 minutes, while Epidemic indicates lower latency when TTL is less than 6 minutes, as shown in Figure 5.23.

5.7.3 Overhead Ratio

It determines how many copies of the same message are sent redundantly to reach a copy successfully to the destination. In the case of overhead ratio, direct-delivery shows zero overhead since the source node directly sends the message to the destination without sharing the message to any node and hence exhibits better performance for varying message copies per minute, the number of nodes, buffer size, and TTL as shown in Figures 5.24 through 5.27. Whereas Epidemic routing shows the higher overhead ratio compared to direct-delivery, first-contact, and PRoPHET routing protocols (Figures 5.24 to 5.27). Therefore, direct-delivery shows good performance in terms of overhead ratio.

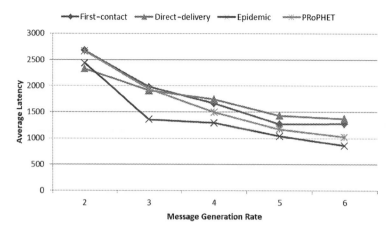

Figure 5.20 Average latency vs. message generation rate.

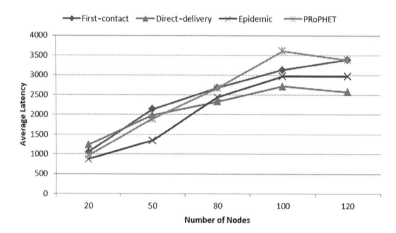

Figure 5.21 Average latency vs. mobile nodes.

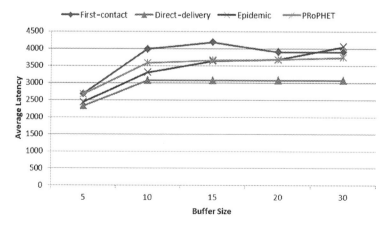

Figure 5.22 Average latency vs. buffer size.

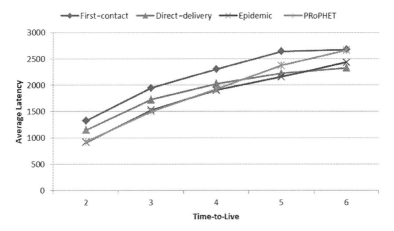

Figure 5.23 Average latency vs. TTL.

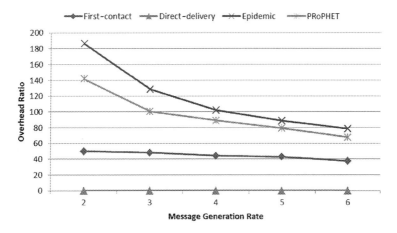

Figure 5.24 Overhead ratio vs. message generation rate.

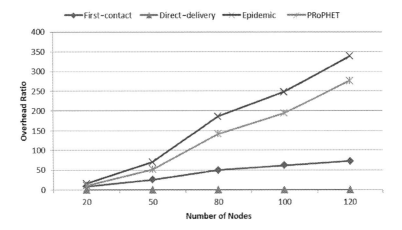

Figure 5.25 Overhead ratio vs. mobile nodes.

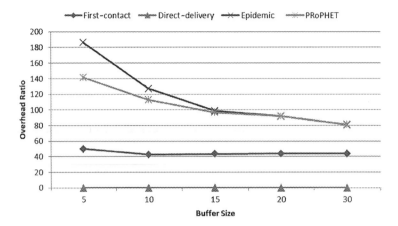

Figure 5.26 Overhead ratio vs. buffer size.

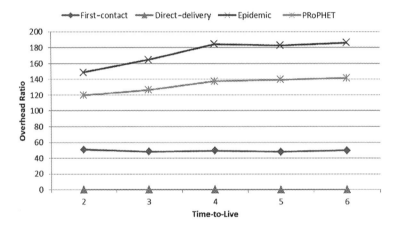

Figure 5.27 Overhead ratio vs. TTL.

5.7.4 Average Hop Count

Average hop count indicates the average number of intermediate nodes to build a successful connection. In the case of average hop count, direct-delivery shows the lowest value, and its value is 1 for varying all cases, i.e., message copies, mobile nodes, buffer size, and TTL as shown in Figures 5.28 to 5.31. First-contact routing shows high overhead ratio among the considered routing techniques (Figures 5.28 to 5.31).

5.7.5 Average Buffer Time

For varying message copies shown in Figure 5.32, the average buffer time of direct-delivery is decreasing but shows value among all routings. With the increase of mobile nodes, the buffer time of direct-delivery is increasing and takes a higher time than all routings as shown in Figure 5.33. As shown in Figure 5.34, when buffer size is greater than 10 MB, direct-delivery shows higher but constant (approximately 18000s) buffer time. With the increase of TTL, the buffer time of direct-delivery is increasing, as shown in Figure 5.35. In all cases, as shown in Figures 5.32 to 5.35, direct-delivery represents a higher average buffer time.

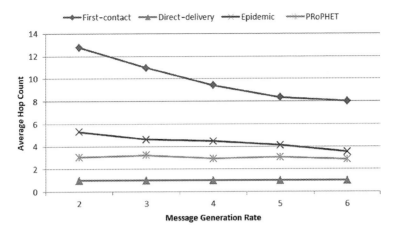

Figure 5.28 Average hop count vs. message generation rate.

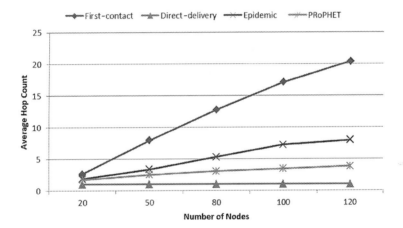

Figure 5.29 Average hop count vs. mobile nodes.

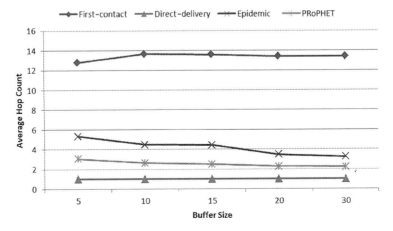

Figure 5.30 Average hop count vs. buffer size.

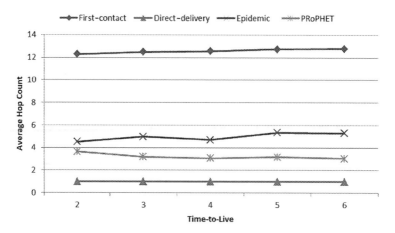

Figure 5.31 Average hop count vs. TTL.

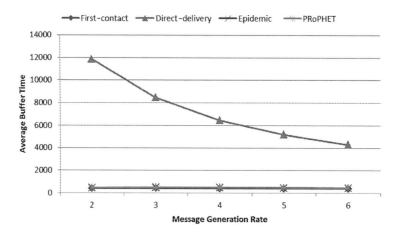

Figure 5.32 Average buffer time vs. message generation rate.

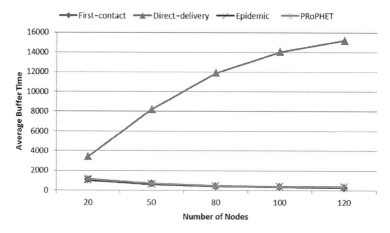

Figure 5.33 Average buffer time vs. mobile nodes.

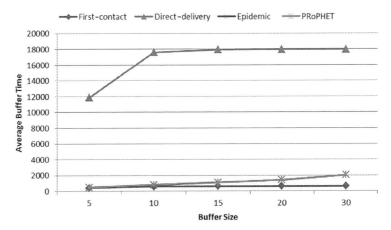

Figure 5.34 Average buffer time vs. buffer size.

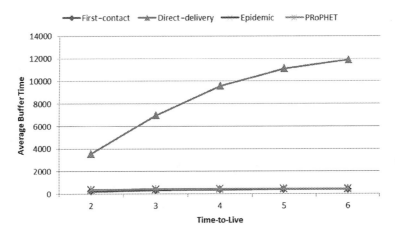

Figure 5.35 Average buffer time vs. TTL.

5.8 CONCLUSION

Delay tolerant network is a kind of ad hoc network having many applications in real life like space communications, wildlife tracking, satellite communication, and so on, where the establishment of an infrastructure-based network is often impossible. Due to the mobility pattern and nodes' behavior of randomly intermittent path selection, the routing in such applications is difficult and challenging. So, the importance of simulating such a scenario is needed where a lot of simulation tools such as NS-2, OMNeT++, ONE, JSim, and OPNET are most usable. This chapter has included the discussion of how to install and integrate the ONE simulator, how to set up the simulation settings in ONE, and how to compare the routing performance using ONE. Hence, the performance of DTN routings, e.g., first-contact, direct-delivery, Epidemic, and PRoPHET has been evaluated by varying the message generation rate, the number of mobile nodes, buffer size, and TTL in terms of delivery ratio, average latency, overhead, average hop count, and buffer time. Hence, direct-delivery routing shows satisfactory performance in terms of the considered metrics.

ACKNOWLEDGEMENTS

The author would like to thank the University Grant Commission of Bangladesh for providing funding at Comilla University for this research.

REFERENCES

[1] M.S. Hossen, "DTN Routing Protocols on Two Distinct Geographical Regions in an Opportunistic Network: An Analysis," *Journal of Wireless Personal Communications*, 108(2), pp. 839–851, 2019.

[2] M.I. Talukdar and M.S. Hossen, "Performance Analysis of DTN Routing Protocols: Single-Copy and Multi-Copy in ICMN Scenario," In *Proceedings of International Conference on Innovations in Science, Engineering and Technology (ICISET-2018), IEEE, International Islamic University Chittagong (IIUC)*, Bangladesh, October 2018.

[3] M.S. Hossen and M.S. Rahim, "On the Optimization of Number of Message Copies for Multi-Copy Routing Protocols in Scalable Delay-Tolerant Networks," In *Proceedings of International Conference on Computer, Communication, Chemical, Materials and Electronic Engineering (IC4ME2)*, pp. 163–167, 2016.

[4] P. Juang, H. Oki, Y. Wang, M. Martonosi, L.S. Peh and D. Rubenstein, "Energy-Efficient Computing for Wildlife Tracking: Design Tradeoffs and Early Experiences with Zebranet," In *Proceedings of ACM ASPLOS*, San Jose, CA, December 2002, pp. 96–107.

[5] M.I. Talukdar and M.S. Hossen, "Selecting Mobility Model and Routing Protocol for Establishing Emergency Communication in a Congested City for Delay-Tolerant Network," *International Journal of Sensor Networks and Data Communications*, 8(1), pp. 1–9, 2019.

[6] T. Spyropoulos, K. Psounis and C.S. Raghavendra, "Efficient Routing in Intermittently Connected Mobile Networks: the Single-Copy Case," In *IEEE/ACM Transactions on Networking*, pp. 63–76, February 2008.

[7] M.S. Hossen and M.S. Rahim, "Analysis of Delay-Tolerant Routing Protocols Using the Impact of Mobility Models," *Journal of Scalable Computing: Practice and Experience*, 20(1), pp. 39–48, 2019.

[8] M.S. Hossen, M.M. Billah and S. Yasmin, "Impact of Buffer Size and TTL on DTN Routing Protocols in Intermittently Connected Mobile Networks," *International Journal of Engineering and Technology*, 7(3), pp. 1735–1739, 2018.

[9] M.S. Islam, S.I. Thaky and M.S. Hossen, "Performance Evaluation of Delay-Tolerant Routing Protocols on Bangladesh Map," *International Conference on Advanced Computing and Intelligent Engineering (ICACIE-2018)*, Springer, 2018.

[10] M.S. Hossen, M.T. Ahmed and M.S. Rahim, "Effects of Buffer Size and Mobility Models on the Optimization of Number of Message Copies for Multi-Copy Routing Protocols in Scalable Delay-Tolerant Networks," *International Conference on Innovations in Science, Engineering and Technology (ICISET-2016)*, IEEE, October 2016.

[11] A. Keränen, J. Ott, and T. Kärkkäinen. "The ONE Simulator for DTN Protocol Evaluation," In *Proceedings of ICSTT*, 2009.

[12] A. Vahdat, and D. Becker, "Epidemic Routing for Partially Connected Ad Hoc Networks," Department of Computer Science, Duke University, Technical Report, April 2000.

[13] A. Lindgren, A. Doria, and O. Scheln, "Probabilistic Routing in Intermittently Connected Networks," *Proceedings of ACM International Symposium on MobiHoc*, vol. 7, no. 3, 2003, pp. 19–20.

[14] J. Burgess, B. Gallagher, D. Jensen, and B.N. Levine, "MaxProp: Routing for Vehicle-Based Disruption-Tolerant Networks," In *Proceedings of IEEE INFOCOM*, Barcelona, Spain, April 2006, pp. 1–11.

[15] A. Balasubramanian, B.N. Levine and A. Venkataramani, "DTN Routing as a Resource Allocation Problem," In *Proceedings of ACM SIGCOMM*, Kyoto, Japan, August 2007, pp. 373–384.

[16] T. Spyropoulos, K. Psounis and C.S. Raghavendra, "Spray and Wait: an Efficient Routing Scheme for Intermittently Connected Mobile Networks," In *Proceedings of ACM WDTN*, Philadelphia, PA, August 2005, pp. 252–259.

[17] T. Spyropoulos, K. Psounis and C. S. Raghavendra, "Spray and Focus: Efficient Mobility-Assisted Routing for Heterogeneous and Correlated Mobility," In *Proceedings of IEEE Per. Com*, White Plains, NY, March 2007, pp. 79–85.

[18] M.S. Hossen and M.S. Rahim, "Impact of Mobile Nodes for Few Mobility Models on Delay-Tolerant Network Routing Protocols," *International Conference on Networking Systems and Security (NSysS), ACM IEEE*, Bangladesh University of Engineering and Technology, Bangladesh, January 2016.

[19] Project page of ONE simulator. Available: http://www.netlab.tkk.fi/tutkimus/dtn/theone, 2013.

[20] F. Bai, N. Sadagopan, and A. Helmy, "IMPORTANT: A Framework to Systematically Analyze the Impact of Mobility on Performance of Routing Protocols for Ad hoc Networks," In *Proceedings of IEEE INFOCOM 2003*, vol. 2, March 2003, pp. 825–835.

[21] T. Issariyakul, *An Introduction to Network Simulator NS2*, New York, NY: Springer.

Chapter 6

Understanding Influencers of Adaptive Social-Aware Opportunistic Forwarding

Soumaia A. Al Ayyat and Sherif G. Aly

American University in Cairo, Cairo, Egypt

CONTENTS

6.1 INTRODUCTION

Social-based opportunistic forwarding algorithms are essential for content dissemination in environments with insufficient network infrastructure or in those that are prone to frequent disruptions [1]. These forwarding algorithms leverage the knowledge of the users' social profiles in ranking the nodes for forwarding decision making [2]. However, many of these algorithms are insensible to both the user's interest in the forwarded content [3] and the limited power resources of the available mobile nodes [4]. Very few of these algorithms [5,6] are aware of these two significant pieces of context information when ranking candidates, yet they are oblivious to the importance of adapting their ranking algorithm to the dynamically changing surrounding environment, as discussed in Section 6.2. For instance, as the message is being transferred from one environment to another, some context information might be inaccessible or may even become less important in the new environment. Due to this change in the availability or the degree of importance of context information, over time, an adaptive approach of ranking candidates for forwarding decision making is much needed.

Adaptive ranking is a framework that dynamically integrates a set of parameters to compose a rank for each node. The forwarding decision making relies mainly on this computed rank in order to adapt to the current context. The main proposed parameters of the rank are weighted attributes which include opportunistic selection of forwarders among the surrounding candidates, the user interest and their mobile node power capability, a measure of how active socially popular users are, and the popularity of the places these users frequent. With such changing rank, the nodes are continuously re-evaluated for the sake of optimizing the achievement in terms of delivery ratio, cost, delay, power consumption, utilization fairness, and f-measure. This framework is defined in Section 6.3.

Space Syntax is a set of techniques that analyze spatial configuration based on theories that link space and society relating spatial configuration to where and how people move and adapt to space. Few contributions have introduced Space Syntax in the domain of opportunistic forwarding [7–9]. In this work, we introduce the integration of Space Syntax into the adaptive ranking framework. Section 6.4 illustrates four proposed implementations of this framework that cater to the adaptive weighing of the attributes. The first version is the basic adaptive algorithm, while the second version additionally introduces opportunistic selection of forwarders. The third and fourth versions of the adaptive algorithm integrate Space Syntax metrics in the adaptive ranking functions of versions 1 and 2, respectively.

Section 6.5 evaluates the performance of the proposed adaptive ranking implementations through simulations based on the SAROS simulator [10], which is introduced in previous research by the authors of this chapter. This simulator includes various environment context-aware simulations that are significant for simulating the forwarding process using the four proposed implementations. The simulator also simulates the forwarding process using the selected related work implementations; namely, the *Epidemic*, and the state-of-the-art social-aware opportunistic forwarding algorithms which include some sort of adaptation, such as the *EBubbleRap*, the *PIPeROp* and the *PISCAROp* algorithms [11]. The simulated context information includes mobility traces from a university campus, power consumption, interest awareness, social profiles, and Space Syntax awareness metrics computation. The results of these simulations are discussed in detail in Section 6.6. The results are quite promising, as they demonstrate improvement in forwarding performance of the proposed algorithms compared to the aforementioned leading-edge social-aware opportunistic forwarding algorithms. For instance, the simulation results illustrate that the first proposed version outperforms the benchmark algorithm, Epidemic, in addition to the state-of-the-art EBubbleRap, PIPeROp, and PISCAROp algorithms by a 40% increase in f-measure, a 96% reduction in the ratio of contacted uninterested users, a 75.6% reduction in cost, and a 9% reduction in power consumption per unit delivery. Comparing the performance of the four proposed versions to one another, it is found that the first version achieves the highest ratio of contacting interested users, while the fourth version contacted the least ratio of interested users. However, this fourth version achieved the highest f-measure among all the algorithms. In addition, the third version incurred the least cost and least power consumption to achieve the highest delivery ratio among the four proposed versions. Finally, we conclude with some recommendations of the most suitable forwarding algorithm implementation(s) for various real-life application environments.

6.2 BACKGROUND AND RELATED WORK

There is significance in dynamically adjusting the weights of the attributes of the ranking function due to the fact that the surrounding context that the nodes encounter along their path in opportunistic networks is continuously varying. Surveying the literature, very few contributions are found to be aware of the importance of dynamically adjusting the weights of the attributes of the ranking function. In this section, we discuss the few approaches that made partial steps towards providing dynamic adjustment of the weights of the ranking function's attributes.

6.2.1 Context-Adaptive Forwarding in Mobile Opportunistic Networks

The research paper [12] has shown that for achieving a more efficient routing in opportunistic networks, it is recommended to compute the weights of the ranking function parameters based on the node's behavior rather than to set predefined weights. The authors proposed setting weights of an opportunistic routing utility function per group of nodes to improve the performance of the network in terms of hit rate, latency, and delivery cost. They also showed through simulations that dynamic adjustment of the weights of the parameters that are similar in value further improves the network performance. However, they propose algorithms that remove the attribute in which two encountering nodes maintain the same value, and thus, the algorithm redistributes the weight of this attribute among the rest of the attributes. They also did not set power-aware attributes to dynamically adapt with the scarceness of power resources in the mobiles nodes.

6.2.2 *EBubbleRap* Algorithm

The community-aware Bubble Rap [13] routing algorithm cares for the fact that nodes' ranking differs as they move within their local community from their global ranking within the whole community. Thus, this algorithm sets two ranking functions. These ranking functions are the local ranking function that ranks the nodes within their local communities, and the global ranking function that ranks the nodes within the global community.

The *EBubbleRap* algorithm [14] is the energy-aware version of the Bubble Rap routing algorithm. This version introduces energy consumption optimization to the socially aware routing [14]. The *EBubbleRap* algorithm introduces an energy-aware utility function into the local and global ranking functions that are used in the *BubbleRap* algorithm. However, being insensible to forwarder's interest in the content, the algorithm contacts uninterested nodes overloading them with content not in their interest domain and also consumes their resources. We would like to mention that throughout the chapter, the term nodes' interest maps to the user's interest.

6.2.3 *PeopleRank* Algorithm

The *PeopleRank* algorithm is a power-oblivious, social-aware message forwarding algorithm based on forwarding content by utilizing the nodes of the socially popular people in a particular place [15]. The PeopleRank algorithm is based on the assumption that socially popular nodes represent better candidates for messages delivery to destinations since these nodes with higher probability will encounter destinations more quickly. Social ranking of nodes is based on their social relationship. Such social relationship is recognized through a declared friendship, or through shared common interests among the nodes.

The PeopleRank algorithm achieves a balance between social-based forwarding and opportunistic forwarding through introducing a damping factor for deciding the ratio of reliance on the social ranking of a node versus the reliance on the opportunistic encounter of carriers. The damping factor (d) ranges from a value of 0, for total reliance on opportunistic forwarding, to 1 for total reliance on social-based forwarding. Values in between represent the relative weight set for each of the two approaches.

The researchers proposed a contact-aware version of the PeopleRank algorithm (CA-PeR) that ranks nodes using both their social rank and their degree of social activeness. Node activeness is measured by how frequent it comes across its social contacts. In effect, a node's social rank is either rewarded or penalized according to a count of the node's contacts with its social contacts. The CA-PeR rank as defined in [15] is computed as follows:

$$CA - PeR(i) = (1 - d) + d * \sum_{j \in F(i)} \frac{CA - PeR(j) * w_{i,j}}{|F(j)|} \qquad (6.1)$$

where d is the damping factor, $CA - PeR(j)$ is the PeopleRank value of the friend j, $F(i)$ is the set of friends of node i who are available in vicinity, and $w_{i,j}$ is the contact-component and computed by the following equation:

$$w_{i,j} = \frac{|encounters_{i,j}|}{\sum_{k \in F(i)} |encounters_{i,k}|} \qquad (6.2)$$

The CA-PeR value consists of a pure opportunistic component $(1 - d)$, a contact-aware component w, and a social-aware ranking component. The social-aware component is based on the concept that the node has a high social rank if, on average, its friends have high social ranks. This component is computed by calculating the sum of the CA-PeR values of all the friends j of the node i that is then averaged by the count of those friends $|F(i)|$.

6.2.4 **SCAR Algorithm**

SCAR [6] is a sensor context-aware opportunistic routing protocol. The SCAR algorithm is an adaptive power-aware version of the *SocialCast* algorithm [16]. SCAR is suitable for environments where mobility patterns can be predicted to a certain extent, such as forwarding sensor data through existing routes among mobile sensor nodes to reach fixed or mobile destination nodes. This protocol is aware of the power consumption and selects the route that consumes less power. The SCAR protocol selects the best candidates for forwarding sensor data based on forecasting the likelihood that they will meet destination nodes. The protocol is distributed and preserves information privacy since each node computes its attributes locally. To improve performance, the SCAR algorithm relies on the anticipated values of the utility attributes instead of on their current values. The forecast mechanism observes three factors: history of co-location of the candidate node with the destination nodes, the change of degree of connectivity of the candidate node with other sensor nodes, and the level of battery of this node. The SCAR algorithm includes a monotonically decreasing function that adapts to the predefined ranges of values of a certain attribute. The utility function weighs all the attributes so that it maximizes the benefit gained from them all, and in order to balance the trade-off between all of these attributes. The route selection process is based on selecting those sensor nodes with the higher utility value from among their neighbor sensors.

The node's utility rank as defined in [6] is shown in Equation 6.3.

$$
\begin{aligned}
Util(i) = {} & Range_{col} * w_{col} * P_{col}(i) + Range_{cdc} * w_{cdc} * P_{cdc}(i) \\
& + Range_{bat} * w_{bat} * P_{bat}(i)
\end{aligned}
\tag{6.3}
$$

where $P_{col}(i)$, $P_{cdc}(i)$, and $P_{bat}(i)$, respectively, are the predicted co-location of node i with any of the destination nodes, the predicted change degree of connectivity of node i, and the predicted remaining battery level of node i. The three predicted attributes are forecasted by applying the Kalman Filter prediction techniques. The weights w_{col}, w_{cdc}, and w_{bat}, respectively, represent the relative importance of the co-location attribute, the change of degree of connectivity attribute, and the battery level attribute. The authors of the SCAR contribution mention that these weights' values depend on the application scenario. Also, $Range_{col}$, $Range_{cdc}$, and $Range_{bat}$ are the adaptive monotonically decreasing range functions for the respective attributes.

6.2.5 PI-SOFA Framework

In another research [11], we presented our vision of a framework that enables integrating interest and power awareness both in ranking the nodes and in the decision process within social-aware opportunistic forwarding algorithms. This framework is applicable for social-aware opportunistic forwarding in one-to-one hop mobile opportunistic networks in urban areas where direct Wi-Fi or Bluetooth connection among mobile nodes enables the message forward for a soft real-time delivery within a couple of hours.

Integration of interest awareness in social-aware forwarding algorithms is simply based on rewarding (or penalizing) the social rank of the nodes based on their potential interest in the forwarded message. The nodes that have potential interest in the forwarded content are those whose interest profiles partial commonality with that of the forwarded message. Both interest profiles can be represented in the form of an interest vector; then both vectors are compared using the Jaccard set similarity Index [17] to coin a new measure, namely, the Similarity Interest (SInt).

To be fair in utilizing the power of the nodes in the place, the social-aware forwarding algorithm has to be aware of two pieces of information, namely, the current power resources of the nodes in the place, and the expected power consumption upon their participation in the message forward process. With such knowledge, the algorithm favors the nodes with higher power capabilities that can sustain the delivery until completion and avoid exhausting the nodes with poor power capabilities.

Within this framework, several implementations of PI-SOFA algorithms that integrate interest and power awareness into the state-of-the-art social-ware opportunistic forwarding algorithms are proposed such as the PIPeROp algorithm, the PISCAROp algorithm, and the PISCast algorithm. In this section, we will detail two of these proposed algorithms.

6.2.5.1 PIPeROp Algorithm

The *PIPeROp* algorithm [5] is an interest and power aware social opportunistic forwarding algorithm within the PI-SOFA framework. The PIPeROp algorithm adds two new dimensions to the PeopleRank algorithm [15] by integrating power awareness and interest awareness into the social-based forwarding process. In order to consider a mobile node for forwarding a message, the PIPeROp algorithm elicits the candidate node's available battery level and the similarity in interest between the node and the forwarded content. The algorithm uses this information as a means of indicating the node's willingness to forward. In addition, it emphasizes power awareness by rewarding the node that is more power abundant; otherwise, it penalizes the node's rank. Accordingly, a candidate node is highly ranked when the following conditions are met: its battery exceeds a pre-defined threshold, its owner is socially connected to popular friends, and both the node owner and their first-order social contacts have interest in the message being sent. Thus, the higher the rank of a mobile node and its social contacts, the higher the chance a node will become a candidate for forwarding the message. However, both algorithms, PeopleRank and PIPeROp, rely on a damping factor that is assigned a predefined value that is static per forwarding application. Such static value does not adapt with the changes in the surrounding environment or the changes in the availability of context information. The node's power-aware rank, *PIPeR*, is formalized by this equation:

$$PIPeR(i) = \mathrm{Re}wardedPower(i) * IPeR(i)$$
$$* \left((1-d) + d * FrInter(i, msg) * \sum_{j \in F(i)} \frac{PeR(j) * w_{i,j}}{|F(j)|} \right) \tag{6.4}$$

where *RewardedPower* is the power of the node being rewarded (or penalized) based on comparing its value against a predefined battery threshold Thr_{bat} as detailed in Equation 6.5. Also, d is the damping factor, and *FrInter(i, msg)* is the cumulative interest of node i and of its friends in the message content (it is calculated as per Equation 6.7), $F(i)$ is the set of friends of node i who are available in vicinity, $PeR(j)$ is the PeopleRank value of the friend j, $w_{i,j}$ is the weight of introducing the friend's PeopleRank value based on the frequency of contact between the two nodes.

$$RewardedPower(i) =$$
$$\begin{cases} (Bat(i) + reward) \ if \ Bat(i) \ge Thr_{bat}, \\ (Bat(i) - reward) otherwise \end{cases} \tag{6.5}$$

Note that the *IPeR* value is the integration of interest-awareness with the PeopleRank algorithm and is computed as per Equation 6.6:

$$IPeR(i) = (1-d) +$$
$$d * FrInter(i, msg) * \sum_{j \in F(i)} \frac{CA\text{-}PeR(j) * w_{i,j}}{|F(j)|} \tag{6.6}$$

$$FrInter(i, msg) = \frac{PSInt(i, msg) + \sum_{j \in F(i)} PSInt(j, msg)}{|F(i)| + 1} \qquad (6.7)$$

The PIPeROp algorithm [11] adds an additional opportunistic component into the candidate selection process. This component favors forwarding the message to any interested forwarder whose battery level is above the fixed battery threshold Thr_{bat}. These favored forwarders need not be socially popular users but should be power-capable users with high interest in the forwarded content. Thus, the next message forwarder will be any candidate whose similarity interest with the message's interest vector exceeds a preset interested forwarder threshold and whose node power is above the given battery threshold. In addition, the algorithm selects candidates with a power-aware rank that is not less than that of the current message holder.

However, both the PeopleRank and PIPeROp algorithms rely on a damping factor that is assigned a predefined value that is static per forwarding application.

6.2.5.2 PISCAROp Algorithm

The PI-SOFA framework integrates interest awareness and power awareness into the SCAR algorithm to propose a new power-and-interest-aware version, the PISCAROp algorithm [11]. The PISCAROp algorithm rewards or penalizes the adaptive SCAR utility value of a node based on the collective interest of both the node's owner and their friends in the forwarded message. It also penalizes or rewards the utility function based on the remaining power of the node compared to a predefined battery threshold. Accordingly, those partially interested nodes that are power capable become favored forwarder candidates.

This version minimizes contact with uninterested ones and favors potentially interested nodes that are socially connected to other potentially interested friends. In addition, through this version, the partially interested nodes that are power capable become favored forwarder candidates.

The new utility function is computed as follows:

$$Plutil(i)\begin{cases} -\big(PSInt(i, msg) * RewardedPower(i)\big) * Util(i) & \text{if } PSInt < 0 \,\&\, RewardedPower < 0, \\ PSInt(i, msg) * RewardedPower(i) * Util(i) & \text{otherwise} \end{cases} \qquad (6.8)$$

where $PSInt(i, Ad)$ is the rewarded/penalized similarity interest of the node with the forwarded content interest vector as detailed in Equation 6.9, $RewardedPower(i)$ is the rewarded/penalized battery level of the node that is computed as per Equation 6.5, and $Util(i)$ is computed as per Equation 6.3 of the SCAR algorithm.

$$PSInt(j, msg) = \begin{cases} (SInt(j, msg) + reward & \text{if } SInt \geq Thr_{Int}, \\ (SInt(j, msg) - reward & \text{otherwise} \end{cases} \qquad (6.9)$$

All the above utility function computations are based on the preset interest and battery thresholds thr_{Int} and thr_{bat}, respectively, that do not adapt with the changes in the surrounding context.

6.2.6 Space Syntax

Space Syntax was initially proposed in the field of architecture by Hillier and Hanxon in 1984 to model natural mobility patterns by analyzing spatial configurations. It accurately predicts natural movement patterns in an area based on how its segments are connected [8, 18, 19]. Space Syntax is comprised of a set of techniques for analyzing spatial configuration, and it is derived from a set of theories linking space and society. Using Space Syntax, one can relate spatial configuration to where people are, how they move, and how they adapt to space [18]. Space Syntax opportunistic forwarding exploits the infrastructure of a particular place and the popularity of specific attraction points in that place in order to predict user mobility patterns, flows, and areas of possible congestion [8, 9].

6.2.6.1 Space Syntax Metrics

In the domain of Space Syntax there are standard metrics to quantify the effect of road maps and architectural configuration on natural movement. Such metrics predict natural movement patterns in an area based on the way of segments connectivity. Space Syntax metrics are pre-calculated once using static information about the environment without a need for recalculation unless the environment changes [8]. There is a set of basic Space Syntax metrics that are commonly used to set the popularity of any street or spot on a given map. We concentrate here on five basic metrics that are detailed as follows:

Integration value [20]: The segment's integration value is the average distance between the segment and all other segments in the map. This value represents the segment's closeness to other segments.

$$
\begin{aligned}
& Integration\,Value(x) = \\
& \quad Average\big(\forall\,y\,in\,Segments\ Distance(x,\,y)\big)
\end{aligned}
\tag{6.10}
$$

Connectivity value [20]: The segment's connectivity value is the number of segments intersecting with this segment. It represents the segment's degree of connectivity.

$$
Connectivity\,Value(x) = \forall\,y\,in\,Segments\,Intersecting\,with\,x\ Count(y) \tag{6.11}
$$

Location index [8]: The location index is the deviation of the segment's integration value from the maximum integration value divided by the range of integration values.

$$
\begin{aligned}
& Location\,Index(x) = \forall\,y\,in\,Segments \\
& \quad \frac{max\big(Integration\,Value(y)\big) - Integration\,Value(x)}{Range\big(Integration\,Value(y)\big)}
\end{aligned}
\tag{6.12}
$$

Attraction value [8]: It is an index that indicates the level of attraction of the spot with respect to the rest of the spots in the area.

$$
\begin{aligned}
& Attraction\,Value(x) = w.r.t.\,all\,the\,Segments \\
& \quad\quad\quad\quad\quad\quad Index(x)
\end{aligned}
\tag{6.13}
$$

Popularity index [8]: The popularity index of a segment is calculated as the normalized attraction value of the segment added to the sum of the normalized length indexes multiplied by the normalized attraction values of all the segments that intersect with this segment. The normalized length of an intersecting segment is the length of the intersecting segment divided by the segment whose popularity index is being calculated.

$$\begin{aligned} Popularity\,Index(x) = &\ Normalize\big(Attraction\,Value(x)\big) \\ &+ \forall\, y\ in\ Segments\ Intersecting\ with\ x \\ &\quad \Sigma\left(\frac{Length(y)}{Length(x)} * Normalize\big(Attraction\,Value(y)\big)\right) \end{aligned} \tag{6.14}$$

6.2.6.2 Space Syntax-Based Forwarding Algorithms

We located a number of studies that looked at Space Syntax specifically in terms of how it affects or controls the social behavior of humans in a particular place and which inferred human mobility and social interaction from these behavior patterns [20–23].

Some studies built on Space Syntax in order to improve the opportunistic forwarding process, including the following:

Space Syntax forwarding by popularity index [8]: Building on the Space Syntax, these authors introduce a new Space Syntax metric, namely, the popularity index of any attraction point on the map. This new metric is computed as a function of the segment's integration value, its location index, and its attraction value. Based on the computed segments' popularity indexes, the popularity index of any attraction point on the map is then calculated. However, the limitation of their proposal is that it relies on the condition of first finding and establishing a route to the destination node before initiating the forwarding process. Satisfying this condition could be difficult in disrupted networks and thus limits the area of application of their proposal. The authors demonstrate the effect of the newly defined metric by proposing the following two Space Syntax-based forwarding algorithms:

- The *SS-Cutoff* algorithm prohibits message forwarding to nodes that are geographically located in areas whose popularity falls between a certain configurable threshold. Also, with a given probability, it forwards messages to nodes in areas whose popularity is above or equal to the threshold.
- The *SS-Sliding* algorithm forwards messages to nodes according to a variable probability that is inversely proportional to the popularity of the area in which these nodes reside. Thus, forwarding becomes less aggressive in areas that are more popular so as to decrease cost.

Select & Spray [19]: This algorithm is a Space Syntax-based opportunistic forwarding algorithm that is applicable on a citywide level. The authors consider Space Syntax information to guide opportunistic forwarding based on the likelihood of a location to attract mobile nodes. The approach also depends on a server/client setup, not an opportunistic network setup. The server consists of a "data collector", a "recommendation engine", and a "select engine" that collect and memorize registered clients' mobility data in addition to a priori knowledge of the city map and the position of attraction points. The algorithm then computes a maximum probability utility function which ranks all nodes in the network based on their probability of reaching a destination, and then it selects the best subset of connected nodes that will be used to disseminate a message with the least potential cost and delay. Based on a citywide range and a server/ client configuration, this application is rather costly in terms of computations, communications, and resources.

The data collector fetches the data of registered clients and the surrounding environment, then inserts or updates them in the databases. The recommendation engine consists of recommendation algorithms to recommend messages to interested clients. It initiates communications by sending the message and the set of destinations to the select engine. The select engine relies on a priori knowledge of the city map, the position of attraction points, and/or the history of node mobility patterns. The select engine has to select the best subset of connected nodes that will be used to disseminate a message with the least potential cost and delay. The client mobile nodes implement a spray engine. This engine consists of an opportunistic dissemination algorithm with forwarding decisions based on social, historical, or mobility data. The select engine detects the probability of divergence between two nodes at time t + 1 to eliminate outliers. There is a maximum probability utility function which ranks all nodes in the network based on their probability to reach a destination.

Exploiting Space Syntax for Mobile Opportunistic Networks: In a study that exploits Space Syntax for Mobile Opportunistic Networks, Mitbaa et al. [9] survey the field of opportunistic forwarding in small-scale urban environments and categorize them according to the assumptions on which the algorithms base the forwarding decisions. The authors illustrate that these algorithms – as they rely on more complex assumptions – achieve more efficient performance, "rendering solutions built on them unusable outside their intended environment categories" [9]. This study proposes three sample algorithms, each of which represents one of their defined categories. The proposed algorithms assume that the forwarded messages may not be fragmented, and that they are designed to ensure there is enough time for nodes to exchange messages. The forwarding decisions are based on static information, such as city map and attraction points, or, alternatively, on local information regarding the node itself. The authors propose the following three sample algorithms:

- *Space Syntax Opportunistic Forwarding* (SOF): This forwarding approach is based on the nodes' awareness of a map layout.
- *Location-Aware Space Syntax Opportunistic Forwarding* (LA-SOF): This forwarding approach is based on the nodes' awareness of a map layout, the nodes' position, and the position of the destination Internet access point locations.
- *Assisted Space Syntax Opportunistic Forwarding* (ASOF): This forwarding approach is based on the nodes' awareness of a map layout, the nodes' position, the position of the destination Internet access point locations, and the nodes' mobility patterns.

Space Syntax and pervasive systems: In a study of how innovative the application of Space Syntax in the domain of pervasive systems can be, the author applied Space Syntax metrics in different phases of the development of pervasive systems [20]. The author proposes three case studies that demonstrate the role of Space Syntax as a simple application module, as an explanatory tool, and as a modeling tool.

The *Weighted PageRank* approach [22] is another study that proves that the underlying space–space topology of an urban environment is far from random. Rather, it exhibits small-world and scale-free structures and proves that the weighted PageRank approach can accurately predict human movement in an urban environment.

All the aforementioned research contributions exhibit the following limitations:

First, the few contributions that combine Space Syntax metrics with other opportunistic forwarding techniques mainly consider expansive areas at a citywide level. They do not consider small scale level application such as a campus-wide or convention center application.

Second, some of these contributions consider social graph metrics (such as closeness centrality and betweenness centrality) in addition to the contact graph, but no proposed algorithm so far considers the forwarder nodes' interest in the forwarded content.

Third, the ASOF approach and its peer algorithms consider destination locations such as streets or static access points [9, 19] – except for the Popularity Index approach which considers mobile destination nodes [8]. Thus, their proposed algorithms cannot be applied to mobile destination nodes.

Fourth, all the proposed Space Syntax-based contributions to date do not take into consideration power consumption, and some complex algorithms consume a great deal of energy. The preservation of energy is crucial in limited-resources mobile opportunistic networks.

Finally, the proposed algorithms by Mitbaa et al. [9, 19] rely on the concept that the mobile node continuously recomputes the weights of its own contact graph to re-compute its utility function. Practically speaking, these computations consume the node's resources especially when this approach is applied on a citywide scope. It is worth noting that the node needs to compute this graph for each forwarded message since each message has its own set of destination locations or destination access points. Given that most of the messages will be forwarded through the betweenness-central mobile nodes, this computational consumption leads to unfairness by exhausting the node's resources.

6.3 ADAPTIVE RANKING

In many cases, there is a frequent dynamic change in the available set of context information needed for ranking nodes in order for the algorithms to properly select the best candidates for forwarding. Thus, dynamic weights of the attributes are needed in order to compensate for that change in context. For this purpose, this research proposes an adaptive ranking algorithm that integrates five context-aware sensors, namely: *interest awareness*, *power awareness*, and *place awareness* with *social awareness* and *user activeness*. The algorithm takes into consideration various factors that affect the rank of the coexisting nodes. These factors include the following:

- The relative interest of the user in the forwarded content;
- The power capabilities of the mobile node the user holds;
- The degree of social popularity of the user;
- The degree of activeness of the user in terms of physically or logically contacting nodes of other users;
- The popularity of the places the user frequents.

This approach dynamically changes the weights of the aforementioned factors to suit the surrounding environment. Based on these changing weights, the factors that control the node's rank are dynamically weighed to adapt the changing environments the node moves across. With such changing rank, the nodes are continuously re-evaluated for the sake of optimizing the achievement of the forwarding algorithm in terms of the following metrics:

- Increasing the delivery ratio;
- Increasing the contact with the interested users;
- Minimizing the contact with the uninterested users;
- Minimizing the cost paid while the targeted delivery ratio is achieved;
- Minimizing the amount of consumed power while the targeted delivery ratio is reached;
- Maintaining fairness in utilizing the resources of the available mobile nodes in the place.

Accordingly, each node will execute this ranking function utilizing its own values of the aforementioned factors. Its rank changes as the weights of these factors change; thus, with each new encounter, the node recomputes its own rank and exchanges its value with the encountered node for decision making on content forwarding.

6.3.1 Adaptive Ranking Framework

In this framework, the forwarding decision relies mainly on the dynamic rank in order to adapt to the current context. The main proposed parameters of the rank are dynamically weighted attributes which include: opportunistic selection; the node's owner interest and the node's power capability; the measure of how active socially popular users are; and the popularity of the place these users frequent. This takes into consideration that as nodes move from one place to another, or as time passes, both the value and the relative importance of their attributes may change. Then, a fixed weight of each attribute within the node's rank might not reflect the current context, leading to an incorrect node ranking approach. Hence, continuously re-evaluating the attributes' weights helps in reflecting the correct current rank of the node. Accordingly, this ranking framework adapts to the current context by changing the relative weights of the attributes that decide the node's rank. The change of weights relies on the values of the aforementioned attributes and whether these current values fall into critical ranges that set such attributes ineffective in comparison to the remaining attributes at the current context. Other changing contexts would take place where any of these attributes' values are unavailable at any certain time due to any reason, or because the user or their mobile node refuses to reveal its value.

Based on this logic, the node ranking function ranks candidates based on dynamically weighted attributes which are enclosed in the following components:

1. The opportunistic selection ($OppComponent$);
2. The user's interest ($IntComponent$);
3. The measure of how active the socially popular users are ($SocActiveComponent$);
4. The power capability of the mobile node ($PowerComponent$);
5. The popularity of the places frequently visited by these users ($PlaceAwareComponent$).

This adaptive ranking function can be formalized as follows:

$$
\begin{aligned}
AdpRank(i) = {} & Oppcomponent + IntComponent \\
& + SocActiveComponent + PowerComponent \\
& + PlaceAwareComponent
\end{aligned}
\tag{6.15}
$$

Let us have a closer look at the various components.

1. *The opportunistic forwarder node selection component, OppComponent,* gives space for a pure opportunistic selection of candidates, *OpportunisticSelection,* without being restricted by any certain criteria. This component is allowed to take place with a w_{Opp} portion of the ranking function components. This component can be formulated as follows:

$$OppComponent = w_{Opp} * OpportunisticSelection \qquad (6.16)$$

2. *The user's interest component, IntComponent,* considers the relative interest of the user in the forwarded content as an important factor in ranking nodes. This component consists of the similarity in interest between the user's interest vector and the forwarded content domain of interest *SInt(i, msg).* This similarity interest metric can be computed using Jaccard similarity between the two interest feature vectors [17]. It is more comprehensive to include the similarity interest of the user's friends *CumInterest(i)* as well. Another interesting component that might be considered in the user's interest is the power capabilities of the mobile node they are carrying, *Power(i),* since the power limitation of some devices might hinder the user's interest in receiving any content even if it is within the user's range of interest. This component can be formulated as follows:

$$IntComponent = w_{IntPower} * CumInterest(i) * Power(i) \qquad (6.17)$$

3. *The social active component, SocActiveComponent,* takes into consideration the social popularity of this user among their friends. It also considers the level at which the mobile user is active in contacting and interacting with other nodes. This component is aware of the active members and the sociable members since this factor can be a good indicator that such members will, with high probability, get in contact giving a higher chance to meet with the destination nodes [24] and to forward the content to anyone of the interested friends. This component consists of the *SocialRank(i)* of the node and the measure of activeness of the node *Activeness(i).* There are several ways to measure these two parameters. Accordingly, this component can have various implementations. The ranking function relies on this component with the relative weight $w_{SocActive}$. Generally speaking, this component is formulated as follows:

$$SocActiveComponent = w_{SocActive} * SocialRank(i) * Activeness(i) \qquad (6.18)$$

4. *The power capability of the mobile node component, PowerComponent,* considers the node's power capability, *Power(i),* as an affecting factor in selecting the next forwarder. In fact, the user whose mobile node's power capability is not high may refrain from participating in the forwarding process, and also the node that has low power capability may not survive until the completion of the delivery process. It is more effective to highly rank the power capable node among all the available nodes in the place. However, to maintain fairness in utilizing the nodes in the place, there is a need to set a threshold below which nodes are forced to drop the forwarded message copy and abandon the participation in the forwarding process. The power capability of a node can be represented by various measures such as the remaining battery charge or by considering the depletion rate of the battery in addition to its capacity. Generally speaking, this component is formulated as follows:

$$PowerComponent = w_{PowerAware} * Power(i) \qquad (6.19)$$

5. *The popularity of the place* is an important component to consider in ranking nodes. This component, namely, *PlaceAwareComponent*, relies on the importance of the Space Syntax metrics on decision making. There are various Space Syntax metrics that can be utilized in computing the place popularity attribute *PlacePopularity(i)*. Some of the popular metrics are the integration value, location index, and attraction value, among others [8]. These metrics consider closeness centrality and betweenness centrality among other social metrics in ranking places in terms of popularity [18]. In a previous work, we have proposed variations of the Space Syntax metrics [7]. The ranking function relies on the place popularity component with the ratio computed by the place popularity weight $w_{PlacePopularity}$. This component is formulated as follows:

$$PlaceAwareComponent = w_{PlacePopularity} * Place\,Popularity(i) \qquad (6.20)$$

where *PlacePopularity(i)* computes the place popularity of node i using Space Syntax metrics such as location popularity of the places frequently visited by node i, or the location popularity of the places node i stays the longest time duration among other Space Syntax metrics. In greater detail, the adaptive ranking function is computed as:

$$\begin{aligned}
AdpRank(i) = {} & w_{Opp} * OpportuniticSelection \\
& + w_{IntPower} * CumInterest(i) * Power(i) \\
& + w_{SocActive} * SocialRank(i)\,Activeness(i) \\
& + w_{PowerAware} * Power(i) + w_{PlacePopularity} * PlacePopularity(i)
\end{aligned} \qquad (6.21)$$

where $w_{Opp} = 1 - (w_{IntPower} + w_{SocActive} + w_{PowerAware} + w_{PlacePopularity})$

$$w_{IntPower} = \left(SInt(i, msg), bat(i) \right)$$

$$w_{SocActive} = f\left(SInt(i, msg) \right)$$

$$w_{PowerAware} = f\left(bat(i), depletionrate(i), capacity(i) \right)$$

$$w_{PlacePopularity} = f\left(Pop(Location(i, now)), Pop(FreqLocation(i)) \right)$$

There are five weights that dynamically change in value to adapt to the current context. These weights are:

1. w_{Opp} is the weight of the opportunistic selection component. Its value is actually the remaining portion after computing all the weights of the other components.
2. $w_{IntPower}$ is a monotonically decreasing function of the similarity interest (SInt) of the node and its power capability *Power(i)*. The *Power(i)* parameter is the remaining power of node i. This power component can be represented as the rewarded predicted power of the node which is computed as *RewardedPower(i)* as per Equation 6.5. The battery

depletion rate awareness can be also introduced in the ranking function where $Power(i)$ is computed as per $PredPower(i)$, detailed in Equation 6.22.

$$\mathrm{Pr}\,edPower(i) =$$

$$\begin{cases} \big(Bat(i) + reward\big) & \text{if } \dfrac{Bat(i)}{DepRate(i)} - TTL(msg) \geq Thr_{dep}, \\ \big(Bat(i) - reward\big) & \text{otherwise} \end{cases} \tag{6.22}$$

where $DepRate(i)$ is the depletion rate of node i, $TTL(msg)$ is the time-to-live of the forwarded message, and Thr_{dep} is the threshold of the battery depletion rate below which $Bat(i)$ is penalized.

In some applications the users' interest in the forwarded content may vary or decrease as time passes. This can be represented by the similarity interest between the interest vector of the user, $IntFV(user)$, and the interest vector of the message content, $IntFV(msg)$, that may not be constant as it may change over time. Accordingly, the $SInt(user, msg)$ becomes a function of time. This is depicted in the following formula:

$$SInt(user, msg) = f\big(time, IntFV(user), IntFV(msg)\big) \tag{6.23}$$

3. $w_{SocActive}$ is the weight of the socio-activeness of the node. It is a function of the $SInt$ of the node. This is attributed to the fact that the ranking function pays more attention to nodes that have interest in the forwarded content. That is why the node's interest in the forwarded content emphasizes its social and activeness components.

4. $w_{PowerAware}$ is the weight that sets the relative effect of the node's power capability in the node ranking process. This weight decreases as the node's power decreases to indicate the decreasing value of this node when selecting it to participate in the forwarding process.

5. $w_{PlacePopularity}$ is the weight of the place popularity $PlacePopularity(i)$ of node i. The weight $w_{PlacePopularity}$ is a function of both the popularity of the current location of node i, $Pop(Location(i, now))$, and the popularity of the most frequent location in which node i stays along its path $Pop(FreqLocation(i))$. By the way, it may be replaced by other space popularity metrics.

Let us clarify with a simple example. Node i will calculate its adaptive rank $AdpRank(i)$ based on these given values: the current remaining portion of the battery of node i is 0.8 (i.e. 80%), its depletion rate is 0.01 per second, and the capacity of this battery is 6000 mAh. Then the weight of power awareness, $w_{PowerAware}$, is a function f(0.8, 0.01, 6000).

Let the user of this mobile node be a sociable person with a social rank 0.9 as per one of the state-of-the-art social ranking functions, and his level of activeness in communicating with others is 0.7. His interest in the content of the forwarded message is 0.75 as reflected by the similarity between the user's interest vector and the interest vector of the forwarded message, and the cumulative interest of this user and his friends in this message is 0.6 as per the equation 1.38 that is detailed in the next section. Let the weight of the socio-activeness of the user, $w_{SocActive}$, be a function $f(0.75, 0.7)$. Thus, the socio-active component will be $f(0.75, 0.7) * 0.75 * 0.7$.

Let the threshold above which the rewarded power of the node is rewarded be 0.5 and let the reward value be 0.5. Then, the $RewardedPower(i)$ is 0.8 + 0.5. From these values, the weight of the interest-and-power awareness component, $w_{IntPower}$, is calculated as a function $f(0.7, 1.3)$, while the interest-power component will be $f(0.7, 1.3) * 0.6 * 0.8$.

Now, let's calculate the place popularity component. If the user frequently visits a place of a popularity $Pop(FreqLocation(i))$ equal to 0.86, while he is currently located in a place of popularity $Pop(Location(i, now))$ equal to 0.5, then, the weight of the place popularity of the node, $w_{PlacePopularity}$, is $f(0.5, 0.86)$. The place popularity of the node is calculated as per one of the Space Syntax metrics detailed in Section 6.2.6.1.

From all of these calculations, the weight of the opportunistic selection is $w_{Opp} = 1 - (w_{IntPower} + w_{SocActive} + w_{PowerAware} + w_{PlacePopularity})$. The opportunistic component itself would be a random value of 1 (to select the candidate) or zero (to ignore this candidate). The summation of all the aforementioned components constitute the adaptive rank of the node as per Equation 6.21.

6.3.2 Adaptive Ranking Versions

The algorithms presented here adapt the ranking function according to the current context. The value of the ranking function relies on several factors that include an interest and power awareness factor, a social and activeness factor, a Space Syntax factor, and an opportunistic forwarder selection factor. The implemented versions of this adaptive algorithm are:

1. The first version is the basic version of the algorithm which relies on the interest and power awareness factor, the social and activeness factor, and the opportunistic forwarder selection factor. This is summarized as:

$$AdpRank(i) \ = \ f\big(IntPower(i), SocActive(i), Opp(i)\big) \tag{6.24}$$

2. The second version implements the adaptive algorithm and inserts an extra explicit interest and power-based opportunistic selection criterion. Thus, the message carrier node i selects node j as a forwarder if the following conditions are satisfied:

$$AdpRank(j) \geq AdpRank(i) \ AND \ SInt(j, msg) \geq Thr_{int} \ AND \ bat(j) \geq Thr_{bat} \tag{6.25}$$

3. The third version relies on the interest and power awareness factor, the social and activeness factor, the Space Syntax factor, and the opportunistic forwarder selection factor. It applies the same factors as the first version but with the Space Syntax factor added to its ranking function. This is summarized as:

$$AdpSpSynRank(i) = f(IntPower(i), SocActive(i), SpaceSyntax(i), Opp(i)) \tag{6.26}$$

4. The fourth version relies on the interest and power awareness factor, the social and activeness factor, the Space Syntax factor, and the opportunistic forwarder selection factor, then inserts an extra explicit interest- and power-based opportunistic selection criterion. It applies the same factors as the second version but with the Space Syntax factor added to its ranking function. Thus, a selected forwarder node j must satisfy the following condition:

$$AdpSpSyn\,Rank\,(j) \geq AdpSpSyn\,Rank\,(i)$$
$$AND\;SInt\,(j,msg) \geq Thr_{int} \tag{6.27}$$
$$AND\;bat\,(j) \geq Thr_{bat}$$

6.4 ADAPTATION PROPOSED IMPLEMENTATIONS

The surveyed works in the field of adaptive ranking and context-awareness seek social awareness, power awareness, and space-popularity awareness, but they lack the ability to adapt the weights of the components that compose these ranking functions. Through the adaptive ranking approach, we present a ranking function that constitutes all the features of social awareness, power awareness, and space-popularity awareness and also provides a set of adaptive weights for all the components in order to be able to adapt continuously to the current context. This section details the implementation of the four proposed versions defined in Section 6.3.2.

6.4.1 Version 1: *Adp*

This is the basic version of the adaptive ranking function. This version basically considers the following parameters in ranking nodes: the pure opportunistic component, the interest-and-power-aware component, the social activeness component, and the node's power component. This ranking function is formulated as follows:

$$
\begin{aligned}
AdpRank\,(i) = {}& w_{Opp} * Opportunistic \\
& + w_{IntPower} * CumInterest\,(i) * Power\,(i) \\
& + w_{SocActive} * SocialRank\,(i) * Activeness\,(i) \\
& + w_{PowerAware} * Power\,(i)
\end{aligned}
\tag{6.28}
$$

where the weights are computed as follows:

$$w_{Opp} = 1 - \left(w_{IntPower} + w_{SocActive} + w_{PowerAware}\right)$$

$$w_{IntPower} = SInt\,(i,\,msg) * bat\,(i)$$

$$w_{SocActive} = \exp ectedusage(i)$$

$$w_{PowerAware} = MonotonicallyDecreasing\left(Bat\,(i)\right)$$

Note that the opportunistic component, *Opportunistic*, can be inclined to those users that may be interested in the forwarded content and as well hold mobile devices with sustainable power charge. It is then computed as follows:

$$Opportunistic = SInt\,(i,\,msg) * \mathrm{Re}\,mainingPower(i) \tag{6.29}$$

where *SInt*(*i*, *msg*) is the similarity in interest between the node *i* and the forwarded message *msg*, and *RemainingPower*(*i*) is the remaining battery level of the node *i* whose function is illustrated in Equation 6.30;

$$Re\,mainingPower(i) = f\left(battery(i), exp\,ectedusage(i)\right) \tag{6.30}$$

where *expectedusage(i)* is the expected amount of power that will be consumed due to the usage profile of the node i which indicates the average time consumed per activity. In addition, the power consumed due to network communications such as Wi-Fi scanning, forwarding, or receiving control messages is factored into the calculation of the expected usage. While we are proposing the aforementioned method, we acknowledge that there may be several methods of computing this component. For the simulation, whose results are illustrated in this chapter, we implement the following version of the expected usage function:

$$ExpectedUsage(i) = KalmanFilter(bat(i),$$
$$Avg\left(Count\left(Contacts(i)\right)\right), TTL(msg), \tag{6.31}$$
$$Consumption\left(WiFi(i), FWD(i), Rcp(i), idle(i)\right))$$

This function predicts the expected amount of battery usage for a specified user – using the Kalman filter predictor [25] – based on three factors. These factors are the node's current battery level $bat(i)$, the expected power consumption due to future contacts as an extrapolation from the average count of contacts recorded so far $Avg(Count(Contacts(i)))$, and the expected power consumption due to Wi-Fi connection $WiFi(i)$ (or Bluetooth connection) and actions of message forwarding or receiving ($FWD(i)$ and $Rcp(i)$) from now until the expiration time of the specified message time-to-live $TTL(msg)$.

The social active component of a node, *SocActive*, is computed as follows:

$$SocActive(i) = w_{SocActive} * Activeness(i) * SocialRank(i) \tag{6.32}$$

where *Activeness(i)* is the measure of activeness of the node i in interacting with other nodes, for which the formula is detailed in Equation 6.33; and *SocialRank(i)* is the social rank of node i, which is detailed in Equation 6.36.

The activeness part of the *SocActive* component is computed as follows:

$$Activeness(i) = \left((1/3) * normalizedContact\left(i, F(i)\right)\right)$$
$$+ \left((1/3) * CDC(i)\right) + \left((1/3) * Col(i, Dist)\right) \tag{6.33}$$

where *normalizedContact* is the normalized count of contacts between node i and its set of friends, $F(i)$. $CDC(i)$ is the change in degree of connectivity of node i, and $Col(i, Dest)$ is the normalized count of co-location of node i with any one of the destination nodes.

In this context, the change in degree of connectivity, CDC, is computed as the intersection of the count of nodes co-located with node i at time t and those co-located with node i at time $t - 1$, divided by the union of the count of those co-located with node i at time t and those co-located with node i at time $t - 1$. As per its definition in the *SocialCast* research [16], CDC is formulated as follows:

$$CDC(i) = \frac{Count\left(col(m, i, t)\right) \cap Count\left((m, i, t-1)\right)}{Count\left(col(m, i, t)\right) \cup Count\left(col(m, i, t-1)\right)} \tag{6.34}$$

where m is any one of the nodes co-located with node i at a certain time instance such that the co-location of a node with any one of the destination node is:

$$Coli(i, DestinationSet(msg)) =$$
$$Normalize(\forall Dest \in DestinationSet(msg) \forall t\ Count(Col(i, Dest))) \quad (6.35)$$

The social rank of node i is computed as the normalized social rank of its friends as follows:

$$SocialRank(i) = \sum_{\forall v \in F(i)} \frac{SocialRank(v)}{|F(v)|} \quad (6.36)$$

Any of the social metrics may be applied in computing the social rank of a node such as centrality or betweenness. In the first version, we combine the interest-aware and the power-aware components to come up with the interest-and-power-aware component which is computed as follows:

$$IntPowerComponent(i) = RewardedPower(i) *$$
$$normalCumInt(i, msg) \quad (6.37)$$

where $RewardedPower$ is computed as per Equation 6.5, while $normalCumInt(i)$ is the cumulative interest of the node which is computed as follows:

$$normalCumInt(i) = \frac{PInt(i) + \forall v \in F(i)\ PInt(v)}{|F(i)| + 1} \quad (6.38)$$

where $PInt(i)$ is the rewarded or penalized $SInt(i)$ when its value is compared to the predefined interest threshold.

The pseudocode of the algorithm is detailed in Algorithm 6.1

ALGORITHM 6.1   ADAPTIVE RANKING ALGORITHM

Require: SInt(source, msg) = 0

1: $\forall$ time t every n seconds, $\forall$ nodes physically in contact
2: **while** node i is physically in contact with node j **do**
3:     $SocialRank(i) = Avg(\forall v \in F(i)\ SocialRank(v))$
4:     **if** $j \in F(i)$ **then**
5:         $Update(SocialRank(i),\ SocialRank(j))$
6:     **end if**
7:     $CalculateAdpRank(i)$
8:     **if** $AdpRank(j) \geq AdpRank(i)\ AND\ TTL(msg) > 0$ **then**
9:         $Forward(msg, j)$
10:     **end if**
11: **end while**

Procedure CalculateAdpRank(*i*)

12: $normalizedContact(i, F(i)) = Avg(\forall v \in F(i) \mid Contact(i,v)\mid)$

13: $CDC(i) = \dfrac{\mid Neighbors(i, now)\mid \cup \mid Neighbors(i, now-1)\mid}{\mid Neighbors(i, now)\mid \cap \mid Neighbors(i, now-1)\mid}$

14: $Col(i, Dest) = Normalize(\forall d \in Dest \mid Contact(i,d)\mid)$

15: $Activeness(i) = ((1/3) * normalizedContact(i, F(i))) + ((1/3) * CDC(i)) + ((1/3) *$
$Col(i, Dest))$

16: $RemainingPower(i) = f(bat(i), expectedusage(i))$

17: $normalCumInt(i, msg) = \dfrac{PInt(i, msg) + \forall v \in F(i)\, PInt(v, msg)}{\mid F(i)\mid +1}$

18: $SocActive(i) = Activeness(i) * SocialRank(i)$

19: $w_{SocActive} = expectedusage(i)$

20: $RewardedPower(i) = bat(i) \pm reward$

21: $IntPowercomponent(i) = RewardedPower(i) * normalCumInt(i, msg)$

22: $w_{IntPower} = SInt(i, msg) * bat(i)$

23: $w_{Opp} = 1 - (w_{IntPower} + w_{SocActive})$

24: $Opportunistic(i) = SInt(i, msg) * RemainingPower(i)$

25: $AdpRank(i) = (w_{IntPower} * IntPowercomponent(i)) + (w_{SocActive} + SocActive(i))$
$+ (w_{Opp} * Opportunistic(i))$

6.4.2 Version 2: *AdpOp*

This second version of the adaptive ranking algorithm, *AdpOp*, includes the option of select-ing the next forwarder node if both its *SInt(i, msg)* and its remaining power *bat(i)* exceed their respective predefined thresholds. This criterion of selection is added to the selection criteria defined in the Adp algorithm detailed in Section 6.4.1.

Accordingly, a node is selected as the next forwarder if it satisfies the following conditions:

$$AdpRank(j) \geq AdpRank(i)\ AND$$
$$SInt(j) \geq Thr_{int}\ AND\ bat(j) \geq Thr_{bat} \tag{6.39}$$

6.4.3 Version 3: *AdpSyn*

This third version, *AdpSpSyn*, introduces space popularity to the first version as a factor in ranking nodes for the purpose of forwarder selection. The ranking function is formulated as follows:

$$AdpSpSyn(i) = w_{Opp} * Opportunistic$$
$$+ w_{IntPower} * CumInterst(i) * Power(i)$$
$$+ w_{SocActive} * Social Rank(i) * Activeness(i) \qquad (6.40)$$
$$+ w_{PowerAware} * Power(i)$$
$$+ w_{PlacePopularity} * Place Popularity(i)$$

where $w_{Opp} = 1 - (w_{IntPower} + w_{SocActive} + w_{PowerAware} + w_{PlacePopularity})$
and $w_{PlacePopularity} = Pop(Location(i, now))$

As for the place popularity of the node, and as per the recommended variations of Space Syntax metrics studied in [7], it is computed as the normalized value of the popularity of the location that this node is most frequently detected in. That is expressed in the following equation:

$$PlacePopularity(i) = Pop\left(FreqLocation(i)\right) \qquad (6.41)$$

6.4.4 Version 4: *AdpSpSynOp*

This fourth version, *AdpSpSynOp*, includes selecting the next forwarder node if both its *SInt(i, msg)* exceeds a predefined interest threshold and its remaining power *bat(i)* exceeds a predefined battery threshold. This criterion of selection is added to the selection criteria defined in the third version which is detailed in Section 6.4.3.

Accordingly, a node is selected as the next forwarder if it satisfies the following conditions:

$$AdpSpSyn(j) \geq AdpSpSyn(i) \ AND$$
$$SInt(j) \geq Thr_{int} \ AND \ bat(j) \geq Thr_{bat} \qquad (6.42)$$

6.5 EVALUATION

Using simulations, this work evaluates the performance of the implemented algorithms in terms of certain metrics while forwarding the message within a defined simulation period to the next candidates. The implemented four versions are compared in performance against the Epidemic forwarding algorithm as a benchmark, the social-aware opportunistic forwarding algorithm PeopleRank, and the adaptive power-aware social-based routing algorithm SCAR. The proposed versions are also compared against three other power-aware social-based opportunistic forwarding algorithms that incorporate some sort of dynamic weighing of the ranking attributes; namely, the EBubbleRap, PIPeROp, and PISCAROp algorithms.

It is worth mentioning that the implementation of the EBubbleRap algorithm in these simulations computes the local and the global ranking functions based on defining the respective community as follows. The local community is defined based on the user's interest in the forwarded content, while the global community is defined based on the social popularity of the user measured by the number of contacts stored in the user's device. In addition, the energy- awareness in the EBubbleRap algorithm is calculated as per the equation defined in their research paper [14].

6.5.1 Simulation Environment

In order to be able to simulate the performance of the adaptive ranking algorithms within a dynamic environment efficiently, this research uses the SAROS simulator [10]. The SAROS simulator is empowered by a set of features and data sets that enable it to simulate social-aware opportunistic forwarding in mobile opportunistic networks. SAROS imports real traces from opportunistic network environments such as university campuses, conferences, and malls, in addition to simulated mobility models such as the SLAW mobility model [26] for human walks in similar environments. From another aspect, for an accurate simulation and for logging of the energy consumption of the mobile devices during the forwarding process, SAROS implements accurate simulation of the power consumption of the mobile devices simulated in the experiments where four mobile brands are accurately simulated in terms of their power consumption per activity, and also implements various usage profiles of these mobile devices. To integrate the social-awareness in the forwarding algorithms, there is a need to simulate the social relationships among the users moving in the simulated environment. For that reason, the SAROS simulator imports several forms of social networks and social profiles. Table 6.1 lists the parameters of the simulation experiment.

The simulations conducted with the SAROS simulator import the mobility traces taken from the American University in Cairo during Fall 2012. The wireless connection traces of 1760 users were collected from 605 Wi-Fi spots located indoors across campus. Figure 6.1

Table 6.1 Simulation environment parameters.

Parameter	Default Value
TTL(msg)	1 hour
Wi-Fi range	50 m
damping factor	0.13 [15]
Thr_{bat}	0.5
Thr_{int}	0.5
Reward	0.5
w_{cdc}	0.25 [6]
w_{col}	0.25 [6]
Initial Battery Distribution	Full battery
Interest Distribution	Uniform Discrete Distribution
SInt(destination, message)	0.8–1
SInt(interested node, message)	0.5–0.79
SInt(uninterested node, message)	0–0.49
Mobile usage profiles	Suspend, Casual, Regular, Business, and Portable Media Device [27]
Percent(Destination nodes)	18% of the sample
Percent(Interested nodes)	36% of the sample
Percent(Uninterested nodes)	46% of the sample
Mobile brands	Samsung i900, HTC Diamond 2, Galaxyi7500, Spica [27]
No. of WiFi Access points	605 Access Points
No. of Traced users	1760 users
No. of Users with social profile	100 users
Source of traces	AUC Campus wireless traces
Duration of simulation	1 hour
Battery model	Kinetic Battery Model [28]

Date\Hour	12AM	1AM	2AM	3AM	4AM	5AM	6AM	7AM	8AM	9AM	10AM	11AM
30-Nov	194	203	185	145	113	100	100	125	119	162	210	262
1-Dec	246	217	177	151	117	108	88	157	286	475	667	992
2-Dec	289	264	215	173	128	118	128	314	0	0	0	0
22-Dec	118	110	91	83	66	63	61	67	72	122	179	231
23-Dec	107	106	85	64	54	62	63	96	321	442	483	577
25-Dec	52	39	37	31	32	21	30	33	36	47	52	47
26-Dec	50	45	48	50	27	39	40	72	271	362	384	348
27-Dec	29	32	22	24	15	18	19	53	266	341	372	413
30-Dec	19	15	17	13	11	12	12	41	208	285	275	309
31-Dec	13	7	7	8	7	5	11	31	183	230	225	257
8-Jan	13	7	3	0	4	3	0	40	256	360	372	491
10-Jan	12	7	6	3	0	6	6	9	134	228	174	249

(a)

Date\Hour	12PM	1PM	2PM	3PM	4PM	5PM	6PM	7PM	8PM	9PM	10PM	11PM
30-Nov	287	327	372	291	257	261	275	247	241	244	244	235
1-Dec	984	1161	1069	1111	695	0	0	0	0	0	0	18
2-Dec	0	0	0	0	0	0	0	0	0	0	0	1
22-Dec	266	263	305	324	283	260	260	215	193	134	124	107
23-Dec	607	641	595	681	572	396	324	292	212	107	94	54
25-Dec	76	94	72	85	106	93	76	73	61	66	58	60
26-Dec	439	472	465	473	404	213	153	126	110	75	52	37
27-Dec	438	419	385	409	354	161	90	83	60	44	46	29
30-Dec	383	346	318	337	292	126	76	74	48	39	31	15
31-Dec	256	259	238	276	224	102	55	43	57	45	51	28
8-Jan	474	442	431	352	287	147	99	73	61	31	22	15
10-Jan	253	218	219	182	163	82	43	24	19	17	10	2

(b)

Figure 6.1 User density heatmap of the AUC wireless traces.

demonstrates a heatmap of a sample of the user density across hours in some high-density working days during the semester. Figure 6.2 illustrates how overlaying the locations of the Wi-Fi spots on the university campus map and the floor maps facilitated the computation of the Space Syntax metrics utilized in the ranking function of the proposed algorithms. Finally, social profiles of 100 users were collected based on consent from these users.

For the sake of the simulation runs held for this research, the simulator classifies the users based on their interest in the forwarded content as per a discrete uniform interest distribution into three categories. The sets are the destination set, in which the users' interest match that of the forwarded content with a Jaccard similarity of 0.8 and above, while the interested

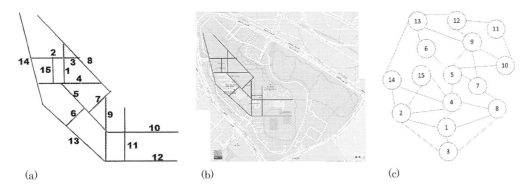

(a)   (b)   (c)

Figure 6.2 Space Syntax graph overlayed on the campus map.

forwarders set includes users with similarity interest between 0.79 and 0.5, and the third set is the set of uninterested users with similarity interest below 0.5. Different interest distributions have been examined, but their simulation results are not shared in this chapter due to space limitation.

As for the initial distribution of the battery level of the simulated nodes, the simulator starts the runs with full battery levels for all nodes in order to extract the pure effect of each algorithm on consuming the nodes' power. Different battery distributions have been examined, but their simulation results are not shared in this chapter due to space limitation.

6.5.2 Evaluation Metrics

To evaluate the performance of the proposed adaptive algorithms, we apply the following categories of evaluation metrics: effectiveness, efficiency, and power-awareness.

6.5.2.1 Effectiveness

The algorithm effectiveness is measured by classifying the contacted nodes as per their interest and by measuring the algorithm's f-measure.

Interest-based effectiveness: An algorithm is effective if it contacts a high portion of the interested users while simultaneously avoiding the uninterested ones. Our simulator measures this effectiveness in terms of the ratio of contacted users classified by their interest. Users are classified as either interested forwarders, destination nodes, or uninterested forwarders.

F-measure: Effectiveness is also measured through f-measure as the harmonic mean of recall and precision [29]. It should be noted here that the targeted true set consists of the interested forwarder nodes in addition to the destination nodes, while the false set consists of the uninterested nodes. The three metrics according to their definitions detailed in another research work [29], we compute to the following equations:

$$Recall = \frac{TP}{TP + FN} \tag{6.43}$$

$$Precision = \frac{TP}{TP + FP} \tag{6.44}$$

$$F-measure = \frac{Recall * Precision * 2}{Recall + Precision} \tag{6.45}$$

where

TP = True Positive = all contacted destination nodes and all contacted interested forwarders.
FP = False Positive = all non-contacted destination nodes and all non-contacted interested forwarders.
TN = True Negative = all non-contacted uninterested nodes.
FN = False Negative = all contacted uninterested nodes.

6.5.2.2 Efficiency

The efficiency of an algorithm is measured in terms of the cost it incurs, the delivery ratio it achieves, and the delay in time that occurs during the delivery process.

Cost is measured by the count of forwarded message replicas that were generated to accomplish this process. We measure the total number of message replicas that have been generated at any given time, and also measure the cost per unit delivery ratio.

Delivery ratio is measured by the portion of successfully reached destination nodes at the end of the simulation run.

Delay is measured as the average amount of time consumed starting from sending each message until it reaches one of the destination nodes. Delay also reflects the degree of user satisfaction. Thus, user satisfaction may be measured by the average delay consumed until a message is delivered to any destination node. The shorter the delay, the higher the degree of user satisfaction.

6.5.2.3 Power Awareness

To measure the level of power awareness of an algorithm, the total consumed power is computed and the degree of fairness in utilizing the nodes' resources is measured.

Power consumption: Algorithm power-efficiency is represented by its ability to conserve the overall power consumption. This metric is measured by computing the total amount of power consumed from all of the nodes' batteries over time. It is also measured by the total amount of power consumed per unit delivery ratio. In addition, a measure of the power consumption due to the exchange of control messages is illustrated to evaluate the overload of exchanging control messages and interest information. The power consumption of control message exchange is represented in the form of a percentage of the total amount of power consumption.

Utilization fairness: A fair algorithm would not exhaust some nodes' batteries in message forwarding while preserving other nodes' power. That is, it seeks to reduce variance among the nodes' battery levels. This is measured through the fairness index of the algorithm.

Borrowing from the fairness index defined in another research work [30], the standard deviation (SD) among the simulated population distribution of batteries' level is divided by their mean value, then subtracted from one. Equation 6.46 represents the calculation of the Fairness Index. The index ranges from 0 to 1, where 1 indicates the highest level of fairness when the SD of the final battery distribution reaches 0.

$$FairnessIndex = 1 - \frac{SD\left(\left\{Bat\left(i\right)|\forall i \in allnodes\right\}\right)}{mean\left(\left\{Bat\left(i\right)|\forall i \in allnodes\right\}\right)} \tag{6.46}$$

Overhead of Exchanged Control Messages: Part of the power consumed is exhorted as an overhead of extra control messages exchanged among nodes. This metric is measured in terms of the number of control messages exchanged among nodes during the forwarding process. It is also measured in terms of the percentage of power consumed due to forwarding these control messages among nodes.

6.5.2.4 Normalized Performance Indices

For a collective performance analysis of the algorithms across various environments, three normalized performance indices are devised by this research; namely, Effectiveness Index, Efficiency Index, and Power-Awareness Index. Each of these indices is computed as the harmonic mean of a group of the aforementioned metrics after normalizing their values. These indices rely on the harmonic mean instead of the arithmetic mean, since the harmonic mean of a list of numbers tends strongly toward the least elements of the list, and it tends (compared to the arithmetic mean) to mitigate the impact of large outliers and to aggravate the impact of small ones [31].

Let us have a look at the detailed calculation of the indices:

Effectiveness Index: This performance index measures the algorithm's effectiveness through the harmonic mean of the f-measure, the ratio of the contacted interested forwarders, and the ratio of the contacted uninterested nodes. Its formula is

$$Effectiveness\ Index = Harmonic\ Mean \begin{pmatrix} Fmeasure, \dfrac{Contacted\ Interested\ Forwarders}{Total\ number\ of\ Interested\ Forwarders}, \\ \dfrac{Contacted\ Uninterested\ Nodes}{Total\ number\ of\ Uninterested\ Nodes} \end{pmatrix} \quad (6.47)$$

Efficiency Index: This performance index measures the algorithm's efficiency in delivery through the harmonic mean of the delivery ratio, the normalized value of the paid cost, and the normalized value of the delay in delivering the message. Since the lower the cost, the higher the efficiency, the formula utilized subtracts the normalized cost from 1. For the same reason, the normalized delay is subtracted from 1. The formula is:

$$EfficiencyIndex = HarmonicMean(Delivery, \\ 1 - Normalize(Cost), 1 - Normalize(Delay)) \quad (6.48)$$

Power-Awareness Index: This performance index measures the algorithm's power awareness through the harmonic mean of the normalized value of the total ratio of consumed power and the Fairness Index. The normalized value of the power consumption is subtracted from 1, since the lower the power consumption, the higher the power awareness of the algorithm. The formula is:

$$PowerAwarenessIndex = HarmonicMean(Fairness\ Index, \\ 1 - Normalize(Power\ Consumption)) \quad (6.49)$$

6.6 RESULTS

This section demonstrates the results of the simulations that are held by the SAROS simulator comparing the performance of the proposed adaptive ranking algorithm versions and the implemented state-of-the-art social and power-aware opportunistic forwarding algorithms that cater for some sort of dynamic weighted ranking attributes.

6.6.1 Interest Awareness

When compared to the simulated algorithms, the adaptive ranking versions show improvement in terms of effectiveness and efficiency as follows.

6.6.1.1 Effectiveness

From the Interest-based effectiveness figure (Figure 6.3a), it is evident that effectiveness of the adaptive ranking versions approaches the effectiveness of the Epidemic benchmark in

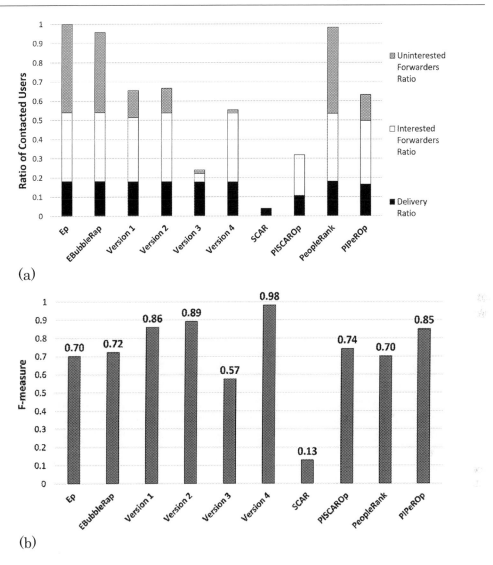

(a)

(b)

Figure 6.3 Effectiveness of the adaptive algorithms.

terms of the delivery ratio and the ratio of contacted interested forwarders (except for the third version, *AdpSpSyn*, which is very conservative in selecting forwarders). The fourth version, *AdpSpSynOp*, significantly reduces the ratio of the contacted uninterested users in comparison to the other algorithms by about 96%, except for the SCAR versions that totally avoid them but on the account of attaining a very low delivery ratio. Moreover, Space Syntax integration within the ranking function (*AdpSpSyn* and *AdpSpSynOp*) maintains the same delivery ratio as that achieved without Space Syntax (*Adp* and *AdpOp*) but, crucially, avoids contact with 86% of the uninterested nodes. It is also noticeable that integrating the interest-threshold opportunistic selection into the ranking function (*AdpOp* and *AdpSpSynOp*) enables contact with a higher ratio of interested forwarder nodes. Compared to the PIPeROp algorithm, the fourth version, *AdpSpSynOp*, achieves a higher delivery ratio and a slight increase in the ratio of the contacted interested nodes and a reduction in the ratio of the contacted uninterested nodes.

From the f-measure figure (Figure 6.3(b)), it is clear that the adaptive algorithm versions achieve the highest level of f-measure (except for the third version, *AdpSpSyn*) where the f-measure of the fourth version, *AdpSpSynOp*, is 32% higher than that of the PISCAROp algorithm and 15% more than that of the PIPeROp algorithm. However, the f-measure of the non-opportunistic third version, *AdpSpSyn*, is much less than the opportunistic fourth version, *AdpSpSynOp*, because it contacts a lower ratio of interested forwarders. The opportunistic version, *AdpSpSynOp*, overcomes this defect through explicit selection of interested forwarders.

From the F-measure sub-figure in Figure 6.3(b), it is clear that integrating the PI-SOFA framework with any of the social-aware forwarding algorithm boosts its f-measure such as the PISCAROp algorithm whose f-measure exceeds the SCAR algorithm by 475.5%.

6.6.1.2 Efficiency

Efficiency is speculated here in terms of the paid cost per unit delivery ratio. If the algorithms are compared in terms of cost without paying attention to the ratio of gained benefit, a false conclusion will be reached. Thus, it is much more accurate to study the cost per unit delivery ratio – as shown in Figure 6.4(a).

The third version, *AdpSpSyn*, is the most cost-efficient algorithm for exerting the lowest cost to attain delivery ratio approaching that of the benchmark with an achievement of 75.6% reduction in cost. Although the SCAR algorithm exerts the least cost, its delivery ratio is the lowest compared to all the other algorithms. Moreover, all four versions of the adaptive algorithm achieve higher delivery ratio when compared to the PIPeROp algorithm, yet only the Space Syntax versions, *AdpSpSyn* and *AdpSpSynOp*, preserve some cost, ranging from 13% to 48% reduction in cost. It is also worth noting that by introducing the threshold opportunistic component to the adaptive ranking with an additional cost, the fourth version, *AdpSpSynOp*, successfully approaches the full delivery ratio.

It is noticeable that the adaptive versions exert a small delay in time while the other PI-SOFA versions incur a long delay in delivery time; thus, the adaptive versions achieve a 93% reduction in delay. It is also worth noting that the fourth version, *AdpSpSynOp*, incurs reasonable delay in order to achieve the highest delivery ratio and the highest ratio of contacted interested forwarders out of the four adaptive versions. Overall, the second version, *AdpOp*, incurs the least delay among the four adaptive versions. Thus, the adaptive versions are able to accomplish the difficult task of achieving a very high delivery ratio with a relatively short delay in delivery time, thus competing with the Epidemic algorithm in terms of the short delay in delivery time on one side, and the high delivery ratio on the other. Over and above, the adaptive versions contact fewer uninterested nodes, something that Epidemic fails to maintain.

6.6.2 Power Awareness

The results of the simulation demonstrate the improvement in power awareness of the proposed algorithms in comparison to the benchmark and the social-based power-aware opportunistic forwarding algorithms. Power consumption and utilization fairness are the two dimensions of metrics measured to demonstrate the power awareness feature of the examined algorithms. Furthermore, a study of the power consumed due to the exchange of control messages in each algorithm is evaluated to reveal the exerted power and communication due to the overhead of the control message exchange.

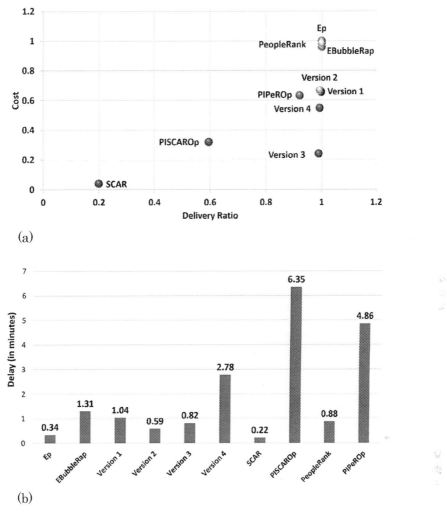

Figure 6.4 Efficiency of the adaptive algorithms.

6.6.2.1 *Power Consumption Awareness*

It is more comprehensive and justifiable to analyze the algorithms' performance in terms of power consumption per unit delivery ratio, as illustrated in Figure 6.5(a). From this figure, it is clear that with a minor increase in power consumption, the first version, *Adp*, succeeds to achieve an exemplary delivery ratio in comparison to the power-and-interest aware PeopleRank algorithm, PIPeROp. It is also clear that integrating Space Syntax in the ranking function, *AdpSpSyn* and *AdpSpSynOp*, significantly reduces the percentage of consumed power per unit delivery ratio by 10%, as it facilitates more effective selection of forwarder nodes that frequent popular places, and thus speeds up the delivery process. On the other hand, the SCAR algorithm and its power-and-interest aware version, the PISCAROp algorithm, conserve power, but they achieve the lowest delivery ratio among all of the compared algorithms. Thus, this power conservation does not meet the desired delivery ratio.

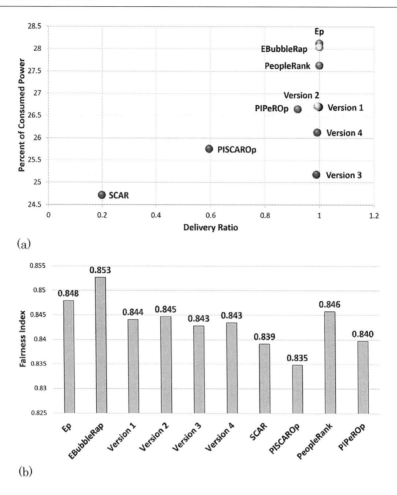

(a)

(b)

Figure 6.5 Power awareness of the adaptive algorithms.

6.6.2.2 Power Utilization Fairness

From the power utilization fairness figure (Figure 6.5(b)), it is clear that integrating the interest-based opportunistic selection in the ranking function slightly improves the level of utilization fairness as exemplified in the performance of the second version, *AdpOp*, compared to that of the other adaptive versions. However, all the adaptive versions achieve a lower level of fairness than that of the EBubbleRap and the PeopleRank algorithms. It is interestingly noticeable how the EBubbleRap algorithm and the Epidemic algorithm satisfy the rule of "equality in injustice amended" by exhausting the power of the majority of the community nodes to attain fairness.

6.6.2.3 Overhead of Exchanged Control Messages

In order to measure the overhead posed by the proposed adaptive algorithms in terms of extra control messages exchanged among nodes, this subsection illustrates the number of control messages exchanged among nodes during the forwarding process. This section also presents the percentage of power consumed due to sending these control messages among nodes.

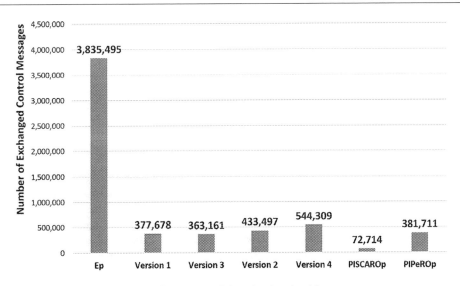

Figure 6.6 Number of produced control messages of the adaptive algorithms.

Figure 6.6 illustrates the number of control messages exchanged among nodes when applying each of the proposed adaptive ranking algorithms within the simulation, while Figure 6.7 illustrates the percentage of power consumed due to the control messages sent by these algorithms. From both figures, it is clear that the fourth version, *AdpSpSynOp*, consumes power for the highest number of control messages, while its non-opportunistic third version, *AdpSpSyn*, consumes the least number of control messages and the least power in forwarding them.

It is worth mentioning that the *Epidemic* algorithm consumes the highest percentage of exchanged control message power consumption due to sending the highest number of such messages. In addition, the PISCAROp algorithm consumes the least number of exchanged control messages, yet it consumes the highest percentage of power compared to all of the other algorithms – except for the Epidemic algorithm – which makes it not power-efficient.

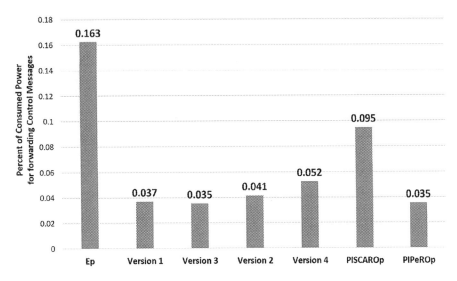

Figure 6.7 Percentage of consumed power for the control messages of the adaptive algorithms.

6.6.3 Normalized Performance Indices

The performance of the proposed algorithms is evaluated through the three normalized performance indices defined in Section 6.5.2.4. From Figures 6.8 through 6.10, we analyze the three performance indices – namely, the effectiveness performance index, the efficiency performance index, and the power awareness performance index – of the simulated algorithms within the defined simulated environment.

6.6.3.1 Effectiveness Performance Index

From the effectiveness performance index figure (Figure 6.8), it is noticeable that the fourth version, *AdpSpSynOp*, maintains the highest effectiveness performance index among all of the compared algorithms with an increase in the index ranging from 19.7% (when compared to the PIPeROp algorithm) to 75.5% (when compared to the SCAR algorithm). On another side, the effectiveness performance index of the third version, *AdpSpSyn*, is the lowest among all of the four versions due to its deficiency in contacting many of the interested forwarders.

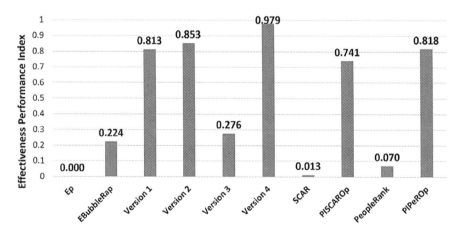

Figure 6.8 Effectiveness performance index.

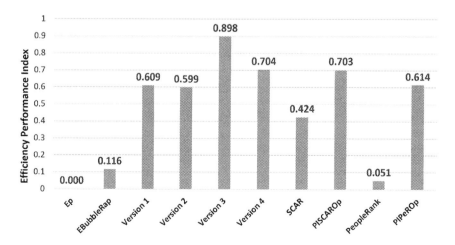

Figure 6.9 Efficiency performance index.

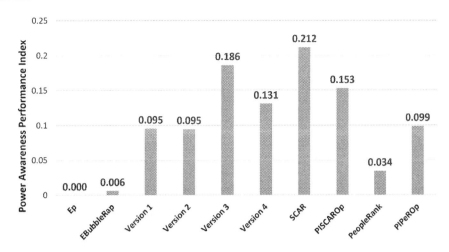

Figure 6.10 Power awareness performance index.

6.6.3.2 Efficiency Performance Index

From the efficiency performance index figure (Figure 6.9), it is evident that the non-opportunistic third version, *AdpSpSyn*, maintains the highest degree of efficiency with an increase in value ranging from 24% in comparison to the PISCAROp algorithm, up to 1660% when compared to the PeopleRank algorithm. The third proposed version of the adaptive algorithm, *AdpSpSyn*, is cost effective as it pays the least cost among the four adaptive versions to achieve a quite comparable delivery ratio.

6.6.3.3 Power Awareness Performance Index

From the power awareness performance index figure (Figure 6.10), it is clear that the algorithms that achieve a low delivery ratio are able to preserve power and maintain a high level of utilization fairness. Thus, the SCAR algorithm achieves the highest power awareness performance index value as it achieves the lowest delivery ratio among the compared algorithms. Note that although the third version, *AdpSpSyn*, scores the second-highest power awareness performance index value while maintaining a very high delivery ratio, it fails to contact many of the interested forwarders, which still leads to a preservation in power consumption. From the power awareness index figure, one deduces that integrating Space Syntax in the ranking function improves the power awareness of the algorithms.

6.6.4 Eight-Metric Performance Comparison

Figure 6.11 illustrates the collective performance comparison among the four proposed adaptive ranking versions. From the figure, we can notice that version 3, *AdpSpSyn*, effectively reduces the paid cost and the delay to maintain the same delivery ratio and to avoid contacting uninterested nodes. However, it achieves the lowest f-measure since it fails to contact the majority of the interested forwarders. On the other hand, version 4 – *AdpSpSynOp* – saves some cost in comparison to the first two versions and avoids contacting uninterested nodes while maintaining the same delivery ratio and the same interested forwarders contact ratio, and thus, it achieves the highest f-measure. All of the four versions maintain the same level of power consumption and utilization fairness.

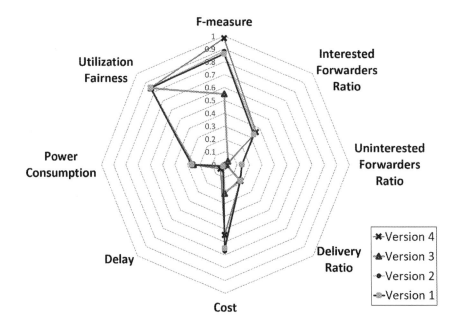

Figure 6.11 Eight-metric performance comparison.

6.7 DISCUSSION AND CONCLUSION

In this chapter, we argue that there are limitations in social-based opportunistic forwarding that is oblivious to the variation in importance of the nodes' ranking attributes as context changes. It may also be unaware of the absence of some context information as the surrounding context varies. More specifically, forwarding algorithms can achieve improvement in performance in terms of effectiveness, efficiency, and power awareness if they leverage knowledge of the current context of the nodes involved in the process. Such context includes the density of users in the place, the distribution of power capabilities of the nodes in proximity, the willingness of the nodes in proximity to share their context information, their willingness to participate in the forwarding process, or the speed of movement of the users which may affect connectivity. This context information can be used to continuously adapt the weights of the ranking function's attributes. Such attributes include the nodes' power capabilities, the user's interest in the forwarded content, the user's social popularity, their activeness in engaging with social contacts, the popularity of the places that are frequently visited, and the popularity of the place the nodes are currently located in.

This research demonstrates that with such usage of adaptation of the ranking function, a forwarding algorithm can achieve greater effectiveness, efficiency, and power awareness.

We thus propose a framework that injects adaptive ranking into opportunistic forwarding algorithms. This framework is illustrated in the proposed four adaptive ranking versions that vary in integrating the opportunistic threshold and the Space Syntax awareness. They are compared to many other interest-and-power-aware social-based opportunistic forwarding algorithms which provide some sort of dynamic weighing of the ranking of a function's attributes. In such comparison, the Epidemic algorithm is selected as a benchmark. Furthermore, we devise performance indices in order to evaluate the performance of all the proposed algorithms.

All the simulations and evaluations presented here were conducted using our own Social-AwaRe Opportunistic forwarding Simulator (SAROS). This simulator provides an implementation of a set of state-of-the-art social-based opportunistic forwarding algorithms, and a variety of devised evaluation metrics. One of the strong features of this simulator is that it is able to simulate users' interest in forwarded content as well as the power capabilities of their devices. The simulator also imports a variety of real traces, such as the gathered mobility traces made by the American University in Cairo (AUC) community and accessed through the AUC wireless connectivity network. The simulation results demonstrate a great improvement in performance in terms of effectiveness, efficiency, power awareness, and utilization fairness.

As a final note, given that these algorithms have numerous real-life applications, it is important to point out that the most suitable algorithm can be determined based on the needs and requirements of the message senders, the nature of the community of users, the frequency and duration of contact, the users' willingness to participate, the capability of the devices, or any other limitations in terms of cost and duration. On the basis of this evaluation, we offer a set of implementation-specific recommendations. One set of useful applications would be the wireless sensor networks where there are power limitations in the sensors used in this type of network. Accordingly, the advisable forwarding algorithms would be the power-aware, depletion-rate-aware versions of the proposed algorithms. From another aspect, for applications that care for forwarding content to interested parties without entirely contacting uninterested ones, the preferable forwarding algorithms would be the interest-aware versions of the proposed PI-SOFA algorithms [11]. From a different aspect, applications that forward content in environments that are characterized by a dynamically changing context, the adaptive forwarding algorithms would fit the most in such environments since these algorithms will adapt the node's ranking function according to the variations in context. In such environments, the third version, *AdpSpSyn*, is cost effective, as it pays the lowest cost and consumes less power to achieve a quite comparable delivery ratio. However, relying on both interest-awareness and Space Syntax, as exemplified in the fourth version (*AdpSpSynOp*), speeds up the delivery process and increases the f-measure since it relies on the interested nodes who frequent popular places.

ACKNOWLEDGMENT

The authors would like to thank Dr. Khaled Harras, Carnegie Mellon University Qatar, for his efforts within the scope of co-supervision of this work within the PhD dissertation of Soumaia Al Ayyat.

REFERENCES

[1] D. Xu, Y. Li, X. Chen, J. Li, P. Hui, S. Chen, and J. Crowcroft. A Survey of Opportunistic Offloading. *IEEE Communications Surveys Tutorials*, 20(3):2198–2236, thirdquarter 2018.

[2] Chiara Boldrini et al. *Modelling Social-Aware Forwarding in Opportunistic Networks*. In *Performance Evaluation of Computer and Communication Systems*, vol. 6821, pp. 141–152. Springer, 2011.

[3] Y. Zhao and W. Song. Survey on Social-Aware Data Dissemination Over Mobile Wireless Networks. *IEEE Access*, 5:6049–6059, 2017.

[4] H. Y. Bohari and Sh. O. Khanna. Energy Efficient Power Aware Routing Algorithm (EEPARA) For Mobile Ad Hoc Network (MANET). *IJERT*, 2(7):1562–1572, 2013.

[5] Soumaia Al Ayyat, Sherif G. Aly, and Khaled A. Harras. *PIPeR: Impact of Power-Awareness on Social-Based Opportunistic Advertising*. In *IEEE WCNC*, pp. 3462–3467, April 2014.

[6] B. Pasztor et al. *Opportunistic Mobile Sensor Data Collection with SCAR*. In *IEEE MASS*, pp. 1–12, 2007.

[7] Soumaia Al Ayyat, Sherif G. Aly, and Khaled A. Harras. *On Integrating Space Syntax Metrics with Social-aware Opportunistic Forwarding*. In *IEEE WCNC*, pp. 1–7, 2019.

[8] Amr Al Jarhi, Hend Adel Arafa, Khaled A. Harras, and Sherif G. Aly. *Rethinking Opportunistic Routing Using Space Syntax*. In *Proceedings of the 6th ACM Workshop on Challenged Networks*, CHANTS '11, pp. 21–26, 2011.

[9] Ahmed Mtibaa and Khaled A. Harras. *Exploiting Space Syntax for Deployable Mobile Opportunistic Networking*. In *IEEE MASS*, pp. 533–541, 2013.

[10] Soumaia Al Ayyat, Sherif G. Aly, and Khaled A. Harras. *SAROS: A Social-Aware Opportunistic Forwarding Simulator*. In *IEEE WCNC*, pp. 1–7, 2016.

[11] Soumaia Al Ayyat, Khaled A. Harras, and Sherif G. Aly. On the Integration of Interest and Power Awareness in Social-Aware Opportunistic Forwarding Algorithms. *Computer Communications*, 71, pp. 97–110, 2015.

[12] Radu-Ioan Ciobanu, Daniel Gutierrez Reina, Ciprian Dobre, and Sergio L. Toral. Context-Adaptive Forwarding in Mobile Opportunistic Networks. *Annals of Telecommunications*, 73, pp. 559–575, 2018.

[13] Pan Hui et al. *Bubble Rap: Social-Based Forwarding in Delay Tolerant Networks*. In *9th ACM MobiHoc Proceedings*, pp. 241–250, 2008.

[14] Cristian Chilipirea, Andreea-Cristina Petre, and Ciprian Dobre. Energy-Aware Social-Based Routing in Opportunistic Networks. *International Journal of Grid and Utility Computing*, 1(3/4):1, 2013.

[15] Abderrahmen Mtibaa, Martin May, and Mostafa Ammar. *Social Forwarding in Mobile Opportunistic Networks: A Case of PeopleRank*. In *Communication and Social Networks*, 58, pp. 387–425. Springer, 2012.

[16] P Costa et al. Socially-Aware Routing for Publish-Subscribe in Delay-Tolerant Mobile Ad Hoc Networks. *IEEE Journal on Selected Areas in Communications*, 26(5):748–760, 2008.

[17] Rares Vernica, Michael J. Carey, and Chen Li. *Efficient Parallel Set-Similarity Joins Using Mapreduce*. In *ACM SIGMOD Proceedings*, pp. 495–506, 2010.

[18] Bill Hillier and Julienne Hanson. *The Social Logic of Space*. Cambridge University Press, 1984. Cambridge Books Online.

[19] Abderrahmen Mtibaa and Khaled A. Harras. *Select & Spray: Towards Deployable Opportunistic Communication in Large Scale Networks*. In *Proceedings of the 11th ACM International Symposium on Mobility Management and Wireless Access*, MobiWac '13, pp. 1–8. ACM, 2013.

[20] Vassilis Kostakos. *Space Syntax and Pervasive Systems*. In *Geospatial Analysis and Modeling of Urban Structure and Dynamics*. Springer Science, 2009.

[21] Peter C. Dawson. *Analysing the Effects of Spatial Configuration on Human Movement and Social Interaction in Canadian Arctic Communities*. In *Proceedings of the 4th International Space Syntax Symposium*, 2003.

[22] Bin Jiang. Ranking Spaces for Predicting Human Movement in an Urban Environment. *International Journal of Geographical Information Science*, 23(7):823–837, 2009.

[23] Alper Unlu, Ozan O. Ozener, Tolga Ozden, and Erincik Edgu. *An Evaluation of Social Interactive Spaces in a University Building*. In *Proceedings of the 3rd International Space Syntax Symposium*, 2001.

[24] Honglong Chen and Wei Lou. Contact Expectation Based Routing for Delay Tolerant Networks. *Ad Hoc Networks*, 36:244–257, 2016.

[25] Chris Chatfield. *The Analysis of Time Series: An Introduction*. Chapman and Hall/CRC, 6th edition, 2004.

[26] Kyunghan Lee Kaist et al. *SLAW: A Mobility Model for Human Walks*. In *INFOCOM*, pp. 855–863, 2009.

[27] Aaron Carroll and Gernot Heiser. *An Analysis of Power Consumption in a Smartphone*. In *Proceedings of USENIX*, 2010.

[28] J. Manwell and J. McGowan. *Extension of the Kinetic Battery*. In *Model for Wind/Hybrid Power Systems*. *Proceedings of EWEC*, pp. 284–289, 1994.

[29] David M. W. Powers. Evaluation: From Precision, Recall and F-Measure to ROC, Informedness, Markedness & Correlation. *Journal of Machine Learning Technologies*, 2(1):37–63, 2011.

[30] Afra J. Mashhadi et al. Fair Content Dissemination in Participatory DTNs. *Ad Hoc Networks*, 10(8):1633–1645, 2012.

[31] E. Henry, T.R. Robinson, J.D. Stowe, and A. Cohen. *Equity Asset Valuation*. CFA Institute Investment Series. Wiley, 2010.

Chapter 7

Performance Analysis of AODV and DSDV Routing Protocols in Mobile Ad Hoc Network Using OMNeT++

Rukhsana Naznin and Md. Sharif Hossen

Comilla University, Cumilla, Bangladesh

CONTENTS

7.1 INTRODUCTION AND RELATED WORKS

A mobile ad hoc network (MANET) is a kind of network where communication among the mobile nodes is established in such a scenario where there exists no fixed infrastructure. Therefore, the need to establish such a network in the areas of emergency communication, wildlife tracking, earthquake, terrorist attack, and so on is dominating. This network is a temporary network where devices communicate with each other with no fixed infrastructure. The nodes are arbitrary. Therefore, there is no centralized administration due to the mobility of the nodes in the network. There is an abrupt change of topology. The devices are self-organized and self-configured because of their dynamic nature. They are connected via wireless links and communicate with each other [1]. To maintain the communication between two nodes requires some rules which will define the route. These well-defined and specified rules are called routing protocols. Routing is the network layer task, which accomplishes the technique of choosing the appropriate path. It will create flooding at the node if broadcasting to all the nodes occurs. So, this should be avoided and also needs an alternative route as a backup [2]. To create routing decisions, the routing protocol is the fundamental need. The routing algorithms for wired networks are not suitable for the wireless networks which are dynamic in nature. The delivery of a packet will not be successful until the topology is stable; that's why the changes in topology require a rapid response. A MANET has limited resources, which is a big challenge for successful communication. So, it requires an efficient routing protocol that can utilize those limited resources properly and can cope with the rapid changes in the network under any circumstances, like the size of the network, traffic density, and mobility of nodes, topology, and even broken routes [3]. The shortest path selection, less power consumption, and less overhead are the main features for a routing protocol. So, the routing will be done in minimal time with the maximum utilization of bandwidth, even after the frequent modification in topology. In this study, we compare the efficiency of routing protocols by the packet delivery ratio, packet loss, average end-to-end delay, and throughput by changing the number of nodes, speed, and packet length [4].

Researchers have examined many protocols that are optimum to implement on a defined network. These researchers have compared performance among some traditional routing protocols such as AODV, CBRP, PAODV, DSDV, and DSR [5,6]. In [7], the authors conducted an analysis between AODV and DSDV on the basis of QoS parameter using NS-2. In [8], the authors review eight routing protocols of MANET, including AODV, DSR, TORA, SSR, ABR, DSDV, CGSR, and WRP, using NS-2. In this chapter, the performance of AODV and DSDV is analyzed using OMNeT++.

The rest of this chapter is organized as follows. Section 7.2 discusses the classification of the MANET routing protocols. Sections 7.3 and 7.4 illustrate the DSDV and the AODV routing protocols, respectively. Section 7.5 represents the discussion of the simulation tool. The result and discussion parts are included in Section 7.6, while Section 7.7 includes the concluding remark with future work.

7.2 MANET ROUTING PROTOCOLS

In MANET, maintaining connectivity is very tough because the topology frequently changes. Therefore, it requires a stronger and more flexible mechanism to maintain routes. Many protocols have been introduced for MANETs. These protocols can be classified into three categories: proactive, reactive, and hybrid, as shown in Figure 7.1.

7.2.1 Proactive Routing Protocol

In this technique, one or more routing tables are maintained by each node in the network that is updated regularly. Each node sends a printed message to the entire network if there is a change within the network. This incurs extra overhead cost because it keeps up-to-date information, which can affect the throughput of the network, but it presents the actual facts to the availability of the network [10]. In this method, all the nodes are involved in creating a route. Such techniques react to topology changes, even if the changes do not affect the traffic. This technique is also called table-driven strategies. Some protocols of this category are WRP, OLSR, and DSDV [11].

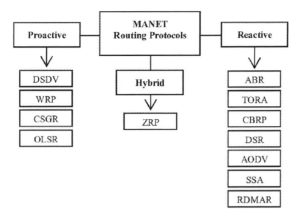

Figure 7.1 The classification of MANET routing protocols [9].

7.2.2 Reactive Routing Protocol

Reactive methods have supported the demand for data transmission and are suitable for ad hoc networks. These types of methods are also known as on-demand methods. In this method, the routing overhead will reduce when the traffic is minimal. Therefore, the changes in the topology are not significant since the router's table is not updated frequently. They find and maintain a route on which there is no traffic.

7.2.3 Hybrid Routing Protocol

Hybrid routing is the amalgamation of proactive and reactive techniques. This routing approach shows good performance by using the advantages of proactive and reactive mechanisms. The most common hybrid routings are ZRP, ZHLS, SHARP, and NAMP [12].

7.3 DESTINATION-SEQUENCED DISTANCE-VECTOR ROUTING (DSDV)

DSDV is one of the proactive routing techniques. It is an improved version of the Bellman-Ford algorithm that guarantees a loop-free routing using a routing table.

7.3.1 Mechanism of the DSDV Routing Protocol

The mechanism of DSDV is very simple. Every entry of the table encompasses a serial number, and when associate nodes send any update message, the sequence number will be increased. The routing table's area unit goes up sporadically once the topology of the community changes, and the area unit is propagated for the length of the network to preserve regular records at some point in the network. Each DSDV node continues two preserved regular records at some points in the network for forwarding and advertising of packets. The active node will send an update packet to its neighbor node if there is any change in the topology of the network. After receiving the updated packet from the neighboring node, the information is extracted by the node, and the router table is updated as follows (Figure 7.2).

    i.   If the current sequence is greater than the previous sequence, then the route consisting of the greater sequence will be chosen. Also, the previous sequence number will be omitted.

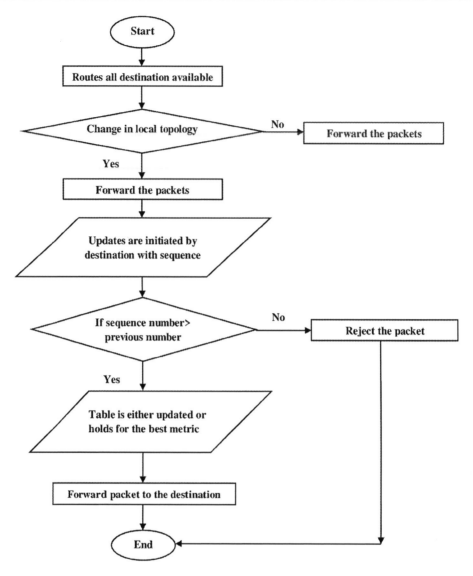

Figure 7.2 The flow chart of the DSDV routing protocol.

ii. A route with the smallest amount fee is chosen if the new and existing sequence selection is the same as this route.

iii. From the newly updated routing table, all the metrics will be chosen.

iv. The process is continued until the whole network units are updated. If a least cost path is detected, it will be maintained and the rest of the path will be omitted.

There are two processes of updating the routing table in the DSDV:

- *Full dump*: In terms of routing, all the information that a node needs is contained by full dump packets. As a result, this full dump approach needs Network Protocol Data Units (NPDUs) when the routing table is full or near full. If there is occasional movement of node in the network, then packets are transmitted occasionally.

- *Incremental*: The incremental packet can choose the best information as they try to get the update of full dump packets dispatched by the nodes. So, this approach utilizes the network resources.

7.3.2 Advantages of the DSDV Protocol

DSDV has some important advantages in MANET. Some of these are as follows [7, 13]:

i.   Loop free paths are guaranteed in the DSDV protocol.
ii. DSDV reduces the count to an infinity problem.
iii. Using incremental updates, extra traffic can be avoided instead of doing a full dump update.
iv. DSDV can reduce the size of the routing table since it maintains not all the paths but only the effective one.
v.   DSDV is a very suitable protocol for an ad hoc network with a lower-size network.

7.3.3 Limitations of the DSDV Protocol

DSDV has some limitations, which are given as follows [7, 13]:

i.   There is unnecessary advertising of routing information even if no changes occur in the network topology. It causes the wastage of bandwidth.
ii. Multipath routing ids are not supported by DSDV.
iii. It is difficult to find out the time delay for sharing the information of routes.
iv. The routing table needs to be updated periodically, which reduces the battery lifetime and also the bandwidth.
v.   Each node needs a table for advertising, so for a large network, the table will be large, which results in high overhead.

7.4 AD HOC ON-DEMAND DISTANCE VECTOR (AODV)

AODV is one of the reactive types of routing protocols in MANETs. As long as the source requests for a route, the AODV protocol builds routes between the nodes. AODV does not produce any additional traffic for communication on links; that's why it is considered an on-demand algorithm. Also, to attach multicast group members, trees are formed. Sequence numbers are used in AODV to make sure the route freshness.

7.4.1 Mechanism of the AODV Routing Protocol

The base of the AODV method is hop-to-hop routing. The node which needs to know the route of any particular destination will send the route request (RREQ). Then, the transitional nodes will forward the route request. At a similar time, those transitional nodes will produce a reverse route to the destination [14]. After receiving the request, a node establishes a Route Reply (RREP), in which a number of hops are used to reach to the destination. Figure 7.3 illustrates the mechanism of AODV routing with RREQ and RREP.

7.4.2 Key Features of AODV

AODV has the following features [7].

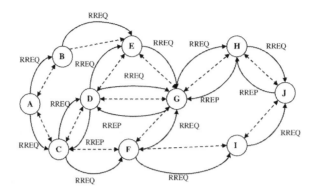

Figure 7.3 Mechanism of RREQ and RREP transmission in AODV [13].

i.   It supports unicast, broadcast, and multicast communication.
ii. It establishes an on-demand route with a small delay.
iii. Link breakage in an active route is solved in an efficient way.
iv. It uses sequence numbers to trace the accuracy of information.
v.   Through the use of sequence numbers, all routes are loop-free.
vi. It uses periodic HELLO messages to trace the neighbors.
vii. A multicast tree connecting cluster members is maintained for a lifespan of the multicast cluster.

7.4.3 Advantages of the AODV Routing Protocol

The advantages of AODV are represented as follows [7, 13]:

i.   Due to the reactive nature of AODV, it can maintain changeable behavior of MANETs.
ii. It supports both unicasts and multicasts.

7.4.4 Limitations of the AODV Routing Protocol

There are some limitations to the AODV routing protocol. These include the following [7, 13]:

i.   The nodes within the network will find each other's broadcasts. It is one of the requirements of this algorithm.
ii. There is no reuse of routing information. The routing information is obtained on demand.
iii. Packets will be distorted due to the insider attackers together with route distraction, and to incursion.
iv. AODV finds a route after the initialization of a flow which results in high delay of a large network.
v.   High overhead on the information measure.

7.5 SIMULATION METHODOLOGY

A simulation is the re-creation of a real-world process during a controlled environment. It involves creating laws and models to represent the world, then running those models to discover what happens. Simulation software is predicted on the method of modeling a true phenomenon with a group of mathematical formulas. It is essentially a program that permits the user to watch an operation through simulation without actually performing that operation. In this project, OMNeT++ is used as the simulation tool.

7.5.1 Mechanism of OMNeT++

OMNeT++ is an event-driven framework to build and implement the emulated scenario of mobile ad hoc networks. It is developed using the C++ language. Figure 7.4 shows the simulation scenario of OMNeT++. The source of the OMNeT++ model usually contains the following files [15]:

i.   The simple module implementations and other codes are included in C++ (.cc and .h) files.
ii. The message definitions which are to be translated into C++ classes are contained by message (.msg) files.
iii. The description of topology and the components are included in NED (.ned) files;
iv. The assignment of parameters of considered model and other settings are included in Configuration (.ini) files.

The process to turn the source into an executable form is the following, in nutshell [15]:

i.   The message compiler translated Message files into C++ (opp_msgc,).
ii. Object form (.o file) are created by compiling C++ sources.
iii. To get an executable or shared library, object files are linked with the simulation kernel and other libraries.

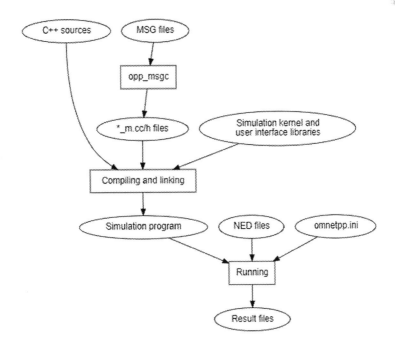

Figure 7.4 Building and running simulation [15].

7.5.2 Simulation Parameters

Some parameters are essential for the simulation. Simulation details are given in Table 7.1.

7.6 RESULT AND DISCUSSION

The simulated results in details with some performance metrics considered in Section 6.1 will be discussed here.

7.6.1 Performance Metrics

Packet delivery ratio: The total number of packets delivered to the destination divided by the total number sent from the source is called the packet delivery ratio.

Packet loss: The subtraction of the number of packets sent from the source and the packets received to the destination is called the packet loss.

Average throughput: The average number of bytes successfully received by the destination is referred to as average throughput.

Delay: It is defined as the average delay of successfully sending a packet from a sender node to the target node.

7.6.2 Performance Evaluation

The simulation is done in different scenarios. The result of the simulation is discussed next.

7.6.2.1 Impact of Changing the Number of Nodes

In this section, the simulation of AODV and DSDV is done for varying the number of nodes, i.e., 10, 20, 30, 40. Here, the speed of mobile node of 25 mps and the packet size of 128 bytes are considered. Then, the analysis is done in terms of delivery ratio, packet loss, throughput, and average delay. When the number of nodes in the network increases, the delivery ratio of both routings also increases, except when the number of nodes = 40 for DSDV. From Figure 7.5, it is clear that AODV exhibits higher packet delivery compared to DSDV.

Table 7.1 Simulation Settings.

Parameter	Value
Simulator	OMNeT++
Traffic model	UDP
Mobile nodes	10, 20, 30, 40
Simulation time	100 s
Node's speed	25 mps, 30 mps, 35 mps, 40 mps
Message length	128, 256, 512, 1024 bytes
Transmission range	250 m
Simulation area	600 × 600
Routing techniques	AODV, DSDV
Transmitter power	1mW

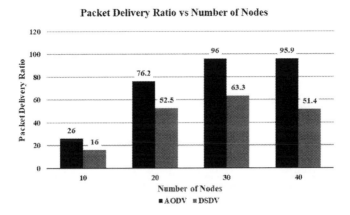

Figure 7.5 Delivery vs. mobile nodes.

For varying the mobile nodes, the packet loss of AODV and DSDV decreases except when the number of nodes = 40, as shown in Figure 7.6. But AODV exhibits the lower packet loss.

With the increase of the number of mobile nodes, the average delay of AODV routing decreases rapidly. But DSDV shows comparatively lower delay compared to the AODV, as shown in Figure 7.7.

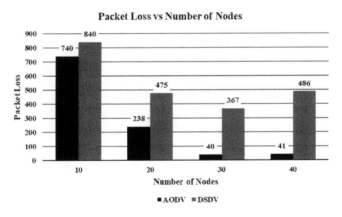

Figure 7.6 Packet loss vs. mobile nodes.

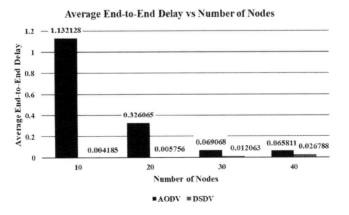

Figure 7.7 Delay vs. mobile nodes.

For increasing the number of mobile nodes, the throughput of both routings increases, as viewed in Figure 7.8. However, the throughput of AODV is significantly higher than the throughput of DSDV routing (Figure 7.8).

7.6.2.2 Impact of Changing the Speed of Nodes

In this investigation, the analysis for varying the speed of nodes on the basis of delivery ratio, packet loss, average throughput, and delay is discussed. It is considered that the packet length is 128 bytes and the number of mobile nodes is 25. The simulation is done for the variation of the speed of the mobile nodes; they are 25 mps, 30 mps, 35 mps, and 40 mps.

For changing the speed of the nodes, the package delivery of both routings decreases overall except the speed of the nodes at 30 mps of AODV routing. As shown in Figure 7.9, the AODV routing exhibits higher delivery ratio compared to DSDV.

For changing the speed of the nodes, the packet loss increases for both the AODV and DSDV routings. The packet loss in AODV is significantly lower than the DSDV routing as shown in Figure 7.10.

For changing the speed of the nodes, DSDV exhibits lower delay than AODV as viewed in Figure 7.11.

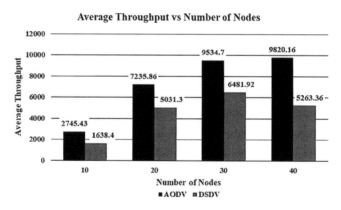

Figure 7.8 Average throughput vs. mobile nodes.

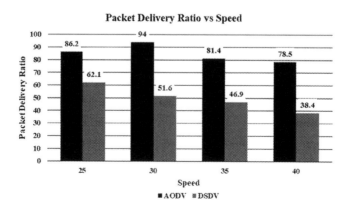

Figure 7.9 Delivery ratio vs. the speed of nodes.

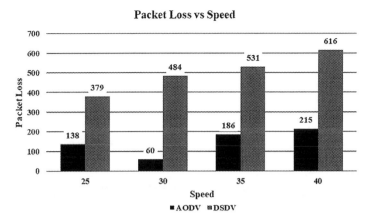

Figure 7.10 Packet loss vs. the speed of nodes.

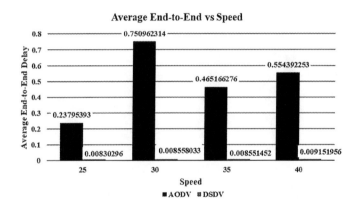

Figure 7.11 Delay vs. the speed of nodes.

The throughput decreases with the increase of the speed of the nodes in the case of AODV, except at 30 mps. But in DSDV, the average throughput decreases for the increase of the speed of the nodes. Hence, the AODV routing exhibits higher throughput compared to DSDV, as shown in Figure 7.12.

7.6.2.3 Impact of Changing the Packet Length

The number of mobile nodes (25) and the speed of the nodes (25 mps) are kept fixed. The simulation is done for varying the packet lengths, i.e., 128, 256, 512, and 1024 bytes. As shown in Figure 7.13, AODV deserves a higher delivery ratio compared to the DSDV routing.

The amount of packet loss increases in both AODV and DSDV when the packet length increases, except when packet length is 256 bytes in AODV and packet length is 512 bytes in DSDV. But DSDV shows higher packet loss compared to AODV (Figure 7.14).

In AODV, the delay increases with the size of packet length, except at 512 bytes. The value of DSDV also increases as the length of packet length increases. But DSDV shows lower delay compared to AODV, as shown in Figure 7.15.

With the increase of packet lengths, the throughput of both routing techniques increases. It is clear from Figure 7.16 that the average throughput of AODV routing is high compared to the average throughput of DSDV.

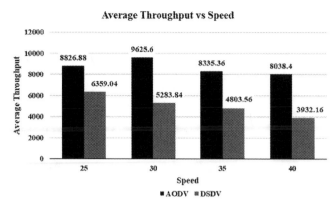

Figure 7.12 Average throughput vs. speed.

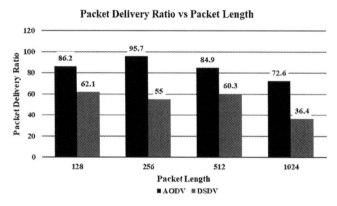

Figure 7.13 Packet delivery ratio vs. packet length.

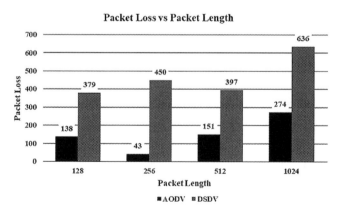

Figure 7.14 Packet loss vs. packet length.

From the discussion of Figures 7.5 through 7.16, it is clear that in all cases, i.e., for changing the number of nodes, the speeds of the nodes, and packet lengths, AODV routing exhibits higher delivery ratio, lower packet loss, and higher average throughput compared to the DSDV routing. However, AODV routing shows higher average delays than DSDV. So, it is a limitation of AODV routing.

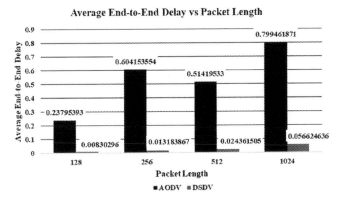

Figure 7.15 Delay vs. packet length.

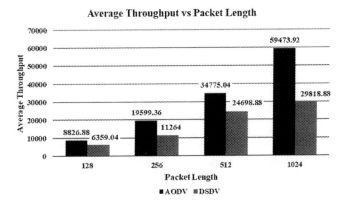

Figure 7.16 Average throughput vs. packet length.

7.7 CONCLUSION AND FUTURE WORK

In this chapter, the functionality of the two routing protocols, namely, DSDV and AODV in MANETs, has been represented elaborately. Here, the performance of these two approaches has been analyzed for changing the number of nodes, speeds, and packet lengths in terms of the delivery ratio, packet loss, delays, and throughput using the OMNeT++ simulator. From the investigated results of the given scenario, it is observed that AODV outperforms DSDV in terms of delivery ratio, packet loss, and throughput, whereas DSDV shows minimum delays compared to AODV. The main limitation of AODV is a high delay. In the near future, AODV routing can be modified to minimize the delays.

REFERENCES

[1] M. Alslaim, H. Alaqel and S. Zaghloul (2014). *A Comparative Study of MANET Routing Protocols*. In *The Third International Conference on e-Technologies and Network for Development*, Beirut.

[2] C. Jayakody, R. Samarasinghe and S. Kodituwakku. (2013). *Routing Protocols in Wireless Mobile Ad-Hoc Network-A Review*. In *The SAITM Research Symposium on Engineering*, Colombo.

[3] C. Wahi and S. Sonbhadra (2012). Mobile Ad Hoc Network Routing Protocols: A Comparative Study. *International Journal of Ad hoc, Sensor & Ubiquitous Computing*, vol. 3, no. 2 pp. 21–31.

[4] M. Khalaf and B. Mohsin. (2019). Analysis and Simulation of Three MANET Routing Protocols: A Research on AODV, DSR and DSDV Characteristics and Their Performance Evaluation. *Periodicals of Engineering and Natural Sciences*, vol. 7, no. 3, pp. 1228–1238.

[5] A. Boukerche. (2004). Performance Evaluation of Routing Protocols for Ad Hoc Wireless. *Mobile Networks and Applications*, pp. 333–342.

[6] T. Salam and M. S. Hossen. (2020). Performance Analysis on Homogeneous LEACH and EAMMH Protocols in Wireless Sensor Network. *Wireless Pers Commun*, vol. 113, pp. 189–222.

[7] M. Vijayalakshmi, A. Patel and L. Kulkarni. (2010). QoS Parameter Analysis on AODV and DSDV Protocols in a Wireless Network. *Indian Journal of Computer Science and Engineering*, vol. 1, no. 1, pp. 283–294 .

[8] M.S. Hossen. (2019). DTN Routing Protocols on Two Distinct Geographical Regions in an Opportunistic Network: An Analysis. *Wireless Personal Communication*, vol. 108, pp. 839–851.

[9] A. Zaballos, A. Vallejo and G. A. A. J. Corral. (2006). *Ad-Hoc Routing Performance Study Using OPNET Modeler*. In *OPNETWORK '2006*, Washington, DC, pp. 1–6.

[10] S. Vanthana and V. S. J. Prakash (2014). *Comparative Study of Proactive and Reactive Ad Hoc Routing Protocols Using Ns2. 2014 World Congress on Computing and Communication Technologies*, Trichirappalli, pp. 275–279.

[11] Harsimrankau and J. Singh. (2017). Comparison of AODV and DSDV Protocols along with Improvement of AODV. *International Journal of Advanced Research in Computer Engineering and Technology*, vol. 6, no. 7, pp. 976–981.

[12] K. Pandey and A. Swaroop. (2011). A Comprehensive Performance Analysis of Proactive, Reactive and Hybrid MANETs Routing Protocols. *International Journal of Computer Science Issues*, vol. 8, no. 6, pp. 432–441.

[13] F. Al-Dhief, N. Sabri, M. Salim, S. Fouad and S. A. Aljunid. (2018). *MANET Routing Protocols Evaluation: AODV, DSR and DSDV Perspective. MATEC Web of Conferences*.

[14] M.D.I. Talukdar and M.D.S. Hossen. (2019). Selecting Mobility Model and Routing Protocol for Establishing Emergency Communication in a Congested City for Delay-Tolerant Network. *International Journal of Sensor Networks and Data Communications*, vol. 8, no. 163, pp. 1–9.

[15] OMNeT++ Discrete Event Simulator. (2020). https://omnetpp.org/

Chapter 8

Message Forwarding and Relay Selection Strategies in Mobile Opportunistic Networks

Mohd Yaseen Mir and Chih-Lin Hu

National Central University, Taoyuan, Taiwan

CONTENTS

8.1 INTRODUCTION

Mobile opportunistic networks (MONs) [1] are sparsely connected mobile ad hoc networks (MANETs), characterized by low node density and lack of long-term and continuous end-to-end connectivity between mobile devices. Unlike traditional connected-initiated message delivery in wired or stationary networks, messages are conveyed through relay nodes by exploiting pair-wise contact opportunities in MONs. Communication by pair-wise contact patterns is multi-hop. In this way, each relay node is considered as a router that stores messages until any next contact opportunity appears to forward the messages. This way of multi-hop communications is named the store-carry-and-forward data delivery model. Note that, in the literature, several terms, e.g. data and message, forwarding and routing, and scheme and strategy, are often used interchangeably, so they are used in the following without ambiguity.

To deal with the inconsistent nature of future contacts, message replication is a naive methodology which has been used to maximize the chances of delivering a message to its final destination [2]. However, the heavy cost of message overhead induced by this sort of naive schemes is unacceptable, particularly in MONs that consist of resource-constrained mobile nodes with intermittent connectivity. Knowing the behavior of target users in network contexts such as the probability of encountering, contact frequency, and inter-contact time can predict future contact opportunities and identify better relays for forwarding the messages towards the destinations in MONs.

To increase the cost-effectiveness of message delivery with high successful delivery rate and low traffic overhead, the design of an appropriate relay selection strategy requires understanding habits in nodal movements and encounter patterns, e.g., average duration, inter-contact time, meeting frequency, periodicity, etc. Due to the diverse nature of MONs, it is difficult and challenging to study the movement patterns and come up with unique designs of efficient message forwarding strategies.

Utilizing social relationships among mobile nodes can help in identifying better relays. Many attributes of social relationships include but are not limited to betweenness, similarity, and tie strength, which can be incorporated into the decision-making process for relay selection. The similarity of two nodes implies the number of neighbors these nodes have in common. Betweenness helps in identifying bridging nodes and carrying messages between two groups. Thus, relay nodes with higher similarity and centrality values have higher probability to be selected for forwarding a message [3, 4]. Similarly, the tie strength of a node reflects its connection strength with the destination node. To further enhance relay selection strategies, recent research like [5] considered frequency and duration of contacts to identify bridge nodes. In addition, [6] considered a node with multiple encounters as an appropriate relay to reflect its ability to forward a message.

Selecting the next relay by understanding the properties of movement behavior of different nodes has been quite investigated in MONs [7–9]. Relay selection decisions are taken by comparing per-node metrics to determine the fitness of the node. Some of the metrics include quality-based, contact-history and frequency, etc. Thus, a relay node with a higher metric value is selected as the next relay node. Historical contact information with respect to any destination node, as studied in [10], is used to form the metrics of contact probability, and a relay node with higher contact probability with the destination node can be treated as the best relay.

In summary, this chapter is organized as follows. Section 8.2 introduces the classical routing protocols used in MONs. Section 8.3 discusses encounter-based routing by first discussing studies based on real traces in Section 8.3.1, and then explains different relay selection strategies used to design routing solutions in Section 8.3.2. Section 8.3.3 explains our proposed regular and sporadic contact-based routing (RSCR) scheme, and results are discussed in Section 8.3.4. Finally, Section 8.4 concludes the chapter.

8.2 MESSAGE FORWARDING/ROUTING

Whereas end-to-end data routing paths are hardly maintained in MONs, conventional routing protocols that adopted the connection-oriented data delivery in MANETs cannot be suitable in MONs. This section gives an overview of routing types and then briefly reviews the classical routing protocols used in MONs.

8.2.1 Background

Routing in MONs is performed by distributing either a single message copy or multiple message copies in the network. The former is done by either direct or first-contact delivery. In the direct-delivery way, the source node itself waits for the destination node and forwards the message. In the first-contact delivery, any node that encountered the source node will receive a message from the source node, and then that encountered node will forward the message to the destination node. However, the single-copy message delivery in either way is not considered feasible in highly dynamic networks, because no reliable message delivery is guaranteed against fragile network topology and constrained resources of mobile nodes in MONs.

Most routing protocols in MONs [1] aim to maximize the message delivery rate, minimize the transmission latency, or minimize the messaging overhead. These performance measures are specified as follows:

- *Delivery rate* means the number of messages created by source nodes are successfully received by the destinations.
- *Transmission latency*, or called delay time, is the time span from the time which a message was sent by the source node to the time which the destination receives this message.
- *Overhead* is usually defined as the number of replicas of an original message created in MONs.

To increase the delivery rate, the same messages are replicated multiple times in the network. Message replication can be either restricted or unrestricted. With a restricted replication-based routing way, only a fixed amount of message copies are distributed to different nodes. By contrast, in an unrestricted routing way, no fixed limit is applied to the number of replicas of a message after the source node created this message. In addition, in order to regulate the network and its storage resources, each message has a time-to-live (TTL) value. When a message expires, it is dropped from the buffer space of a node. To improve the delivery rate, any routing scheme should attempt to deliver messages to their destinations before their associated TTL expires.

8.2.2 Classical Routing Protocols

Following the store-carry-and-forward data delivery model in MONs, source nodes attempt to distribute messages to destinations with the help of relay nodes. Two neighbor nodes can exchange messages kept in local buffer whenever they are in contact with each other. A contact is defined as a period when two nodes temporarily stay in a reciprocal transmission range during their movement in a network. In general, a node may be in contact with several other nodes at the same time, so messages can be handed over across neighbor nodes quickly to expedite the message distribution. To provide a background knowledge of design and development of mobile opportunistic routing protocols, the following briefly mentions some of the classical routing protocols.

8.2.2.1 Epidemic

One of the earliest mobile opportunistic routing protocols was designed to forward a message whenever two nodes contact with each other [2]. Epidemic belongs to the unrestricted type, since this scheme floods messages into the network by duplicating a message copy to each encountered node. As long as the encountered node does not keep the same message copy, a message is forwarded to it as the source node. Any node carrying the message copy will continue to forward more copies to other encountered nodes if they did not receive that message before. In this way, a great number of message copies of the original message from the source node can be broadly distributed in a network. Eventually, the destination node can have the maximum possibility of receiving the message eventually. Admittedly, it is known that the Epidemic protocol consumes considerable resources of nodes, such as buffer space and energy battery, and causes heavy message overhead in a network. To maintain the system performance, Epidemic can be used fairly subject to the provision of sufficient buffer space and definite message lifetime, i.e., TTL.

8.2.2.2 PRoPHET

PRoPHET [10] belongs to the unrestricted type of routing. In PRoPHET, a node forwards a message copy to another node only if the latter node has higher delivery predictability (DP) than the former node. The value of DP implies the probability of one node meeting with other nodes in MONs. Each node maintains a table of DP values with different nodes. Whenever two nodes meet, their corresponding DP values are updated according to Equation 8.1. In Equation 8.1, $P_{(x,y)}$ and $P_{(x,y)old}$ represent current and previous DP of node x with node y, respectively. The value $P_{init} \in [0,1]$ is an initial constant.

$$P_{(x,y)} = P_{(x,y)old} + \left(1 - P_{(x,y)old}\right) \times P_{init} \tag{8.1}$$

$$P_{(x,y)} = P_{(x,y)old} {}^{*} \gamma^{k} \tag{8.2}$$

$$P_{(x,z)} = P_{(x,z)old} + \left(1 - P_{(x,z)old}\right) \times P_{(x,y)} \times P_{(x,z)} \times \beta \tag{8.3}$$

If x and y do not contact each other for a certain period, their DP values will decay accordingly, and the updated DP value is calculated by Equation 8.2. In Equation 8.2, $\gamma \in [0,1]$ is an aging constant, and k denotes the number of time units that have expired since the last update of this DP value. Equation 8.3 updates a transitive DP value of x with node z where β is a parameter with value $\in [0,1]$.

8.2.2.3 Spray and Wait

Spray and Wait (SNW) [11] belongs to the restricted type of routing where a fixed number of messages copies are replicated in the network. SNW consists of two phases, the Spray phase and the Wait phase, as follows.

In the Spray phase, SNW distributes at most L copies of a message to different relay nodes. In source spraying, a source node itself spreads L message copies in MONs whenever it comes into contact with other relay nodes. However, if the source node contacts the destination node in between, the message is directly forwarded, and no message copy is distributed further. Otherwise, the relay nodes that received the message copy from the source node waits for the destination node to transfer the message. In binary SNW, a source node initially transfers $n = L/2$ messages where $L > n > 1$, copies to the first encountered relay node, and keeps $n = L/2$ to itself. Later, both the source node and the encountered node transfer half of the remaining $n/2$ copies to other relays, which is the same as the previous iteration. This spraying process continues until each node retains a copy with $n = 1$ and waits for the destination node to transfer the message directly.

Provided that the destination node is not encountered during the Spray phase, a relay node goes into the Wait phase. After that, a relay node carrying a message copy will wait to send that message directly to the destination node once they come into contact with each other.

8.2.2.4 Bubble Rap

Bubble Rap [4] exploits social relationships among different nodes to forward a message. Such social relationships are built by following the interaction among different people, recorded by using their mobile devices. This interaction helps in predicting the next contact time between mobile nodes, and thus enhances the delivery rate. Social metrics such as centrality and community based on examination of a node's past interactions are often jointly considered for forwarding a message in social-based MONs.

- *Local centrality*: This relative rank is allocated to a node and is used to forward a message within its group/community.
- *Global centrality*: This rank allocated to a node reflects its ranking in the entire MON. This rank is utilized across the whole MON for forwarding a message.

Initially, a node with a high global rank is given a larger priority to be a message transfer. This message is replicated before it reaches a node belonging to the same community as the destination of the node. Afterward, the local ranking within a community is used to further disseminate the message until it reaches the destination node.

8.3 ENCOUNTER- AND CONTACT-BASED ROUTING

Encounter-based routing techniques, a.k.a. contact-based routing, deal with contact histories between mobile devices. Because some of the nodes in MONs may contact frequently, such nodes may help to increase data delivery rate cost-efficiently rather than other nodes that do not have apparent contact relationship. For example, [8] uses the past encounter rate as a metric to decide the number of replicas of a given message which would be sent to other nodes. Note that the terms *encounter* and *contact* are used interchangeably hereinafter, while they are commonly presented in the literature.

Our goal in this section is to mention recent efforts of encounter-based routing studies in MONs. Encounter patterns in the context of nodal moment behavior correspond to the manners which nodes meet with each other during movement in MONs. By exploiting encounter patterns, mobile nodes can be guided to select the next relay for message forwarding. Some practical mobility features, such as repeated appearances at the same location, repetitive patterns, and strong correlation between mobile nodes, can be exploited to obtain either periodicity between node pairs or differentiate sporadic and regular contacts among nodes. The next subsection describes previous studies on real-life traces and then explains different routing algorithms designed by reference to various encounter patterns in MONs.

8.3.1 Study on Real-Life Traces

Meeting time between node pairs can be considered a key element to identify the repetitive contacts in the MONs. Many studies on human mobility have rendered a thorough understanding of nodal encounter patterns [12–17]. Various techniques, e.g., auto-correlation function, Fourier transform, correlation analysis, time-variant, etc., are employed to examine some measurement results based on wireless and mobile trace data sets.

The sort of trace data sets in wireless and mobile environments include Bluetooth device proximity, T-mote sensor devices carried by users, and Wi-Fi access points (AP) placed inside the university campus and at different locations for data collection. The data collected include the node's address, associated AP, and the timestamp for start and end contact time between mobile devices. The processed information on such trace data includes duration, location, and timestamps of encounters. Some of these traces in the research community can be found here [18]. In what follows, we will go on to understand mobility patterns in reference to the studies of [12, 14–16] which reported some measurement results based on these traces.

8.3.1.1 Periodicity/Pattern between Node Pairs

Given a specific time period, a contact between any two nodes occurs frequently is referred to as a periodic contact. Finding periodicity between node pairs can help in predicting the next contact time and can identify a relay with a shorter inter-contact time with the destination node.

The study in [14] investigated the periodicity in encounter patterns using power spectral analysis. They used the auto correlation function (ACF) to identify repetitive patterns and also used the discrete Fourier transform (DFT) to find distinct periodicity. Encounter rate was used as a metric to group encountered pairs and individual nodes to highlight any periodicity involved in each of them. To analyze the encounter pattern in a trace data, different variables such as encounter frequency $F_d(i, j)$, daily encounter $E_d(i, j)$, hourly encounter $E_h(i, j)$, and duration of encounter $L_d(i, j)$, where i and j represent encountered nodes, are used. ACF is then applied to this time-domain encounter trace to find the repeated periodical patterns.

ACF measures the correlation between observations at different lags or distances apart. At the lag of $k = 0$, the auto-correlation is the maximum, resulting in a variance δ. If λ represents the mean rate with respect to a pair (i, j) in a total trace time of T days, the auto-correlation coefficients for each lag of $k(1 < k < T)$ with respect to (i, j), which is denoted as $r_k(i, j)$ and can be calculated by Equation 8.4:

$$r_k(i,j) = \frac{\sum_{d=0}^{T-k}(E_d - \lambda)(E_{d+k} - \lambda)}{\sum_{d=0}^{T-1}(E_d - \lambda)^2} \qquad (8.4)$$

The resulting series gives a repetitive pattern for certain lags. DFT is then applied to find distinct/hidden periodic properties by transforming autocorrelation coefficients of the time series data to the frequency domain as:

$$y_c(i,j) = \sum_{k=1}^{T-1} r_k(i,j)e^{-\frac{2\pi i}{T}kc}, \qquad (8.5)$$

where $y_c(i, j)$ is a power spectrum of the pair (i, j) for each frequency component $c(1 \leq c < T)$. To further observe the periodicity of an encounter pair, the average auto-correlation coefficient of each lag is calculated and then transformed to the frequency domain.

Similarly, to derive periodic patterns, [12] examined skewed and periodic reappearances at the same places. They observed from different traces and found that, on average, 65% time is spent by a node with one AP, and more than 95% of online time at as few as five APs. The finding of periodical re-appearance signifies the re-appearance of a node at the same AP after some fixed periods. To capture skewed locations, they defined some popular locations for a node by defining some communities for each node. Thus, a node visits its community more often than others.

Re-appearance of a group of people in "only a specific period" is known as a transient community (TC). To detect TCs, [16] proposed a Contact-burst-based Clustering Method (CCM) by utilizing contact pairs. Each pairwise process is formulated to be a regular presence of contact bursts, which is defined as a period when contacts regularly occur between two node pairs. TCs are detected by clustering the node pairs with similar contact bursts. Members of a node pair may have frequent contacts with each other in the same duration. Thus, a set of contact bursts with a similar duration will form a TC.

To detect frequently occurring contacts and predict the next contact time between any node pair, [15] used the correlation analysis method, called "Auto Regression Moving Average." To detect a pattern, each node stores a contact history list with a fixed length, including contact time (up and down), meeting amount to obtain encounter history, contact frequency, and contact interval. Using these variables, the meeting probability can be calculated to determine nodes with definite patterns. Correlation analysis is used with derived nodes to obtain a confidence degree, which is then used to improve prediction accuracy. This process is briefly described next.

First, a transaction set is characterized as a sequence of events, where an event is a meeting of two nodes at any time. The support analysis and confidence analysis of the transaction set, as will be described later, are used to reveal the pattern of a mobile node. For any node with the encounter history, $(Z_1, ..., Z_n)$ is termed as an event set S. For any two consecutive encounters (Z_i, Z_{i+1}) in S, it is defined as an event. For any recent event, (Z_{n-a}, Z_n) denoted A, its support degree is calculated by Equation 8.6:

$$support(A) = \frac{N(A)}{N(All_2)},$$ (8.6)

where $N(All_2)$ represents the occurrence number of all two consecutive encounters event in S, and $N(A)$ represents the occurrence number of A in S. For each node Y, let B be an event for two consecutive encounters (Z_n, Y). Then, the support degree of event $A \Rightarrow B$ can be calculated by Equation 8.7:

$$support(A \Rightarrow B) = \frac{N(A \cup B)}{N(all_3)},$$ (8.7)

where $A \cup B$ represents the three consecutive encounters event (Z_{n-1}, Z_n, Y), $N(A \cup B)$ represents the occurrence number of $A \cup B$ in S, and $N(all_3)$ represents the occurrence number of all three consecutive encounters event (Z_{n-1}, Z_n, Y) in S. Confidence degree is then defined to denote the conditional probability of the event $A \Rightarrow b$:

$$confidence(A \Rightarrow B) = \frac{support(A \cup B)}{support(A)}.$$ (8.8)

To determine the strong associated pattern for the contact pattern $A \Rightarrow B$, a minimum confidence threshold $min_{sup} \in [0,1]$ is used. If $confidence(A \Rightarrow B) > min_{sup}$, then $A \Rightarrow B$ has a strong association. Here, the Time Series Analysis Model (TSAM) is used to estimate the next contact time. TSAM is a statistical method for discovering the dynamic patterns in a dynamic data set. With the obtained strong association, a dynamic time series is extracted, and then ARMA takes it as an input to predict the next contact time.

8.3.2 Relay Selection

Having established the presence of pattern periodicity between mobile nodes in MONs, we now focus on some of the forwarding strategies by further exploiting these patterns to enhance delivery rate cost-effectively. Upon prior studies in the literature [6, 17, 19–23], we describe some strategies in regard to appropriate relay selection methods in MONs.

Using social network concepts such as centrality and communities, [21] proposed an analytical model for multicast relay selection to minimize cost in terms of the number of relays used. The centrality metric measures the ability of a node to communicate with other nodes. With a single data source and N nodes in a network, this work proposed the cumulative contact probability (CCP) for a node i with a mean rate $\lambda_{i,j}$. Then, a centrality metric is defined in Equation 8.9. A node with a higher CCP implies its effectiveness to deliver data to destinations.

$$C_i = 1 - \frac{1}{N-1} \sum_{j=1, j \neq i}^{N} e^{-\lambda_{i,j} T}. \tag{8.9}$$

In Equation 8.9, C_i represents the average probability that a node randomly contacts i within T. Data forwarding to all the destinations will start according to a specific threshold value: delivery ratio p. To forward data, a minimum number of relays are selected to satisfy both p and T constraints. With multiple data sources, a community-based approach is utilized to forward data. A node keeps track of nodes that have contacts with destinations within its social community and selects relays accordingly.

To optimize delay and improve performance, [19] proposed the conditional intermeeting time (CIMT) metric. Using past encounter history, CIMT calculates the average inter-contact between two nodes relative to a meeting with a third node. To take forwarding decisions, CIMT is used to the compare meeting time with the destination node whenever two nodes contact each other. If i, carrying a message, meets j and learns j's CIMT is less than itself, it selects j as the next relay to forward the message.

To improve performances of data forwarding in MONs, [20] exploited the transient social contact patterns. These patterns, discussed later, are formulated based on experimental studies on data traces. Further, the centrality of a node is evaluated using the transient social patterns. In the following, we first describe the transient social patterns, and then data forwarding and centrality.

- *Transient contact distribution*: At different periods, the contact distribution is highly skewed. For example, during the daytime, nodes frequently contact other nodes, in contrast to the situation at nighttime. Nonetheless, at nighttime, some nodes may still be able to transfer data using transient contact distributions.
- *Transient connectivity*: This includes contacts that occur merely in some specific periods. For example, students in a class remain connected with each other only for that specific time.
- *Transient community structure*: A node belongs to different communities during specific periods. For example, a student at some university may be connected with classmates during the daytime and with roommates during the nighttime.

Data forwarding is mentioned below. If node i meets node j, then i will forward the data to j in the following cases only:

1. If j is the destination node, or both the destination node and j belong to the same community.
2. If the first condition is false, then data forwarding metrics for both i and j are calculated by their transient contact patterns. If j has a higher metric value than i, then i forwards the data.

The centrality value is used as a metric for data forwarding. It measures the expected number of nodes that a node can meet within the given time. At time t_c, the centrality metric C_i of i is defined by $C_i = \sum_{j \in N} c_{ij}$, where N indicates the set of nodes, and c_{ij} implies the number of nodes expected by i to encounter within some time constrains by contacting j.

The authors of [17] investigated social contact patterns from the temporal contact patterns in order to predict future contact patterns, and thus improved the cost-efficiency of data forwarding in MONs. A new metric, called temporal closeness (TCL), is defined to capture the presence of social contact patterns. Average separation, which includes both the frequency and duration of contact between two nodes, is utilized to deduce TCL. Let i contact j at time t_1, and let the duration of contact be t_{d1}. Now, if both i and j contact again at t_2 such that $t_2 > t_1$, then the separation time is equal to $t_2 - (t_1 + t_{d1})$. A shorter average separation time reveals a closer relationship, and the variance of the separating time reflects irregularity. To derive a temporal centrality metric value, the following procedure is adopted.

Let θ_{uv} and δ_{uv} respectively denote average separation (AV_{uv}) and variance of separation time (VA_{uv}) between two nodes u and v in a certain window (w). The exponential function is used to normalize both the average separation time and average separation in each time window. Then, given with $AV_{uv} = e^{1-\frac{\theta_{uv}}{w}}$ and $VA_{uv} = e^{1-\frac{\delta_{uv}}{w}}$, TCL between u and v is measured as $TCL(uv) = g \times AV_{uv} + (1 - g) \times VA_{uv}$, where $g \in [0,1]$. Finally, taking all V nodes in the MONs into consideration, the temporal centrality (TC) in N time windows is used as a centrality metric based on TCL of pairwise nodes, expressed in Equation 8.10.

$$TC(u) = \frac{1}{N-1} \sum_{v \in V, v \neq u} TCL(uv). \tag{8.10}$$

The Recent Weighted Method (RWM) is proven to be best for predicting future TC. Herein, we skip the description of the prediction method; readers are encouraged to see [17] for further details. With some time constraints on the destination node, a node with a higher TCL is selected as the next relay node. Note that if there is no relationship with the destination node, then the future TC is utilized.

Using the concept of social energy metric, [6] proposed a Social Energy Based Routing (SEBAR) protocol. A node with multiple encounters will have a higher social energy and can be considered a better relay. SEBAR uses the concept of community to calculate social energy. Let $N_{k,l}$ be the amount of energy generated when nodes k and l encounter each other. If k belongs to one community, then it contributes $p \times \frac{N_{k,l}}{2}$ to its community and keeps $(1-p) \times \frac{N_{k,l}}{2}$, where p represents percentage. Else, if it belongs to n_k communities, then $p \times \frac{N_{k,l}}{2n_k}$ is shared with each community. By this process, the community gains the energy from its members and distributes this energy to its members based on the node's community centrality, which is defined in Equation 8.11.

$$c_j(j) = \frac{\sum_{i=1}^{m_k} D_k(i)}{\sum_{any \ v_x \in C_j} \sum_{i=1}^{m_k} D_k(i)}. \tag{8.11}$$

In Equation 8.11, $c_k(j)$ represents the community centrality of k in community C_j and with a total number of m_k encounters of k, $\sum_{i=1}^{m_k} D_k(i)$ is the total contact duration, where $D_k(i)$

represents the duration of the ith contact. Similarly, $\displaystyle\sum_{any\ v_x \in C_j} \sum_{i=1}^{m_x} D_x(i)$ represents the total contact duration of all member nodes in C_k. The energy $(\mathbb{E}_k)$ gained by k consists of two parts: the reserved energy during collision $(\mathbb{E}\_\mathbb{N}_k)$ and the gained energy from communities $(\mathbb{E}\_\mathbb{C}_k)$, as calculated in Equation 8.12.

$$\mathbb{E}_k = \sum_{i=1}^{m_k} \mathbb{E}\_\mathbb{N}_k(i) + \sum_{i=1}^{n_k} \mathbb{E}\_\mathbb{C}_k(j), \tag{8.12}$$

$$\mathbb{E}\_\mathbb{N}_k(i) = (1-p) \times \frac{N_{k,l}(i)}{2},$$

$$\mathbb{E}\_\mathbb{C}_k(j) = c_k(j) \times \sum_{any\ v_x \in C_j} \sum_{i=1}^{m_x} p \times \frac{N_{x,l}(i)}{2n_x}.$$

However, if a node does not encounter other nodes for a long time, its energy decays. An energy decay coefficient is defined by $\epsilon(0 \le \epsilon \le 1)$, so that after each τ, the energy of a node will decay by ϵ. So, Equation 8.12 is defined again and rewritten in Equation 8.13.

$$\mathbb{E}_k^t = \mathbb{E}\_\mathbb{N}_k^t + \mathbb{E}\_\mathbb{C}_k^t, \tag{8.13}$$

$$\mathbb{E}\_\mathbb{N}_k^t = (1-\epsilon) \times \mathbb{E}\_\mathbb{N}_k^{t-\tau} + \sum_{i=1}^{m_k^t} \mathbb{E}\_\mathbb{N}_k(i),$$

$$\mathbb{E}\_\mathbb{C}_k^t = (1-\epsilon) \times \mathbb{E}\_\mathbb{C}_k^{t-\tau} + \sum_{any\ C_j \ni k} \left(c_k(j) \times \sum_{any\ v_x \in C_j} \sum_{i=1}^{m_x^t} p \times \frac{N_{x,l}(i)}{2n_x} \right),$$

where $\mathbb{E}_k^t$, $\mathbb{E}\_\mathbb{N}_k^t$, and $\mathbb{E}\_\mathbb{C}_k^t$ denote the energies gained at time t, and m_x^t represent a number of encounters of node x in the period $[t - \tau, t]$. Now, consider k carries q message copies for the destination d and encounters l. Node k forwards half of the copies in two cases: (1) k does not belong to d's community, and $\mathbb{E}_k < \mathbb{E}_l$; (2) both k and l belong to d's community and $\mathbb{E}\_\mathbb{C}_k < \mathbb{E}\_\mathbb{C}_l$.

In [22], the design of selecting the next relay for data forwarding with the shorter inter-contact time to the destination was considered. A utility value for each relay node, based on the features such as periodicity and similarity, was designed to select the next relay.

The authors of [23] developed a new comprehensive metric to build an accurate social relationship between different nodes in a social graph. In particular, they considered several attributes, including average contact duration, shortest separation time, contact frequency, and average separation time. Two nodes with frequent connections create stronger relationships, which means that the weight, a.k.a. strength, between them is higher. Using this new metric, they also designed a method to detect the hierarchical community structure. Finally, by exploiting the overlapping community structure, they proposed a routing strategy in social-based MONs.

8.3.3 Proposed Routing with Relay Selection

Our study efforts specify a novel decision-making process for contact-based relay selection in MONs. To explore the use of contact patterns to increase delivery rate cost-effectively, we examine the intrinsic properties of contact patterns and clarify two contact types, i.e., regular and sporadic contacts. Contacts occurring periodically are defined as regular, and occasional contacts are termed as sporadic. Readers can refer to [24] for further details. The content that follows first explains the finding and differentiation of these contact patterns in Section 8.3.3.1. In Section 8.3.3.2, we describe the proposed routing scheme, named Regular and Sporadic Contact-Based Routing (RSCR).

8.3.3.1 Link Definitions on Regular and Sporadic Links

To discover contact patterns in a time scale, we first intend to determine the link type corresponding to the contact behavior. Let a link define a contact between two nodes n_x and n_y, expressed as $L_{x=y}$ where $L_{x \leftarrow y}$ indicates n_y contacts n_x. To be a regular link, we say that $L_{x \leftarrow y}$ is a regular link $R_{x \leftarrow y}$ if n_x and n_y contact each other multiple times over a particular threshold value in MONs. Oppositely, a sporadic link $R_{sp(x \leftarrow y)}$ is formed when n_y contacts n_x only sometimes in a random time period T_r. If there exists a regular link between n_z and n_y, then a link between n_x and n_y is a transient link and is defined as

$$L_{x \leftarrow y \leftarrow z}^{tr} \mid if \exists \{R_{x \leftarrow y} \mid \{\exists R_{y \leftarrow z}\}\}.$$

We now plainly use the mean inter-contact time as a specific time period to define any occurrences of either regular or sporadic patterns by considering the following two cases:

1. If n_x contacts n_y at times $t_0, t_1, t_2, ..., t_M$, then we obtain M inter-contact time samples as $\{(T = t_1 - t_0), (T_2 = t_2 - t_1), ..., (T_M = t_M - t_{M-1})\}$. Then, the mean of inter-contact time is calculated as $T^{x \leftarrow y} = \dfrac{\sum_{i=1}^{M} T_i}{M}$. Node n_x can contact other nodes, and we can obtain different means of inter-contact time of n_x. Given $\{T^{x \leftarrow 1}, T^{x \leftarrow 2}, ..., T^{x \leftarrow Z}\}$ with respect to Z distinct nodes encountered by n_x during M time periods, we will average these means to be

$$T^x = \frac{\sum_{y=1}^{Z} T^{x \leftarrow y}}{Z}. \tag{8.14}$$

2. In this case, we update $T^{x \leftarrow y}$ incrementally with $T_{M_s}^{x \leftarrow y}$ in the previous number of M_s samples and the current sample $T_{M_s} + 1$, i.e. $t_{M_s+1} - t_{M_s}$. Thus, $T^{x \leftarrow y}$ can be updated as

$$T^{x \leftarrow y} = \frac{M_s \times T_{M_s}^{x \leftarrow y} + T_{M_s+1}^{x \leftarrow y}}{M_s + 1}. \tag{8.15}$$

Thus, $T^x/T^{x \leftarrow y}$ is used as the specific time period to distinguish regular and sporadic links associated with n_x.

With the link type definitions and mean of inter-contact time, we briefly describe how to use $T^x/T^{x \leftarrow y}$ to extract any link types between two nodes. We consider $L_{x \leftarrow y}$ as an initial state (s_0) in a Markov chain. The change of a state occurs whenever n_x or n_y meets another node to form a new link and form a new state. If n_x and n_y again contact each other in the near future, we call it a recurrence $f_{L_{x \leftarrow y} L_{x \leftarrow y}}$ to s_0. If $f_{L_{x \leftarrow y} L_{x \leftarrow y}}^{(n)}$ denotes the first recurrence to $L_{x \leftarrow y}$ at the nth step, then its probability is defined as $f_{L_{x \leftarrow y} L_{x \leftarrow y}}^{(n)} = P\{s_n = L_{x \leftarrow y} | s_0 = L_{x \leftarrow y}, s_1 \neq L_{x \leftarrow y}, ..., s_{n-1} \neq L_{x \leftarrow y}\}$, where s_n denotes the nth state. The probability of recurrence to $L_{x \leftarrow y}$ can be written as

$$F_{l_{x \leftarrow y}} = \sum_{n=1}^{\infty} f^{(n)}_{L_{x \leftarrow y} l_{x \leftarrow y}}. \tag{8.16}$$

Then, s_0 is a recurrent state, if $F_{l_{x \leftarrow y}} = 1$. Thus, $L_{x \leftarrow y}$ is regular if there is at least one recurrent state in T^x or $T^{x \leftarrow y}$. Otherwise, $L_{x \leftarrow y}$ is sporadic, and may occur at T_r.

8.3.3.2 RSCR: Scheme Design

The RSCR scheme is divided into two parts:

- *Link type determination*: Given that each node records contacts encountered in time t, the link determination process determines the link type between n_x and n_y by $T^x / T^{x \leftarrow y}$ together with the fixed time length T. Given that $T < t$, T is used to get a certain number of time periods with respect to T^x by $\frac{T}{T^x}$. Then, if a $L_{x \leftarrow y}$ i.e., $F_{l_{x \leftarrow y}} = 1$ is formed within t in each T^x, $L_{x \leftarrow y}$ is regular. If $T^x > T$, $\frac{T}{T^x}$ is set as 1 by default. Later, $T^{x \leftarrow y}$ in Equation 8.15 is used to decide the link type for $L_{x \leftarrow y}$ formed after t. Let n_x contact n_y at time t_1, t_2, and t_3, such that $t_1 > t$ and $t_2 > t_1$. The first two inter-contact samples are calculated as $T_1^{x \leftarrow y} = t_1 - (t_0 = 0)$, and $T_2^{x \leftarrow y} = \left(T_1^{x \leftarrow y} + (t_2 - t_1) \right) / 2$. Until now, $L_{x \leftarrow y}$ is considered as sporadic. Now, If $T_2^{x \leftarrow y}$ is greater than $t_3 - t_2$ and $t_3 - t_1$ is less than T, $L_{x \leftarrow y}$ is taken as regular. If this proposition is false, a link is sporadic and the determination process stops here. Otherwise, the same steps are run to check the new contacts between n_x and n_y.
- *Message forwarding policy*: To select the next relay node to forward a message, let n_x contain a set of messages $\{m_x\}$ in local buffer for the destination node n_d. As with contacting n_y, n_x applies the RSCR scheme to take a decision by considering the link type of $L_{x \leftarrow y}$, a threshold $\alpha > 0$, and the remaining TTL value of every m_x in $\{m_x\}$, denoted as TTL_R^x. Node n_x takes a decision to select a next message in m_x to send it to n_y by referring to the following cases:
  - If either n_y is the destination node or $L_{x \leftarrow y}$ is $R_{sp(x \leftarrow y)}$, $R_{y \leftarrow d}$, or $L_{x \leftarrow y \leftarrow d}^{tr}$, n_x forwards m_x to n_y.
  - If the value of $\frac{TTL_R^x}{T_{M_s}^{x \leftarrow y}}$ is smaller than α, n_x forwards m_x to n_y without considering $L_{x \leftarrow y}$. Note that when $L_{x \leftarrow y}$ is updated either regular or sporadic, this condition is not followed further.
  - If $L_{x \leftarrow y}$ is $R_{x \leftarrow y}$ and the value of $\frac{TTL_R^x}{T^x \mid T^{x \leftarrow y}}$ is greater than α, n_x retains m_x. Otherwise, it forwards m_x.

8.3.4 Performance Results

In this section, we present the simulation settings and compare the performances of RSCR with Epidemic [2] and PRoPHET v2 [10].

8.3.4.1 Simulation Settings

We use the opportunistic network simulator ONE [25] with two mobility benchmarks, i.e., the SLAW [13] mobility model and real trace-based dataset Infocom '05 [26]. FIFO dropping policy is adopted in all the schemes to drop a message whenever a node's buffer operates at its limit. Table 8.1 lists the simulation parameters.

Following two performance metrics are used.

Table 8.1 Simulation parameters.

Parameters	Infocom '05	Slaw
Number of nodes	41	100
Map size	–	2000 × 2000 m²
Simulation time	24 hours	12 hours
Contact record collection time (t)	12, 24 hours	9, 12 hours
Time-to-live (TTL) duration	1 to 10 hours	1 to 5 hours
Message size	50 KB	50 KB
Buffer size	20 MB	10 MB
Transmission radius	–	10 m
Number of sources	1	1
Message creation intervals	100 seconds	50 seconds
Transmission speed	2 MB/s	2 MB/s

- *Delivery rate*: The ratio of the number of messages delivered successfully to the total distinct messages created by all source nodes.
- *Overhead ratio*: The ratio of number of messages relayed between relay nodes to the distinct messages created by all source nodes.

8.3.4.2 Results and Discussion

This section describes the practical usage of α and T, and then relative performance among RSCR, PRoPHET v2, and Epidemic.

Experiential values of α and T are gathered from experimental trails according to the mean inter-contact time and number of nodes in the network. In order to decide appropriate values of α and T to achieve good performance of RSCR, we intend to find fit values using SLAW and Infocom '05. According to the comparison between $\dfrac{TTL_R^x}{T^x \mid T^{x \leftarrow y}}$ and α, a larger α means a message will be forwarded earlier, and a smaller means keeping a message in a buffer for a longer time. Similarly, a larger T implies the number of time periods by $\left\lceil \dfrac{T}{T^x} \right\rceil$ comes out more as compared with a smaller T. In general, the variance of the following factors can influence the performance:

1. *Number of regular and sporadic links*: RSCR greedily forwards messages when sporadic links are encountered, which results in an increase in overhead. With more time periods, the condition for a regular link is not true for each time period, which results in more sporadic links. With a smaller T, more regular links are produced. As shown in Figure 8.1, the cases as $T = 3$ hours cause more overhead than those as $T = 0.5$ hours.
2. *Node density*: In Infocom '05 with low node strength, we take a large value of α to disseminate messages. SLAW has higher node density, so a smaller value of α is used to wait for a better relay in the near future.
3. *Time length of TTL*: With a sufficient buffer size, a message with a longer TTL value has more chances of reaching the destination. TTL_R^x also becomes one of the dominating factors in the message forwarding decision.

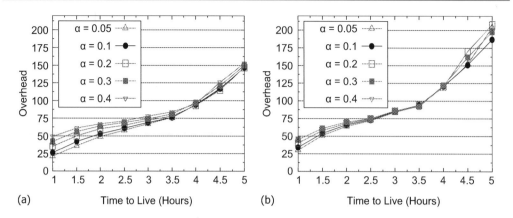

Figure 8.1 Overhead ratio by RSCR with different α, T, and *TTL* under SLAW (baseline: (a) $T = 30$, and (b) $T = 180$).

Performance results are shown by varying TTL = [1,5] and [1,10] hours in SLAW and Infocom '05, respectively. RSCR is implemented by considering three variants, i.e., Cases 1, 2 and 3, as described next.

- Case 1 – $RSCR_{base}$: We calculate the mean inter-contact and link type information between each node pair in a network in advance. To measure the performance of RSCR, we vary TTL from 1 to 5 hours and use T = 1 hour. Epidemic and PRoPHET v2 are sensitive to number of nodes in a network, encounter frequency, and message generation. With particular values of α and T, a longer TTL makes RSCR to keep messages longer on nodes and thus reduces the overhead as shown in Figure 8.2. A shorter TTL results in shorter TTL_R^x and invokes message forwarding quickly. In $RSCR_{base}$, messages staying longer with regular nodes will not get replicated multiple times. Nonetheless, messages with a lower TTL value are replicated as soon as contact opportunity occurs. For a smaller TTL value, Epidemic attains a high delivery rate, and it decreases after an increase in TTL.
- Case 2 – $RSCR_{pred}$: $RSCR_{pred}$ can predict the link type and mean inter-contact time for any forthcoming contact. We use Equation 8.15 to obtain the mean inter-contact time and use it to determine the link type, as explained in Section 8.3.3.2 for any new contact after time t. During the link determination process after t, initially, a link formed between any two nodes is treated as sporadic, which increases the overhead. As Figure 8.3 shows, $RSCR_{pred}$ increases the delivery rate with less overhead as compared with Epidemic, PRoPHET v2, and $RSCR_{base}$.
- Case 3 – $RSCR_{pred}$: $RSCR_{pred}$ adjusts the α and T as a result of Case 2 to improve the delivery rate further. With SLAW, $RSCR_{pred}$ alleviates the overhead with new values of α and T. In Infocom '05, the overhead by $RSCR_{pred}$ becomes less than PRoPHET v2's and Epidemic's overheads under a higher TTL, as shown in Figure 8.3(b).

8.4　SUMMARY AND FUTURE DIRECTIONS

This chapter described the encounter- and contact-based routing methodology in MONs. A brief review of many typical routing schemes was given, followed by a study of how those schemes exploit the contact periodicity among mobile nodes to enhance the delivery

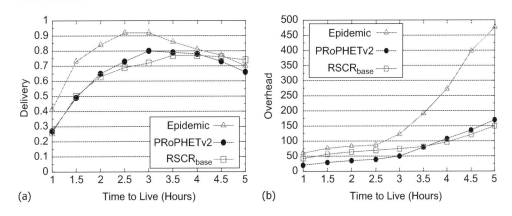

Figure 8.2 Performance of RSCR$_{base}$ (baseline: α = 0.3, T = 1 hour, t = 12 hours in SLAW).

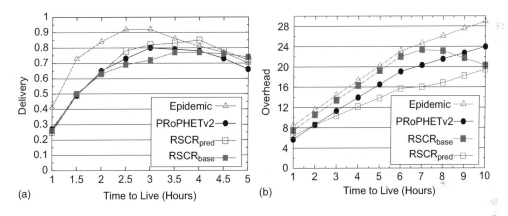

Figure 8.3 Performance of RSCR$_{pred}$ in comparison with RSCR$_{base}$ and PRoPHETv2 (baseline: (a) α = 0.2, T = 2 hours, and t = 9 hours in SLAW, and (b) α = 4, T = 0.5 hour, and t = 12 hours in Infocom '05).

Table 8.2 A summary of the topics of interest

Research Area	Related Work	Key Points
Social-based patterns	[3, 4, 21]	• Use of social relationship to build nodes' *betweenness* centrality with other nodes, social *similarity,* and tie strength with the destination node. • Nodes with higher centrality and similarity values are given priority to receive a message.
Social-based contact patterns	[5, 6, 16, 20, 23]	• Use of transient social contact patterns such as transient contact distribution, transient connectivity, and transient community structure. • Messages are forwarded using community structure and transient contact patterns. • Social-based contacts within community/communities to calculate the centrality value and the use of frequency and duration with respect to contacts.

(*Continued*)

Table 8.2 (Continued)

Research Area	Related Work	Key Points
Encounter-based routing	[8, 9, 15, 19, 24]	• Use of past encounter histories and inter-contact time to find different patterns between two nodes. • Use properties of movement behavior of different nodes to select the next relay. • Finding patterns using the correlation analysis to improve the accuracy of prediction.
Periodicity-based routing	[12, 14, 17, 22]	• Use of some techniques like the auto-correlation function (ACF) to find repetitive patterns and discrete Fourier transform (DFT) to find distinct patterns. • Capturing of skewed location, periodical reappearances of a node. • Use of social-based contact patterns from the temporal perspective to predict future patterns. • Use of inter-contact time to the destination node to select the next relay node.

rate cost-effectively. This chapter explained the notion and finding of contact periodicity and contact patterns between any node pair in a network. Contact patterns were then used to determine the centrality metric of a node for its facility to communicate with other nodes. Particularly, social and periodic features are used to calculate the centrality value of each node or to find the strength with respect to a node pair. In addition, this chapter also presented our recent study efforts on designing a new cost-effective contact-based routing scheme, named RSCR, for message forwarding. The RSCR design can determine a link type regular or sporadic between two nodes in a network. The proposed relay selection strategy can use such link types together with TTL of a message to enhance the delivery rate with low message overhead. Future directions in MONs attempt to integrate new trends, including machine learning to predict future contacts, opportunistic offloading, device-to-device communication, etc. With integration of new techniques and paradigms, the research community in MONs will come up with novel solutions to resolve the uncertainty of contact predictability and relay selection for cost-effective data delivery in MONs (Table 8.2).

REFERENCES

[1] Yue Cao and Zhili Sun. Routing in delay/disruption tolerant networks: A taxonomy, survey and challenges. *IEEE Communications Surveys Tutorials*, 15(2):654–677, 2013.

[2] Amin Vahdat and David Becker. Epidemic routing for partially-connected ad hoc networks. Technical report, Duke University, Durham, NC, 2000.

[3] E. M. Daly and M. Haahr. Social network analysis for information flow in disconnected delay-tolerant MANETs. *IEEE Transactions on Mobile Computing*, 8(5):606–621, 2009.

[4] P. Hui, J. Crowcroft, and E. Yoneki. Bubble Rap: Social-based forwarding in delay-tolerant networks. *IEEE Transactions on Mobile Computing*, 10(11):1576–1589, 2011.

[5] K. Wei, D. Zeng, S. Guo, and K. Xu. On social delay-tolerant networking: Aggregation, tie detection, and routing. *IEEE Transactions on Parallel and Distributed Systems*, 25(6):1563–1573, 2014.

[6] F. Li, H. Jiang, H. Li, Y. Cheng, and Y. Wang. Sebar: Social-energy-based routing for mobile social delay-tolerant networks. *IEEE Transactions on Vehicular Technology*, 66(8):7195–7206, 2017.

[7] Vijay Erramilli, Mark Crovella, Augustin Chaintreau, and Christophe Diot. *Delegation forwarding*. In *Proceedings of the 9th ACM International Symposium on Mobile Ad Hoc Networking and Computing*, pp. 251–260, Hong Kong, China, May 2008.

[8] S. C. Nelson, M. Bakht, and R. Kravets. *Encounter-based routing in DTNs*. In *Proceedings of IEEE INFOCOM '09*, pp. 846–854, Rio de Janeiro, Brazil, April 2009.

[9] Sujata Pal and Sudip Misra. *Contact-based routing in DTNs*. In *Proceedings of the 9th International Conference on Ubiquitous Information Management and Communication*, pp. 1–6, Bali, Indonesia, January 2015.

[10] Samo Grasic, Elwyn Davies, Anders Lindgren, and Avri Doria. *The evolution of a DTN routing protocol – PRoPHETv2*. In *Proceedings of ACM CHANTS '11*, pp. 27–30, Las Vegas, NV, September 2011.

[11] Thrasyvoulos Spyropoulos, Konstantinos Psounis, and Cauligi S. Raghavendra. *Spray and wait: An efficient routing scheme for intermittently connected mobile networks*. In *Proceedings of the ACM SIG-COMM Workshop on Delay-Tolerant Networking (WDTN)*, pp. 252–259, Philadelphia, PA, August 2005.

[12] Wei-Jen Hsu, T. Spyropoulos, K. Psounis, and A. Helmy. *Modeling time-variant user mobility in wireless mobile networks*. In *Proceedings of IEEE INFOCOM '07*, pp. 758–766, Barcelona, Spain, May 2007.

[13] S. Hong Kyunghan Lee, S. J. Kim, I. Rhee, and S. Chong. *Slaw: A new mobility model for human walks*. In *Proceedings of IEEE INFOCOM '09*, pp. 855–863, Rio de Janeiro, Brazil, April 2009.

[14] Sungwook Moon and Ahmed Helmy. *Understanding periodicity and regularity of nodal encounters in mobile networks: A spectral analysis*. In *Proceedings of IEEE GLOBECOM '10*, pp. 1–5, Miami, FL, December 2010.

[15] Weitao Wang, Y. Bai, P. Feng, Y. Gu, S. Liu, W. Jiang, and J. Huang. *DTN-KNCA: A high throughput routing based on contact pattern detection in DTNs*. In *Proceedings of IEEE COMPSAC '18*, vol. 1, pp. 926931, Tokyo, Japan, July 2018.

[16] Xiaomei Zhang and G. Cao. Transient community detection and its application to data forwarding in delay tolerant networks. *IEEE/ACM Transactions on Networking*, 25(5):2829–2843, 2017.

[17] Huan Zhou, Victor C. M. Leung, Chunsheng Zhu, Shouzhi Xu, and Jialu Fan. Predicting temporal social contact patterns for data forwarding in opportunistic mobile networks. *IEEE Transactions on Vehicular Technology*, 66(11):10372–10383, 2017.

[18] CRAWDAD Dataset. Mobile opportunistic network trace. Online; accessed October 25, 2020.

[19] Eyuphan Bulut, S. C. Geyik, and B. K. Szymanski. Efficient routing in delay tolerant networks with correlated node mobility. In *Proceedings of IEEE MASS '10*, pp. 79–88, San Francisco, CA, November 2010.

[20] W. Gao, G. Cao, T. La Porta, and J. Han. On exploiting transient social contact patterns for data forwarding in delay-tolerant networks. *IEEE Transactions on Mobile Computing*, 12(1):151–165, 2013.

[21] Wei Gao, Qinghua Li, Bo Zhao, and Guohong Cao. *Multicasting in delay tolerant networks: A social network perspective*. In *Proceedings of ACM MobiHoc '09*, pp. 299–308, New Orleans, LA, May 2009.

[22] Yu-Feng Hsu, C. Hu, and H. Hsiao. *On exploiting temporal periodicity for message delivery in mobile opportunistic networks*. In *Proceedings of IEEE COMPSAC '18*, vol. 1, pp. 809–810, Tokyo, Japan, July 2018.

[23] X. Meng, G. Xu, T. Guo, Y. Yang, W. Shen, and K. Zhao. A novel routing method for social delay-tolerant networks. *Tsinghua Science and Technology*, 24(1):44–51, 2019.

[24] M. Y. Mir and C. Hu. *Exploiting mobile contact patterns for message forwarding in mobile opportunistic networks*. In *IEEE Wireless Communications and Networking Conference (WCNC)*, pp. 1–7, Seoul, Korea (South), May 2020.

[25] Ari Keränen, Jörg Ott, and Teemu Kärkkäinen. *The one simulator for DTN protocol evaluation*. In *Proceedings of Simutools '09*, pp. 55:1–55:10, Rome, Italy, March 2009.

[26] James Scott, Richard Gass, Jon Crowcroft, Pan Hui, Christophe Diot, and Augustin Chaintreau. CRAWDAD dataset Cambridge/haggle (v. 2009-05-29), May 2009.

Chapter 9

Routing Techniques for Opportunistic Network

Sonam Kumari
Raj Kumar Goel Institute of Technology, Ghaziabad, India
Itu Snigdh
Birla Institute of Technology, Mesra, India

CONTENTS

9.1 INTRODUCTION

The mobile opportunistic network (MON) [5] is a challenged network as it is characterized by long and variable delays, high latency, frequent disconnections, long queuing times, limited longevity, and limited resources. The connection is unstable between the nodes and breaks as swiftly as it is discovered. The opportunistic network exploits human social behavior such as mobility patterns and daily routine to accomplish routing of messages and sharing data. With no fixed end-to-end path, the communication takes place only when opportunistic contacts are established among mobile nodes. Therefore, it is tightly coupled with human social behavior. Since the connection is highly dynamic, and since even the routes built here are highly dynamic, traditional routing strategies in an opportunistic network cannot be used efficiently.

With the abundance of mobile devices equipped with Wi-Fi, Bluetooth, sensors, and cameras, several domains of research are being done on exploiting the advantages of the opportunistic network. Moreover, modern vehicles are also employing such facilities and are installed with communication interfaces and sensors. This situation is advantageous for communication as the number of available mobile devices in an environment, and hence the contact opportunities, increase, which makes it useful in establishing an opportunistic network [6]. An MON carries a message created by a source node, with the help of intermediate nodes that forward the message in such a way as to bring the message closer to the destination. The communication model used in an MON is based on "store-carry-forwarding". Due to the variability in link existence and stability, nodes store the message on their buffer until they establish contact with the next hop node and forward the message after discovering the intermediate node [6].

The Internet we are using currently has certain assumptions on which it works, like end-to-end connectivity, low round-trip times and two-way symmetry of link bandwidth, i.e. uplink and downlink. Considering the case of an interplanetary communication taking place, the propagation delay would be more if we use the usual Internet. There are no end-to-end connectivity guarantees at a different instance of time – i.e. at a certain instance of time, two nodes might be connected, but at another instance, they might not be connected, and a long duration of time may elapse before those two nodes are connected again. The link bandwidth for uplink and downlink might not be similar in such a network, and these networks are highly error-prone.

9.2 CLASSIFICATION OF ROUTING ALGORITHMS

As we know that in an MON, a complete path between the sender and destination node probably does not exist most of the time, we are not able to use traditional routing protocols in such an environment. Therefore, designing an efficient routing protocol for opportunistic networks is complex, as information about the network topology and the contact information is insufficient. The existing routing strategies for the opportunistic network are broadly categorized into two routing schemes: the context-oblivious approach, and the context-aware approach. In a proactive protocol [7], every node in a network maintains a routing table that is updated periodically and will broadcast the same to all other nodes available in the network. The reactive routing protocol [7] is an "on-demand routing protocol". In this case, by flooding the Route REQuest (RREQ) packets on the network, every node dynamically discovers its route to destination node.

9.2.1 Context-Oblivious Approach

The context-oblivious routing classification contains flooding-based routing protocols. These routing protocols indiscriminately follow either flooding or controlled flooding strategies. In the flooding-based strategy, a source node forwards packets to every neighboring node. Each neighbor gets a packet and advances it to every one of its neighbors apart from the one from which the packet was received. The destination node thus may receive many copies of that packet through various nodes at an alternate time. The routing protocols of this category do not exploit any form of context information of nodes, and the message is transmitted to the destination by disseminating it as widely as possible. So, it is the simplest approach, as no other prior information is calculated here for forwarding purposes. However, the flooding-based strategy disperses an enormous number of packets in the network, which creates a traffic issue, and it is also expensive as far as memory and energy utilization is concerned. Epidemic routing and, Spray and Wait routing belong to this category.

- *Direct transmission*: Spyropoulos et al. [8] proposed a simple single-copy routing called direct transmission routing. Direct transmission is a simple, single-copy routing protocol. In this routing protocol, the node which generates a message will hold that message in its buffer until the time it reaches the destination node. In this scheme, data transfers are minimized, but the delay in message delivery is high for this routing strategy and is less robust.
- *Epidemic routing*: Vahdat and Becker [9] proposed epidemic routing for forwarding data packets in an opportunistic network. It ensures that every node receives the packet

by a sufficient number of random exchanges of packets in the network. The dissemination process is bounded by providing hop count limits for each message.

- *Spray and Wait*: Spyropoulos et al. [10] proposed a technique called Spray and Wait, which limits the flooding of packets, and thus the bandwidth requirement degrades in this strategy. It basically consists of two phases, namely, the spray phase and the wait phase. In the Spray phase, N copies of the message are created and are distributed to N number of intermediate nodes randomly. Then, if the message is not delivered to the destination node, $N/2$ numbers of copies are created and spread randomly. This process is repeated until and unless one copy of the message is left out. Then in the Wait phase, direct transmission is performed, and in this way, the packet is delivered to the destination node.

- *Binary Spray and Wait based on Average Delivery Probability (ADPBSW)*: Xue et al. [11] proposed another variation of the Spray and Wait routing protocol, Binary Spray and Wait, which limits the flooding of packets by calculating average delivery probability. Based on this, the next hop is selected for forwarding the packet towards the destination node. Whenever a node is encountered, it updates the delivery probability value. In Spray and Wait, forwarding is done randomly, whereas in ADPBSW, forwarding is based on delivery probability. If a message with destination node D is sent by node A which possess N copies of this message and it encounters node B such that $Pavg(b,d) > Pavg(a,d)$, then node A will forward $\dfrac{N}{2}$ messages to node B. If $Pavg(b,d) \leq Pavg(a,d)$, A will not forward messages to B.

- *Network Coding protocol*: Widmer and Boudec [12] proposed a network coding routing convention which is like flooding-based routing or probabilistic routing, except it utilizes network coding to restrict message flooding. The overhead of probabilistic routing protocol is reduced in the Network Coding protocol. Here, each message is linked with an encoding vector, and the message vector is transmitted along with the encoding vector. A decoding matrix is used to store the message and encoding vector, as well as the original message vector in case it is a source node. So, in this protocol, the message is transformed by the nodes in another format before transmission. The performance of this protocol is higher than the probabilistic-based protocol and is suitable for the extreme environment such as sparse mobile networks.

- *Erasure-coding based routing*: Wang et al. [13] proposed this protocol where forwarding of data is based on erasure codes. The longer delays are reduced, as it allows to use a larger number of intermediate nodes while maintaining the overhead. The message is converted into code blocks with the end goal that any adequately enormous subset of the created code blocks can be utilized to reproduce the original message. It takes as information a message of size M and a replication factor r. The calculation produces $M \times r/b$ equivalent measured code blocks of size b, with the end goal that any $(1 + \epsilon) . \dfrac{M}{2}$ code blocks can be utilized to recreate the message. Here, ϵ is a constant and differs relying upon the specific algorithm utilized; for example, Reed-Solomon codes or Tornado codes. The determination of the algorithm includes compromises between coding/decoding effectiveness and the minimal number of code blocks required to remake a message.

- *Hybrid erasure-coding based routing* (H-EC): Chen et al. [14] proposed the H-EC protocol which is intended to completely consolidate the robustness of the erasure coding-based protocol while saving the performance of replication techniques. In this methodology, two duplicates of eradication-coded blocks are forwarded to intermediate nodes. The first duplicate of coded blocks is sent by the Erasure coding way such that each relay node gets the coded block, and the second duplicate of erasure coded blocks is communicated by the Aggressive Erasure Coding protocol.

9.2.2 Context-Aware Approach

Context is any relevant information that describes the environment of any objects (i.e. a person, groups, and physical and computational devices) that interact. In the knowledge-based approach [15–17], based on certain information related to the environment, a node is selected as an intermediate node such that it increases the delivery probability and makes a faster delivery possible. The information may be related to either network topological order, network traffic, or node location.

- *Randomized routing*: Spyropoulos et al. [8] proposed this single copy routing protocol where it uses intermediate nodes to deliver the message to the destination node. Here, a node transmits the message to the intermediate node it encounters with probability $p > 0$ only if its probability of delivering the message to the destination node is greater than the present node holding the packet. The time of contact information is used to calculate the delivery probability.
- *Utility-based routing*: Spyropoulos et al. [8] proposed this single copy routing protocol where the position data regarding each node is updated in the last encounter timer and is diffused in the network though the mobility process of other nodes. This position data is not absolute, but rather relative to the position of another node. If a node is seen at some time instant having a low timer value for another node, then this other node is expected to be somewhere nearby. This protocol is the variation of the randomized routing protocol, as the speed, mobility pattern, and last encountered time information are used to calculate the delivery probability.
- *Seek and Focus routing protocol*: Spyropoulos et al. [8] proposed the Seek and Focus protocol. A utility-based protocol has the disadvantage of a slow-starting initial stage, which can be more prevalent in larger networks. Particularly in a huge network, where anticipated separation between a source and a destination node is greater, it will take the source node quite a while to find a higher-utility next hop at the start. The Seek and Focus protocol consists of two stages. In the Seek phase, if the utility around the hub is low, perform randomized forwarding with boundary p to rapidly look for close-by hubs; and in the Focus phase, when a high utility hub is found, change to a utility-based protocol.
- *Prioritized Epidemic routing*: Ramanathan et al. [18] proposed *PR*ioritized *EP*idemic routing (PREP), where the priority of the packet is dependent on cost to destination, source, and expiry time. The value of cost is dependent on the per-link "average availability" information that is spread in the epidemic manner. Here every node executes a neighbor discovery algorithm to create and maintain a set of bidirectional links with neighboring nodes. Each message is allocated with drop priority pd and transmit priority pt. The packet whose transmit priority is high will be forwarded first to the node who comes in its contact. When the buffer is full than the high drop priority packet will be deleted first than the packet with low drop priority. PREP uses expiry time information and topology awareness to take a decision of deleting or holding the packet in a scenario where resources (buffer, bandwidth) are fully consumed.
- *Spray and Focus routing protocol*: Spyropoulos et al. [15] proposed the Spray and Focus protocol in which packet replication is reduced and distributed among a few relay nodes. Further, the relay node will forward the packet using a single copy utility-based scheme and carry the packet until time to live expires. Since many relay nodes are in parallel looking for the destination node, it becomes an efficient protocol in the sparse connectivity environment. In the spray phase, the number of copies of the message generated "L" is such that the average delay is α times the optimal one where M number of nodes are present, and it is calculated using the following equation:

$$\left(H_M^3 - 1.2\right)L^3 + \left(H_M^2 - \frac{\pi^2}{6}\right)L^2 + \left(a + \frac{2M-1}{M(M-1)}\right)L + = \frac{M}{M-1} \tag{9.1}$$

In the focus phase, the message can be forwarded to a different intermediate node based on forwarding criteria which is based on a set of timers that record the time since two nodes last encountered each other. So, here, every node keeps up a vector with IDs of all messages that it has stored, and at whatever point two nodes encounter one another, they exchange their vectors and check which messages they have in common and their time to live (TTL). If the TTL expires, the message is discarded.

- *Motion vector (MoVe)*: LeBrun et al. [19] introduced a scheme using the motion vector (MoVe) of mobile nodes to predict their future location. In this strategy information of nodes, i.e. relative velocity, is calculated, and using this information, a node's future location is predicted and this gives its neighboring nodes information such that they bring the message closer to the destination. The data success rate of MoVe as a function of the number of nodes is less than Epidemic routing, and an end-to-end delay is longer than epidemic routing.

- *Bubble Rap routing protocol*: Hui and Crowcroft [20] proposed the Bubble Rap protocol, which is based on the concept of community such that the effectiveness of forwarding strategy is improved. It follows two significant observations. First is the popularity of the node in the community, so the packet is forwarded to the most popular node. Second is the community to which the relay node belongs; if the relay node is of the same community as that of destination node, there is a greater chance that the interaction between the nodes will take place. There are a few assumptions of the protocol: Each node must belong to at least one community, and a community can consist of a single node. Every node has two types of rankings: global ranking and local ranking. Global ranking represents the popularity of the node across the network, whereas local ranking indicates only the popularity of the node within the community to which it belongs. The source node constructs the hierarchical ranking tree based on global ranking until it reaches a node which belongs to the destination node community. After that, it uses a local ranking tree to deliver the message to the destination node.

- *Context-Aware Routing (CAR)*: Musolesi et al. [21] presented Context-Aware Routing (CAR). In CAR, the message is transmitted to a node that has the highest delivery probability. The delivery probability is calculated and predicted using some context information with the help of the Kalman filter. Delivery probabilities are exchanged periodically so that the best carriers are computed based on the nodes' context information.

- *MaxProp*: Burgess et al. [22] proposed a protocol called MaxProp for effective routing of messages. In this routing strategy, a node schedules the packets for transmission to its neighboring nodes based on path likelihoods which are predicted using the history of contacts. This scheme also uses few mechanisms, such as providing an acknowledgment, a head-start for new packets, and lists of previous intermediaries used in order to enhance the prediction

- *Shortest expected path routing (SEPR)*: Kun et al. [23] proposed a shortest expected path routing (SEPR). SEPR estimates the probability of link forwarding using the historical information. Effective path length (EPL) is estimated in this scheme, and whenever two nodes meet, they exchange this information with each other and update its EPL value in the link probability table. The smaller the EPL value, the higher the chance of successful message delivery. Nodes recalculate their probability to meet other nodes when they get connected. The probability is calculated basically using the time those two nodes stay connected divided by the sampling time window. The Dijkstra shortest path algorithm is used to get the expected path length for each node in the network.

- *Estimate/prediction routing*: Lindgren et al. [24] proposed a probabilistic routing protocol called PRoPHET (Probabilistic Routing Protocol using History of Encounters and Transitivity). PRoPHET is based on the estimation of delivery probability. This routing strategy works similar to the Epidemic routing. In real life, humans mostly move predictably according to their repeating behavioral patterns. For example, if a user has visited a spot ordinarily previously, it is likely that the user will visit that place once more. In view of these perceptions, the convention utilizes the recurrence of contacts between the nodes as setting data to improve the routing performance. Whenever nodes meet, they exchange tables containing the delivery probability value. It is also observed that if two nodes meet each other frequently, they have a high delivery probability to each other. Hence, the delivery probability ages with time.
- *History-based prediction routing (HBPR)*: Dhurandher et al. [25] proposed HBPR, which uses the behavioral information of the nodes to find the next node for forwarding the packet. A few assumptions are taken into consideration:
  - The movement of the nodes are based on the Human Mobility pattern
  - The nodes are cooperative
  - The simulated area is termed as cells, and they can be locations that nodes visit during simulation process.

  The protocol is divided into three phases:
- Initializing the Home Locations
- Message generation and Home Location update
- Next hop selection.

  Each node has a certain region that it would visit frequently, some regions that it would visit but not often, and so forth. Based on this, one could predict a node's future location quite effortlessly while using its past history. Every node floods their past location in the network and keeps updating their history after performing a re-scan process.
- *kROp*: Deepak et al. [26] proposed the k-Means clustering based routing protocol. kROp utilizes unsupervised machine learning in the form of an optimized k-Means clustering algorithm to train on these features and make next-hop-selection decisions. This protocol is based on Feature identification, a Clustering model, and an Evaluation function. In the Feature identification phase, the nodes' information, such as encounters of the node with the destination node, distance of the node from the destination node, available buffer space of the node, and number of messages successfully delivered by that node until that point of time, are used for next-hop prediction. An evaluation function is defined that helps to select the best cluster among the cluster centers generated.

Table 9.1 compares the context oblivious routing protocols, and Table 9.2 compares context-aware protocols on the basis of buffer management, estimation of forwarding probability, reliability, energy efficiency, and types of routing whether reactive or proactive.

9.3 FACTORS AFFECTING ROUTING PROTOCOL PERFORMANCE

The key problem in the opportunistic network is to choose to the appropriate relay nodes to transfer the message such that the message is delivered with minimum delay latency and maximum probability of successful delivery of the message to the destination node. The parameters that affect the overall performance of routing protocol are:

- Social reference
- Delivery probability
- Node speed
- Node remaining energy

Table 9.1 Comparison between context oblivious routing protocols.

Protocols	Buffer Management	Estimation of Forwarding Probability	Reliability	Energy Efficiency	Reactive or Proactive
Direct transmission	Infinite	No	More reliable (high traffic)	More bandwidth is required	Reactive
Epidemic	Infinite	No	Less reliable (high traffic serves high contention)	More bandwidth is required	Reactive
Spray and Wait	Infinite	No	Less reliable (similar to Epidemic)	Less bandwidth than Epidemic	Reactive
Binary Spray and Wait based on Average Delivery Probability	Infinite	No; spreads copies of message in controlled way	Less than epidemic	Less bandwidth compared with Epidemic	Reactive
Network Coding protocol	Infinite	No, merges several messages into one to reduce number of transmissions	Less than epidemic	Less bandwidth compared with Epidemic	Reactive
Erasure-coding based routing	Infinite	No, converts message into several code blocks	Less than epidemic	Less bandwidth compared with Epidemic	Reactive
Hybrid erasure-coding based routing	Infinite	No, combination of EC and A-EC	Less than epidemic	Less bandwidth compared with Epidemic	Reactive

Table 9.2 Comparison between context-aware routing protocols.

Protocols	Buffer Management	Estimation of Forwarding Probability	Reliability	Energy Efficiency	Reactive or Proactive
Randomized routing	No	Last encountered time	Less reliable	Less bandwidth	Reactive
Utility-based routing	No	Last encountered time with speed and mobility pattern	Less reliable	Less bandwidth	Reactive
Seek and Focus routing protocol	No	Last encountered time with speed and mobility pattern	Less reliable than epidemic	Less bandwidth than epidemic	Reactive

(Continued)

Table 9.2 (Continued)

Protocols	Buffer Management	Estimation of Forwarding Probability	Reliability	Energy Efficiency	Reactive or Proactive
Prioritized Epidemic routing	Yes	Use expiry time information and topology awareness	Less reliable than epidemic	Less bandwidth than epidemic	Reactive
Spray and Focus routing protocol	Infinite	Last encountered time	Reduces delay of Spray and Wait Routing up to 20 times in some scenarios	Less bandwidth than epidemic	Reactive
MoVe	Infinite	Yes, using the motion vector	Less than epidemic	Less bandwidth with substantially less buffer space compared with epidemic	Reactive
Bubble Rap routing protocol	Yes, infinite	Social community information	Better forwarding efficiency compared to context-oblivious routing and PROPHET routing.	Less bandwidth than epidemic	Reactive
CAR	Infinite	Yes, using Kalman filter	More reliable than epidemic if the buffer size is small	Less bandwidth than an epidemic	Reactive
MaxProb	Infinite	Yes, estimating the delivery likelihood	More reliable than CAR (higher delivery rate)	Less bandwidth than CAR	Reactive
SEPR	Yes, removes those packets with smaller EPL	Yes	More reliable than epidemic (almost 35 percent)	Less bandwidth than epidemic (almost 50 percent)	Reactive
PRoPHET	Infinite	Yes, using delivery probability vector	More reliable than epidemic	Less bandwidth than an epidemic	Reactive
History-based prediction routing	Infinite	Yes, using nodes mobility pattern	More reliable than epidemic	Less bandwidth than epidemic	Reactive
kROp	Infinite	Yes, utilizes unsupervised machine learning	More reliable than HBPR	Less bandwidth than HBPR	Reactive

Social reference: The MON mimics human social behavior, and this is estimated on the value of betweenness centrality [27]. Betweenness centrality of a node is the ratio between the numbers of the shortest path that passes through that node to the total number of the shortest path:

$$\text{Betweenness}(v) \sum_{s \neq v \neq t} \frac{\sigma_{st}(v)}{\sigma_{st}} \qquad (9.2)$$

where σ_{st} is the total number of shortest paths from node s to node t and $\sigma_{st}(v)$ is the number of those paths that pass through v.

Delivery probability: The delivery ratio (DR) is directly proportional on the two factors, contact duration (CT) and inter-contact time (ICT) [3]. Contact duration is the time period for which the devices are in contact with each other. Intercontact time is the time interval after which the two devices will come in contact with each other.

$$DR \alpha \frac{Contact\ Duration}{Inter\ Contact\ Time} \qquad (9.3)$$

Node speed: Normally, the user carries the mobile device with them, so the node speed parameter is dependent on transportation mode of the user. In general, it has been observed that the speed of a walking person varies from 0.5 to 1.5 meters per second, the speed of person traveling by a vehicle is assumed to vary from 8.3 to 30 meters per second and of a cyclist is assumed to vary from 4.2 to 11 meters per second.

Node remaining energy: Node remaining energy (NRE) inversely depends on the number of packets it transmits and receives and the inter-probe time [28]. As the NRE value decreases, the likeliness of the node suffering downtime increases and the probability of node being used as the next hop decreases.

$$NRE \alpha \frac{1}{No\ of\ packets\ received\ and\ transmitted} \qquad (9.4)$$

$$E\big(Send\,(k,d)\big) = Eelec * k + \varepsilon amp * k * d_{ij}^{2} \qquad (9.5)$$

$$E\big(Recv\,(k)\big) = Eelec * k \qquad (9.6)$$

where *Eelec* is the coefficient of radio frequency, *εamp* is the amplification coefficient of sending the device, d_{ij} is the data transmission distance of the node, and *k* is the number of bits being transferred.

Inter probe is the time interval after which the node will start the neighbor discovery process again:

$$Inter\ probe\ time(\tau) = \frac{1}{\alpha f_s} \qquad (9.7)$$

$$\text{Where } f_s = \frac{2B}{R}$$

where B is the speed at which the node travels in meter per second. R is the transmission radius of the node in the meter. The variable α lies within the range $0 < \alpha \leq 1$ and it defines the degree of closeness of nodes to be classified as its neighbors. Generally, the transmission range of Bluetooth is 30 meters and that of a Wi-Fi device is 50 meters. Therefore, let's say a person is carrying a Bluetooth-enabled mobile phone in his pocket and traveling in a vehicle at a speed of 25 meters per second. The inter probe time calculation is:

$$fs = \frac{2B}{R}$$

After substituting, R = 30 meters, B = 25 meters/second, and fs = 1.67 per second. Therefore, Inter probe time (τ) = 1/(αfs) = 0.59 second (assuming α = 1).

9.4 QoS CONSTRAINTS OF MONs

Though the opportunistic network seems to be the available solution to the incessant mobility-based communication, deploying an opportunistic network in the real environment poses several challenges. The two major challenges of an opportunistic network are:

- Opportunistic contact
- Storage constraint

    Contact: Since in such environments, nodes are mobile, the contact between nodes becomes unpredictable. The contact duration and inter-contact time are important factors through which we can determine the capacity of the opportunistic network. Contact duration is the time period for which the two devices are in contact with each other and the inter-contact time is the time interval after which the contact between the same pair of devices takes place.

    Storage constraint: In order to avoid packet drop in an opportunistic network, the packets must be stored in the intermediate nodes buffer space for the time duration the packet is delivered to the appropriate destination node. As the number of messages in a network increases, the buffer requirement increases. We can say that the buffer requirement is directly proportional to a number of messages created in the network; for example, the epidemic routing protocol, which, based on the flooding approach, needs unlimited relay buffers. Routing can be improved if we are able to gather more information such as network expected topology and contact information. Unfortunately, such data are not easily available.

9.5 ONE SIMULATOR-BASED PERFORMANCE EVALUATION OF ROUTING ALGORITHMS

There exist several simulation tools for the delay tolerant network, such as dtnsim, dtmsim2, ns2, and OMNeT for MANET, through which various approaches can be tested and evaluated by performing simulation under a certain scenario. Similarly, for the opportunistic network, a simulation environment called the Opportunistic Network Environment (ONE) simulator is available. The ONE simulator not only focuses on routing but also combines various mobility model and visualization in one package that is easily extensible and provides a rich set of reporting and analyzing modules. The ONE simulator incorporates various types of movement models proposed for the opportunistic network, such as map-based movement, Random Waypoint, Random Walk, and shortest path map-based movement.

9.5.1 Case Study: Comparison of Average Latency

Latency is the time delay of a packet to reach from the source to the desired destination node. Simulation could be performed for several existing routing protocols. In a case study, Epidemic routing, Spray and Wait routing, and PRoPHET routing simulations were performed. The observation was based on different numbers of nodes, and the average latency of these routing protocols has been compared and plotted in the graph shown in Figure 9.1.

From the graph, it has been observed that the average latency of PRoPHET routing is less than that of the other two routing protocols.

9.5.2 Analysis of the Routing Algorithms in View of Open Issues

Opportunistic networks present a lot of challenges owing to their mobility and disconnected behavior. As in the previous section, simulation was performed for comparing the average latency of different protocols, in a similar way we can also evaluate other factors (i.e. average buffer time, overhead ratio) based on which of several routing protocols can be compared.

Buffer space is the memory space required to store the packet into nodes memory. Packets are stored in an intermediate nodes memory until it transmits the packet to another node. Packet buffering is important for faster delivery and to reduce the packet loss problem. The buffer management algorithm is used in order to schedule the packet and to take a decision whether to accept or reject the packet. For the average buffer, time simulation was performed.

In Figure 9.2 a graph of average buffer time versus a number of nodes for the PRoPHET, Epidemic, and Spray and Wait routing protocols has been plotted. It is observed that the average buffer time of PRoPHET is less than that of Epidemic and Spray and Wait as the number of nodes increases.

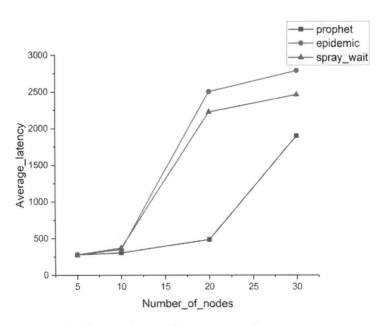

Figure 9.1 Comparison graph of average latency of routing protocols.

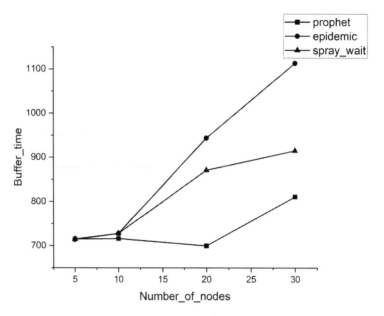

Figure 9.2 Comparison graph of average buffer time of routing protocols.

9.6 CONCLUSION

We have outlined the frequently used routing protocols for opportunistic networks. We now understand that the success of any network is the capability to efficiently and reliably communicate data. The requirements of an MON are different from those of the regular wireless networks wherein we can leverage the mobility of nodes to our advantage. Also, in wireless networks, the increase in node density cause interference and packet drops. On the contrary, the performance of MONs improves with the increase in the number of nodes. So, with the increase in the number of communicating nodes, the chances of forwarding increase, and hence the delay and the required buffer time decreases.

REFERENCES

[1] Lindgren, Anders, and Avri Doria. "Experiences from deploying a real-life DTN system." In *IEEE Consumer Communications and Networking Conference: 11/01/2007-13/01/2007*, pp. 217–221. IEEE Communications Society, 2007.

[2] Huang, Chung-Ming, Kun-chan Lan, and Chang-Zhou Tsai. "A survey of opportunistic networks." In *22nd International Conference on Advanced Information Networking and Applications-Workshops (AINA Workshops 2008)*, pp. 1672–1677. IEEE, 2008.

[3] Pelusi, Luciana, Andrea Passarella, and Marco Conti. "Opportunistic networking: Data forwarding in disconnected mobile ad hoc networks." *IEEE communications Magazine* 44, no. 11 (2006): 134–141.

[4] Santos, Rodrigo M., Javier Orozco, Sergio F. Ochoa, Roc Meseguer, and Daniel Mosse. "Providing real-time message delivery on opportunistic networks." *IEEE Access* 6 (2018): 40696–40712.

[5] Conti, Marco, Silvia Giordano, Martin May, and Andrea Passarella. "From opportunistic networks to opportunistic computing." *IEEE Communications Magazine* 48, no. 9 (2010): 126–139.

[6] Kumari, Sonam, and Itu Snigdh, "Analyzing impact of factors on routing decisions in opportunistic mobile networks," Conference Publications, ICRTCST – 2018.

[7] Kumar, Manish, Itika Gupta, Sudarshan Tiwari, and Rajeev Tripathi. "A comparative study of reactive routing protocols for industrial wireless sensor networks." In *International Conference on Heterogeneous Networking for Quality, Reliability, Security and Robustness*, pp. 248–260. Springer, Berlin, Heidelberg, 2013.

[8] Spyropoulos, Thrasyvoulos, Konstantinos Psounis, and Cauligi S. Raghavendra. "Single-copy routing in intermittently connected mobile networks." In *2004 First Annual IEEE Communications Society Conference on Sensor and Ad Hoc Communications and Networks, 2004. IEEE SECON 2004*, pp. 235–244. IEEE, 2004.

[9] Vahdat, Amin, and David Becker. "Epidemic routing for partially connected ad hoc networks." (2000): 2019.

[10] Spyropoulos, Thrasyvoulos, Konstantinos Psounis, and Cauligi S. Raghavendra. "Spray and wait: An efficient routing scheme for intermittently connected mobile networks." In *Proceedings of the 2005 ACM SIGCOMM Workshop on Delay-Tolerant Networking*, pp. 252–259. 2005.

[11] Xue, Jingfeng, Xiumei Fan, Yuanda Cao, Ji Fang, and Jiansheng Li. "Spray and wait routing based on average delivery probability in delay tolerant network." In *2009 International Conference on Networks Security, Wireless Communications and Trusted Computing*, vol. 2, pp. 500–502. IEEE, 2009.

[12] Widmer, Jörg, and Jean-Yves Le Boudec. "Network coding for efficient communication in extreme networks." In *Proceedings of the 2005 ACM SIGCOMM Workshop on Delay-Tolerant Networking*, pp. 284–291. 2005.

[13] Wang, Yong, Sushant Jain, Margaret Martonosi, and Kevin Fall. "Erasure-coding based routing for opportunistic networks." In *Proceedings of the 2005 ACM SIGCOMM Workshop on Delay-Tolerant Networking*, pp. 229–236. 2005.

[14] Chen, Ling-Jyh, Chen-Hung Yu, Tony Sun, Yung-Chih Chen, and Hao-hua Chu. "A hybrid routing approach for opportunistic networks." In *Proceedings of the 2006 SIGCOMM Workshop on Challenged Networks*, pp. 213–220. 2006.

[15] Spyropoulos, Thrasyvoulos, Konstantinos Psounis, and Cauligi S. Raghavendra. "Spray and focus: Efficient mobility-assisted routing for heterogeneous and correlated mobility." In *Fifth Annual IEEE International Conference on Pervasive Computing and Communications Workshops (PerComW'07)*, pp. 79–85. IEEE, 2007.

[16] Verma, Anshul, K. K. Pattanaik, Aniket Ingavale, "Context-Based Routing Protocols for OppNets", *Routing in Opportunistic Networks*, New York, Springer, chapter 3, pp. 69–97, 2013. doi:10.1007/978-1-4614-3514-3_3

[17] Verma, Anshu, K. K. Pattanaik, "Routing Protocols in Opportunistic Networks." *Opportunistic Networking: Vehicular, D2D and Cognitive Radio Networks*, CRC Press (Taylor & Francis Group), chapter 5, pp. 125–166, 2017. doi:10.1201/9781315200804.

[18] Ramanathan, Ram, Richard Hansen, Prithwish Basu, Regina Rosales-Hain, and Rajesh Krishnan. "Prioritized epidemic routing for opportunistic networks." In *Proceedings of the 1st International MobiSys Workshop on Mobile Opportunistic Networking*, pp. 62–66. 2007.

[19] LeBrun, Jason, Chen-Nee Chuah, Dipak Ghosal, and Michael Zhang. "Knowledge-based opportunistic forwarding in vehicular wireless ad hoc networks." In *2005 IEEE 61st Vehicular Technology Conference*, vol. 4, pp. 2289–2293. IEEE, 2005.

[20] Hui, Pan, and Jon Crowcroft. *Bubble Rap: Forwarding in small world DTNs in ever decreasing circles*. No. UCAM-CL-TR-684. University of Cambridge, Computer Laboratory, 2007.

[21] Musolesi, Mirco, and Cecilia Mascolo. "CAR: Context-aware adaptive routing for delay-tolerant mobile networks." *IEEE Transactions on Mobile Computing* 8, no. 2 (2008): 246–260.

[22] Burgess, John, Brian Gallagher, David D. Jensen, and Brian Neil Levine. "*MaxProp: Routing for vehicle-based disruption-tolerant networks*." In *Infocom*, vol. 6, pp. 1–11. 2006.

[23] Tan, Kun, Qian Zhang, and Wenwu Zhu. "Shortest path routing in partially connected ad hoc networks." In *GLOBECOM'03. IEEE Global Telecommunications Conference (IEEE Cat. No. 03CH37489)*, vol. 2, pp. 1038–1042. IEEE, 2003.

[24] Lindgren, Anders, Avri Doria, and Olov Schelén. "Probabilistic routing in intermittently connected networks." *ACM SIGMOBILE Mobile Computing and Communications Review* 7, no. 3 (2003): 19–20.

[25] Dhurandher, Sanjay K., Deepak Kumar Sharma, Isaac Woungang, and Shruti Bhati. "HBPR: history based prediction for routing in infrastructure-less opportunistic networks." In *2013 IEEE 27th International Conference on Advanced Information Networking and Applications (AINA)*, pp. 931–936. IEEE, 2013.

[26] Sharma, Deepak Kumar, Sanjay Kumar Dhurandher, Divyansh Agarwal, and Kunal Arora. "kROp: k-Means clustering based routing protocol for opportunistic networks." *Journal of Ambient Intelligence and Humanized Computing* 10, no. 4 (2019): 1289–1306.

[27] Liu, Tong, Yanmin Zhu, Ruobing Jiang, and Bo Li. "A sociality-aware approach to computing backbone in mobile opportunistic networks." *Ad Hoc Networks* 24 (2015): 46–56.

[28] Orlinski, Matthew, and Nick Filer. "Neighbour discovery in opportunistic networks." *Ad Hoc Networks* 25 (2015): 383–392.

Chapter 10

Blockchain Leveraged Node Incentivization in Cooperation-Based Delay Tolerant Networks

Souvik Basu

Heritage Institute of Technology, Kolkata, India

Soumyadip Chowdhury

University of Engineering and Management, Kolkata, India

Siuli Roy

Heritage Institute of Technology, Kolkata, India

CONTENTS

10.1 INTRODUCTION

A DTN, which is essentially an opportunistic peer-to-peer network, involves the transmission of messages from a sender to the receiver in multiple hops through forwarder nodes. Thus, the success of such a network is highly dependent on the cooperation of these forwarder nodes, and selfish activities contribute to low delivery ratio and long latency of delivery [1–6]. Intentions behind such selfish behavior may be either non-malicious (concerns about energy and bandwidth consumption, etc.) or malicious (attacks like denial of service, collusion, impersonation, etc.) [5]. One solution to this problem is to reward forwarder nodes with real money or credit to stimulate them to engage in forwarding activities. The characteristics of an opportunistic peer-to-peer network, however, bring challenges to the creation of an incentive mechanism. On one hand, how can a sender be convinced that the message, containing shelter-needs, is actually delivered to the receiver? If a sender rewards a selfish node beforehand, the node will not faithfully store-carry-forward the message to the receiver and will indulge in a "dine and dash" behavior [7]. On the other hand, a sender does not know who the forwarder nodes will be, and how many such nodes will be involved in transmitting a message; thus, it is hard to know who will be rewarded and how much reward is to be paid [8]. Finally, forwarder nodes may have different constraints on consumed resources and different expectations about the incentive. A node will not diligently store-carry-forward the message to the receiver and engage in "dine and dash" behavior if a sender rewards a greedy node beforehand. On the other hand, a sender does not know who the forwarder nodes will be and how many such nodes will be involved in the transmission of a message; it is therefore difficult to know who will be rewarded and how much reward will be paid [8]. Finally, forwarder nodes can have distinct resource constraints and different expectations.

Several credit-based approaches [9–14] to developing effective incentive structures for DTNs have been published, where rewards are given to the cooperative forwarder nodes through paying credit or virtual money. Most credit-based reward schemes, however, either rely on or do not have an explicit digital currency framework that is proven to be secure [8]. Blockchain [15–18] is a distributed digital public transaction leader of cryptocurrencies such as Bitcoin or Ethereum [17, 18] which, in lieu of a central trusted authority, can verifiably and permanently record transactions between two parties. The most realistic digital currency, Ethereum, makes it possible for P2P networks to formulate practical credit-based reward schemes. However, the usage of blockchain is restricted by the user's access to an end-to-end Internet connection. This limitation restricts the use cases to access blockchain and prevents its adoption in intermittently connected network environments like DTNs. In fact, reliance on the Internet is becoming a critical concern for blockchain developers, since network interruptions can cause serious harm to the blockchain ecosystem. Thus, it becomes imperative to look for new avenues through which blockchain can be used even in intermittently connected network environments. To the best of our knowledge, this domain is much less investigated.

This chapter first proposes a mechanism that enables usage of blockchain technology in an intermittently connected network such as DTN based on the Ethereum platform using smart contracts in Solidity. The mechanism develops an alternative way of broadcasting and authorizing blockchain transactions without having to rely on Internet connectivity. It exploits the store-carry-forward feature of DTN nodes in conjunction with stationary relay nodes (e.g., smartphone, laptop), called DropBoxes to transmit and authorize transactions, hence enabling the usage of blockchain in DTN. The mechanism successfully integrates blockchain with DTN to exploit the immutability and accessibility of blockchain in an intermittently connected network. The chapter next proposes BlockCent, a blockchain-based node incentivizing scheme for a DTN-leveraged, smartphone-based, peer-to-peer network that runs on the blockchain-DTN integrated environment. In this scheme, forwarder nodes who help in successful transmission of messages are given rewards, in terms of Ethereum. Exploiting the Ethereum cryptocurrency enables us to design the incentive scheme at low cost, alleviating concerns about the reliability of virtual coin rewards in DTNs. Unlike the conventional blockchain system, the parties that make transactions, i.e. the forwarder nodes, do not need direct access to the blockchain network in our system. The scheme uses a novel reward model to bring rationality to the incentivizing process.

The proposed incentivizing scheme is applied to the disaster management use case. When a disaster strikes, collection of the victims' needs at different relief shelters and their transmission to the emergency operation center (EOC) is crucial. These need information are used by different stake holders (e.g. relief agencies, government agencies, NGOs, and disaster managers) appropriately coordinate and manage relief activities [1–5]. Since large portions of communication facilities get incapacitated as a result of the disaster, it gets hard to gather and communicate the needs from various remote and inaccessible shelters. The networking research group has strongly suggested the use of delay tolerant networks (DTNs) to set up a post-disaster communication infrastructure using mobile phone, Wi-Fi Direct, or Bluetooth interfaces carried by volunteers operating in DTN mode [1–6, 19]. Such networks can be used to collect and transmit post-disaster shelter needs to the EOC on a peer-to-peer basis through the volunteers carrying smartphones. The proposed BlockCent scheme, running on the blockchain-DTN integrated environment, helps in fast transmission of relief shelter needs to the EOC by enforcing cooperation among the forwarding nodes. This leads to efficient disaster relief, even in an intermittently connected environment.

The rest of this chapter is organized as follows. Section 10.2 summarizes related work in this field. In Section 10.3, we provide an overview of the rudimentary elements of the blockchain system. Section 10.4 describes the system model of the proposed scheme. We elaborate the mechanism for integrating blockchain with DTNs in Section 10.5 and illustrate the blockchain-based incentive scheme in Section 10.6. A detailed security analysis is given in Section 10.7. Section 10.8 presents the experimental results. We conclude the chapter with a direction towards future work in Section 10.9.

10.2 LITERATURE REVIEW

Incentive schemes for DTN applications can be classified into three categories: reputation-based schemes, barter-based schemes, and credit-based schemes. Since the proposed work deals with incentivizing DTN nodes for cooperation, we review a few prominent works on credit-based incentive schemes for DTNs only.

Zhu et al. [9] propose SMART, a secure, multilayer, credit-based incentive scheme to stimulate bundle forwarding cooperation among DTN nodes. To address the selfishness problem in DTNs, Lu et al. [10] propose a practical incentive protocol, called Pi, such that when a source node sends a bundle message, it also attaches some incentive on the bundle, which is not only attractive but also fair to all participating DTN nodes. Shevade et al. [11] propose the use of pair-wise tit-for-tat (TFT) as a simple, robust, and practical incentive mechanism for DTNs. Authors also develop an incentive-aware routing protocol that allows selfish nodes to maximize their own performance while conforming to TFT constraints. Chen et al. [12] propose MobiCent, a credit-based incentive system for DTN. While MobiCent allows the underlying DTN routing protocol to discover the most efficient paths, it is also incentive compatible. Ning et al. [13] present effective schemes to estimate the expected credit reward, and they formulate nodal communication in a DTN as a two-person cooperative game, whose solution is found by using the Nash Theorem. MobiID, a user-centric and social-aware reputation based incentive scheme for DTNs, proposed by Wei et al. [14], allows a node to manage and demonstrate its reputation.

All of these schemes either (i) rely on central trusted authorities that do not exist in a post disaster communication network, or (ii) use no explicit virtual digital currency system. These issues are addressed in the works proposed by He et al. [8], where a blockchain-based incentive mechanism for distributed P2P applications uses Bitcoin to incentivize forwarders, and Park et al. [7], where a Bitcoin-based incentive scheme incentivizes forwarders in vehicular DTNs. However, both [7] and [8] assume that all forwarders are equipped with Internet connectivity and directly access the Blockchain network. Such assumption is impracticable in several intermittently connected network environments, such as the post-disaster communication network, where forwarder nodes are DTN-enabled smartphones communicating over Wi-Fi Direct but having no Internet connectivity. The proposed blockchain-based incentive scheme is inspired by the existing research but is very different. Major contributions of this work are:

- A mechanism that enables usage of blockchain technology in an intermittently connected network such as DTN
- A fully decentralized blockchain-based node incentivizing scheme for DTNs
  - without relying on central trusted authorities
  - without requiring end-to-end Internet connectivity
- A novel reward model to make the incentivizing process more rational
- Extensive security analysis to justify security of the proposed incentivizing scheme
- Exhaustive simulations on the ONE simulator and the Ethereum platform substantiate the efficiency of the incentivizing scheme

10.3 RUDIMENTARY ELEMENTS OF BLOCKCHAIN

Blockchain is a system in which a record of transactions made in Bitcoin or any other cryptocurrency are maintained across several computers that are linked in a peer-to-peer network. It is a distributed database at the same time. A blockchain is constantly growing as new sets of records/blocks are added to the chain [7, 18]. The overview of blockchain preliminaries relevant to our proposal is given next.

10.3.1 Blocks

In a blockchain, transactions are grouped into blocks so that they can be efficiently verified. Once a block is verified, they are added to the previous block. In a blockchain, each block is chained to its previous block through the use of a cryptographic hash. A change to a block forces a recalculation of all subsequent blocks, which requires enormous computation power. This makes the blockchain immutable, a key feature of crypto currencies like Bitcoin and Ethereum.

10.3.2 Transaction

A blockchain transaction can be defined as a small unit of task that is stored in a block. A typical transaction [7] from a sender to a recipient is shown in Figure 10.1. A transaction has a unique identifier (TX) and consists of a set of inputs and outputs. Inputs to a transaction include previous transaction identifier (TX') and signature of the user. Outputs from a transaction include the amount transferred to a recipient and the recipient's public key. To authorize spending the coins in the transaction input, the user should present its signature for the transaction and corresponding public key. Output of a transaction is used as input of the next transaction for verification. Transaction verification is distributed to P2P network nodes, and only the valid transactions are recorded.

10.3.3 Mining

Mining involves creating a hash of a block of transactions that cannot be easily forged, protecting the integrity of the entire blockchain without the need for a central system. Some computers storing the entire blockchain are called miners. When a transaction is made over a blockchain, the transaction information is thus put on a block and gets recorded. All the miners crack a cryptographic riddle so as to locate the correct cryptographic hash. When a miner creates the hash, it is added to the blockchain and must be confirmed by different nodes on the system in a cycle known as consensus. At the point when a miner effectively confirms the hash, it is given a reward [18].

10.3.4 Ethereum and Smart Contracts

Ethereum [18] is an open-source, public, blockchain-based distributed computing platform. Ether is the cryptocurrency generated by the Ethereum platform as a reward to mining nodes for computations performed and is the only currency accepted in the payment of transaction fees. Ethereum provides a decentralized virtual machine, the Ethereum Virtual Machine (EVM), which can execute scripts using an international network of public nodes. "Gas" is the fee, or pricing value, required to successfully conduct a transaction or execute a contract on the Ethereum blockchain platform.

A smart contract is a piece of code that runs on Ethereum to execute credible transactions without any third parties. It is a self-executing contract between network participants, without the need for traditional legal contracts. The contracts are high-level applications that are compiled down to EVM bytecode and deployed to the Ethereum blockchain for execution.

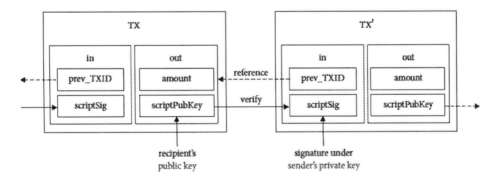

Figure 10.1 Overview of blockchain transaction structure and spending [7].

An Ethereum user creates a contract and pushes the data to that contract so that it can execute the desired command.

10.4 INTEGRATION OF BLOCKCHAIN WITH DTNS

In this section, we propose a mechanism for integrating blockchain technology with DTNs. The major challenges of such integration are establishing the connection between DTN and the Ethereum server, converting DTN messages to blockchain-compatible blocks, and deploying blocks to the blockchain network.

We develop an application program interface (API) and an application binary interface (ABI) that play the pivotal role in deploying the blocks containing DTN messages, in the form of smart contracts on the Ethereum network. Our specific contributions are:

- Developing an API for establishing the connection between a DTN and the Ethereum server
- Developing an ABI for converting the blocks, containing DTN messages, to Ethereum blocks using Solidity contracts
- Deploying the converted blocks on the blockchain network using JSON-RPC protocol

In this work we use programming languages like Java and Solidity; frameworks like Spring Boot, Spring Rest, Spring MVC, Spring Data, and Spring Security; and technologies like Maven-Building tool, Metamask-Crypto wallet, and gateway to blockchain wallet and Rinkeby-Ethereum Test Network. Initially, a Maven & Spring Boot project is created which adds all dependencies (Spring Rest, MVC, Spring Data, Spring Security, JSON-RPC, ABI, Lombok, Web3J, Slf4j, etc.) and creates the pom.xml and application.properties files. After creating the project successfully, we develop the API that works as the bridge between the local blockchain data and the Ethereum network using the Rest API Controller. The Rest API Controller creates a link ("/accept-local blocks-from-dropbox") with support of Get Request and Post Request. The DropBox can send data to the API control using this API.

After receiving the blocks, containing DTN messages, the API control frames the blocks as Pojos (Plain old Java Objects). All Pojos are then converted into Solidity contracts using the developed ABI. These Solidity contracts are framed as Ethereum blocks. Thus, DTN blocks are converted into Ethereum blocks. These blocks are deployed to the Ethereum server in real time, using the JSON-RPC protocol. Figure 10.2 elucidates the functioning of the developed API control.

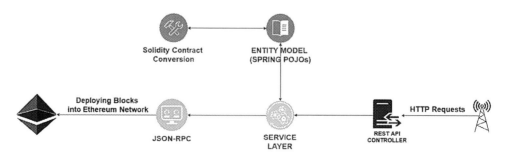

Figure 10.2 The designed API control.

After the blocks are deployed on the Ethereum server, the server mines all blocks using its own miners and pays a certain amount of gas to the miners as payment for mining the blocks. After every successful mining, all DTN blocks are deployed into the blockchain network, which is completely immutable.

10.5 SYSTEM MODEL

In this section, we provide the system model, consisting of the network architecture and security issues, pertaining to the disaster management use case of the proposed BlockCent scheme.

10.5.1 Network Architecture

During and after disasters, victims normally take shelter in nearby safe areas. An EOC is set up at nearby resource-rich areas [1–5, 19]. Our proposed scheme maps the earlier scenario where the smartphones carried by the volunteers communicate over a DTN and create a post-disaster communication network. In this scheme, the entire disaster-affected area is virtually divided into a number of non-intersecting zones, having many shelters. Network architecture used in our scheme, illustrated in Figure 10.3, consists of six major components, whose functions are described next.

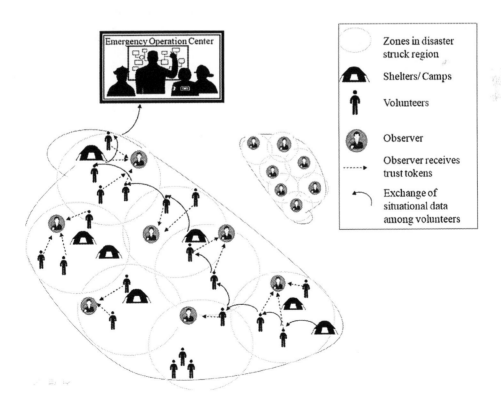

Figure 10.3 Network architecture of the proposed scheme.

Shelter node: Each shelter has a shelter node (for example, a laptop having Wi-Fi Direct and LTE/5G Internet connection). Emergency managers in a shelter upload the shelter requirements to the shelter node, which transmits this data to the volunteers working in and around it.

Control node: The EOC has a control node (e.g. a workstation fitted with an LTE/5G Internet link and Wi-Fi Direct) that gathers situational information and distributes resources to the remote shelters.

Forwarder node: Volunteers carrying Wi-Fi Direct-equipped smartphones are forwarder nodes that travel around the disaster-stricken region, collect shelter node situational messages, communicate these messages with each other through the PRoPHET [18] DTN routing protocol, and eventually transmit them to the control node.

Observer node: Disaster management agents equipped with Wi-Fi Direct smart devices are observer nodes [20] that travel at a pace greater than that of the volunteers around the disaster-stricken region. The zone does have a designated observer node which collects and transmits to the nearest DropBox all transactions created by forwarder nodes therein.

DropBox: Stationary relay nodes (e.g., smartphone, laptop having LTE/5G Internet connectivity, possessing sufficient storage and computation capacity) that receive and store transactions submitted by the observer nodes passing by it and forward them to the API control, running on a cloud platform, for further processing. DropBoxes are deployed at high-priority locations, based on the uDBdep scheme [21] to maximize data transfer.

Ethereum server: The Ethereum server runs the open-source, public, blockchain-based distributed computing system. It mines blocks and uploads them to the global blockchain.

Each of these components runs different modules of the proposed scheme to accomplish message forwarding and appropriate incentivizing.

10.5.2 Security Model

In this work, we assume the control node to be trusted, and all security issues, described next, arise from selfish shelter nodes, forwarder nodes, and observer nodes.

- A shelter node may refuse to pay back the cooperating forwarder nodes and observer node in its zone when the message finally reaches the control node.
- A forwarder node may not:
  - forward a message;
  - send acknowledgment to the previous forwarder node.
- An observer node may pose as a forwarder node and claim incentives for forwarding messages.

These selfish behaviors can be further complicated by the collusion of two or more nodes. However, such collusion-based security issues are out of the scope of the current work.

10.6 BLOCKCENT: BLOCKCHAIN-BASED NODE INCENTIVIZING SCHEME

In this section, we present BlockCent, a blockchain-based node incentivizing framework for a smartphone-powered DTN that runs on the integrated blockchain-DTN setting. In this scheme, when transmitting a message to the control node through one or more forwarder nodes, a shelter node belonging to a specific zone gives a certain amount of Ethereum as a reward to all cooperative forwarder nodes that assist in transmitting the message and the observer node inside it. Each intermediate forwarder node collects from the next-hop forwarder a digitally signed acknowledgment (trust-token) to which it forwards the message as a sign of cooperation. If and only if it has a digitally signed acknowledgment from its successor, then a forwarder node is called cooperative. It should be remembered that each reward is actually a promise, and the forwarder nodes can reclaim rewards only after the shelter node recognizes that the message is successfully transmitted to the control node. This helps in thwarting the forwarder nodes from taking refuge in "dine and dash" behavior.

10.6.1 Setup Phase

In our scheme, we assume N forwarder nodes, K shelter nodes, P observer nodes (one per zone) and one control node. In the setup phase, each shelter node is assigned a shelter id SN_j, $j = 1, 2, ..., K$, and each forwarder node a forwarder id FN_i, $i = 1, 2, ..., N$, by the control node. Each shelter node and each forwarder node is also assigned a private and public key pair (PR_{SNj}, PU_{SNj}), $j = 1, 2, ..., k$ and (PR_{FNi}, PU_{FNi}), $i = 1, 2, ..., N$, respectively, by the control node. The control node has its own private and public key pair (PR_{CN}, PU_{CN}). All public keys are distributed in the network. Whenever a new volunteer wants to join relief operations, he/she has to register with EOC. On registration, the message transmission and incentive redemption modules of the proposed blockchain-based incentive application is installed on the volunteer's smartphone along with necessary details like private and public key pair of its own and public key of the other stakeholders.

10.6.2 Incentivizing Process

A reward is known as a transaction in the proposed scheme, and a shelter node uses the blockchain method to enable cooperative forwarder nodes with sufficient quantities of Ethereum. Three modules comprise the incentivizing process. First, a certain deposit that is used to facilitate the cooperative forwarder nodes is made by the shelter node and generates a transaction. Second, the shelter node transmits the message through one or more forwarder nodes to the control node. Finally, the forwarder nodes that cooperate in forwarding messages redeem their incentives.

To illustrate the incentivizing process, we assume that a shelter node SN_j in zone Z sends message M to the control node CN through the cooperative forwarder nodes FN_1, FN_2, and FN_3. The observer node ON_Z moves around Z for collecting transactions generated by FN_1, FN_2, and FN_3 and broadcast them to the blockchain network for necessary validation and inclusion in the blockchain. As mentioned in Section 10.5.1, the forwarder nodes exchange messages using the PRoPHET protocol [17]. This protocol uses history of previous encounters and transitive property to estimate delivery predictability (DP), $P(A,B)$, of each node A for all known destinations B. $P(A,B)$ depicts the probability of A meeting B. As nodes meet, they exchange DP tables and update their own DP table according to certain protocols.

10.6.2.1 Transaction Creation

SN_j generates two nonce values N_1 and N_2; N_1 is to be used for proving that FN_1, the next hop forwarder node receives the message correctly and N_2 is to be used for proving that CN receives the message successfully. SN_j publishes $E_{PU_{SNj}}(N_1)$ and $E_{PU_{SNj}}(N_2)$ to the blockchain network over the Internet. SN_j then makes a deposit to commit that it will give incentives to FN_1 if FN_1 cooperates to transmit the message to CN. SN_j constructs a time-locked deposit transaction $TX_{SN_j}^{Deposit}$ and broadcasts it to the blockchain network. The time-lock condition restricts SN_j from withdrawing the deposited Ethereum earlier than the locking period. The locking period could be the maximum time required by the miners to verify the transaction based on information provided by FN_1, FN_2, FN_3, and CN.

10.6.2.2 Message Transmission

SN_j composes and sends the message bundle $msg1 := \{M\|E_{PU_{CN}}(N_2)\|Sig_{PR_{SN_j}}(Hash(M)\|E_{PU_{CN}}(N_2))\|Sig_{PR_{SN_j}}(N_1)\}$ to its next-hop forwarder node FN_1 over DTN, using the DTN routing protocol. SN_j constructs a reward transaction $TX^{Reward}_{SN_j \to FN_1}$ and broadcasts it to the blockchain network for validation along with delivery predictability $P(FN_1, CN)$, which depicts the probability of FN_1 meeting CN, final destination of M.

On receiving $msg1$, FN_1 extracts N_1 and saves it to be used for incentive redemption. It also verifies the signature $\{Sig_{PR_{SN_j}}(Hash(M)\|E_{PU_{CN}}(N_2))\}$ and sends SN_j a digitally signed and encrypted acknowledgment $E_{PU_{SN_j}}(Sig_{PR_{FN_1}}(ACK_{FN_1}))$, encrypted with SN_j's public key. When FN_1 opportunistically meets FN_2, it forwards the message bundle $msg2 := \{M\|E_{PU_{CN}}(N_2)\|Sig_{PR_{SN_j}}(Hash(M)\|E_{PU_{CN}}(N_2))\}$ to its next-hop forwarder node FN_2 over DTN and constructs a reward transaction $TX^{Reward}_{FN_1 \to FN_2}$. Since FN_1 is a normal forwarder node with less speed, it carries $TX^{Reward}_{FN_1 \to FN_2}$ along with $P(FN_2, CN)$ until it meets the high-speed observer node ON_Z opportunistically. On receiving these values, ON_Z transfers them to the DropBox, which in turn transmits them to the API control, running on a cloud server, over the Internet.

A similar process is executed between FN_2 and its next-hop forwarder node FN_3. Finally, when FN_3 reaches CN, it forwards the message bundle $msg4 := \{M\|E_{PU_{CN}}(N_2)\|Sig_{PR_{SN_j}}(Hash(M)\|E_{PU_{CN}}(N_2))\}$ to CN. CN receives the message, verifies the signature $\{Sig_{PR_{SN_j}}(Hash(M)\|E_{PU_{CN}}(N_2))\}$, and sends FN_3 a digitally signed acknowledgment $E_{PU_{FN_3}}(Sig_{PR_{CN}}(ACK_{CN}))$, encrypted by FN_3's public key.

Transactions $TX^{Reward}_{SN_j \to FN_1}$, $TX^{Reward}_{FN_1 \to FN_2}$, and $TX^{Reward}_{FN_2 \to FN_3}$ are commitments that coins will be transferred only after the miners have verified them.

10.6.2.3 Incentive Redemption

Once the message M is successfully delivered to CN, CN provides $\{E_{PU_{CN}}(N_2), E_{PU_{CN}}(ACK_{CN})\}$ and SN_j provides $E_{S_{N_j}}(ACK_{F_{N_1}})$ to the blockchain miners. The Ethereum server collects from the API control the proofs of cooperation of FN_1, FN_2, and FN_3 and submits them to the blockchain miners. In particular, the Ethereum server collects:

$\{E_{PU_{FN1}}(N_1), E_{PU_{FN_1}}(ACK_{FN_2}), PU_{SN_j}, PU_{FN_2}\}$ from FN_1,
$\{E_{PU_{FN_2}}(ACK_{FN_2}), E_{PU_{FN_2}}(ACK_{FN_3}), PU_{FN_3}\}$ from FN_2,
$\{E_{PU_{FN_3}}(ACK_{FN_3}), E_{PU_{FN_3}}(ACK_{CN}), PU_{CN}\}$ from FN_3.

The incentives could be redeemed only if the miners validate the transactions using these values and the previously encrypted nonce values broadcast by SN_j. The conditions for validity are as follows, all encryption functions being commutative:

- The encrypted nonce provided by CN can be verified with the encrypted nonce provided by SN_j as:

$$E_{PU_{CN}}\left(E_{PU_{SN_j}}(N_2)\right) = E_{PU_{SN_j}}\left(E_{PU_{CN}}(N_2)\right)$$

- The encrypted nonce provided by FN_1 can be verified with the encrypted nonce provided by SN_j as:

$$E_{PU_{FN1}}\left(E_{PU_{SN_j}}\left(N_1\right)\right) = E_{PU_{SN_j}}\left(E_{PU_{FN_1}}\left(N_1\right)\right)$$

- There is a delivery path from SN_j to CN verified by the acknowledgments:

$$ACK_{FN_1} \rightarrow ACK_{FN_2} \rightarrow ACK_{FN_3} \rightarrow ACK_{CN}$$

- Digitally signed acknowledgments provided by FN_1, FN_2, and FN_3 can be verified as:

$$E_{PU_{FN1}}\left(E_{PU_{FN2}}\left(ACK_{FN2}\right)\right) = E_{PU_{FN2}}\left(E_{PU_{FN1}}\left(ACK_{FN2}\right)\right)$$

$$E_{PU_{FN2}}\left(E_{PU_{FN3}}\left(ACK_{FN3}\right)\right) = E_{PU_{FN3}}\left(E_{PU_{FN2}}\left(ACK_{FN3}\right)\right)$$

$$E_{PU_{FN3}}\left(E_{PU_{CN}}\left(ACK_{CN}\right)\right) = E_{PU_{CN}}\left(E_{PU_{FN3}}\left(ACK_{CN}\right)\right)$$

Incentives are rewarded to cooperative forwarder nodes and the observer node by the blockchain network through the Ethereum server, API control, and the DropBoxes. The BlockCent scheme, running on the blockchain-DTN integrated environment is illustrated in Figure 10.4.

10.6.3 Reward Model

A popular rewarding method in incentive schemes is paying per message, where, for a successfully transmitted message, each of the n cooperative forwarder nodes receives x amount of credit and the sender needs to pay $x \times n$ in total. However, in an opportunistic network like ours, it is not possible to predict how many forwarder nodes will get involved in the message transmission process. We therefore follow the approach of benefit sharing, where a portion of the profit is shared by each of the n cooperative forwarder node, i.e. the amount deposited by the sender. Assuming that sender SN_j has made the deposit transaction $TX_{SN_j}^{Deposit}$

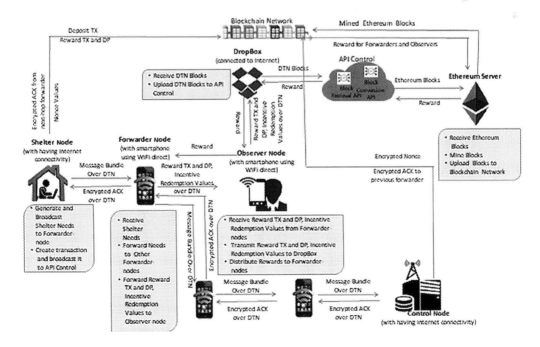

Figure 10.4 BlockCent scheme running on the blockchain-DTN integrated environment.

and deposited an amount of X at the beginning of transmission, and there exists n cooperative forwarder nodes FN_1, FN_2, ..., FN_n that are involved, the reward is:

$$R_{U=} \begin{cases} \alpha, if\, U = ON_Z \\ \dfrac{X-\alpha}{n} \times P(FN_i, CN), if\, U = FN_i, i =, 2, ..., n \end{cases}$$

Incorporating delivery predictability $P(FN_i, CN)$ in the reward model makes the scheme more reasonable. First, nodes having a higher DP with the control node are given higher rewards to ensure faster message delivery. Second, in order to reduce the number of hops to optimize their benefit, forwarder nodes try to forward messages to forwarders with the highest DP. This in turn results in the delivery of quick messages.

10.7 SECURITY ANALYSIS

We provide a thorough security analysis of the proposed incentivizing scheme with respect to the security issues mentioned in Section 10.5.2.

10.7.1 Shelter Node Refuses to Pay Back

A shelter node SN_j makes a deposit at the beginning of message transmission, committing to pay back all stakeholders. SN_j prepares a time-locked deposit transaction $TX_{SN_j}^{Deposit}$ and broadcasts to the blockchain network. The time-lock restricts SN_j from withdrawing the deposit before the expiration of the locking period and refuses to pay back. This guarantees fairness to cooperating forwarders and the observer.

10.7.2 Forwarder Node Refuses to Forward Message

If a forwarder node FN_i does not forward a message and to the next-hop forwarder node FN_{i+1}, FN_i will not receive the digitally signed acknowledgment $E_{PU_{FN_i}}(Sig_{PR_{FN_{i+1}}}(ACK_{FN_{i+1}}))$ from FN_{i+1}. Consequently, FN_i is unable to provide $E_{PU_{FNi}}(ACK_{FN_{i+1}})$ to the observer node ON_Z at the time of incentive redemption. Thus, FN_i cannot get incentive without forwarding a message.

10.7.3 Forwarder Node Refuses to Send ACK

If a next-hop forwarder FN_{i+1} refuses to send acknowledgment to its previous forwarder FN_i, FN_i is unable to provide $E_{PU_{FNi}}(ACK_{FN_{i+1}})$ to ON_Z during incentive redemption. However, if FN_{i+1} provides $E_{PU_{FNi+1}}(ACK_{FN_{i+2}})$ to O_Z, the miners will fail to verify the delivery path ... $\rightarrow ACK_{FN_i} \rightarrow ACK_{FN_{i+1}} \rightarrow ACK_{FN_{i+2}} \rightarrow$ ... and will invalidate the transaction. Thus, FN_{i+1} cannot get incentive in spite of forwarding the message.

10.7.4 Observer Node Poses as Forwarder Node

If an observer node ON_Z intends to pose as a forwarder node FN_i, it has to provide $\{E_{PU_{Oz}}(ACK_{ONZ}), E_{PUOz}(ACK_{FN_{i+1}}), PU_{ONz}\}$ to the miners. However, ON_Z has no ACK_{ONZ} to its credit, nor can it decrypt ACK_{FNi+1} with PR_{FNi} to create $E_{PUOz}(ACK_{FN_{i+1}})$. Thus, ON_Z cannot pose as FN_i and claim incentives for forwarding messages.

10.8 EXPERIMENTAL RESULTS

In this section we evaluate the performance of the proposed BlockCent scheme through extensive experiments conducted in a realistic simulation environment.

10.8.1 Simulation Environment

We use a real disaster scenario for setting up the simulation environment, based on the 2015 Nepal earthquake. The Google Map of water, food, shelter, and medical resources for the Nepal earthquake [22], shown in Figure 10.5, marks the shelters and medical relief centers set in Kathmandu and its adjoining districts like Nuwakot, Sindhupalchowk, and others. We set up our simulation environment based on Figure 10.5.

10.8.2 Simulation Setup

The three main objectives of the experiments are (i) to observe how far BlockCent succeeds in achieving major design goals, (ii) to evaluate network performance of BlockCent, and (iii) to observe the overheads introduced by BlockCent for blockchain activities. We use the ONE simulator [23] optimization tool for accomplishing the first two objectives and the Ethereum [24] blockchain platform for the third.

10.8.2.1 ONE Setup

Our simulation setup in the ONE simulator [23] is based on information given by the map [22] concerning the post-disaster relief operation carried out in the Kathmandu region of Nepal. We set up 11 shelters covering about 100 sq. km across the disaster-hit city. We set up one EOC as well. We consider 88 forwarder nodes, that is, mobile volunteers carrying smartphones. Such nodes share and relay situational messages from the shelters to the EOC. The entire region affected by the earthquake is divided into five virtual areas. To collect the transactions of the forwarder nodes in that zone, each zone has one observer node. There are stationary shelter nodes and the control node. The forwarder nodes and observers, however, are mobile nodes and adopt the Post Disaster Mobility (PDM) model

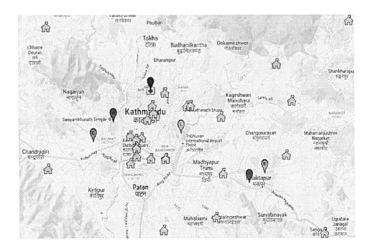

Figure 10.5 Google Map for the 2015 Nepal earthquake [22].

suggested in [25] by Uddin et al. A shelter node produces and transmits situational messages to the forwarder nodes at a rate of one per hour. We use real situational messages originating from the WhatsApp chat log of an NGO's "Doctors For You" volunteers [26], who provided medical assistance during and after the earthquake. These messages are 120 bytes in size, containing the real emergency resource specifications at various shelters. In our network, we consider cooperative forwarder nodes (CFN) and selfish forwarder nodes (SFN). Cooperative forwarder nodes take part in the PRoPHET [27] routing protocol routing operation, while selfish forwarder nodes drop all messages sent to them (destined for other nodes) that were supposed to be redirected to other forwarder nodes. The findings presented here are the 20 independent run average. The relevant parameters used in simulation are described in Table 10.1.

10.8.2.2 Ethereum Setup

We implement and evaluate a prototype of our proposed BlockCent scheme using Ethereum [24] as the blockchain platform. Our entire code is approximately 500 lines, consisting of smart contracts written in Solidity. The rationale behind using Ethereum is that it provides a broader range of validation strategies compared to the other blockchain platforms, with the power of smart contracts. We deployed our implementation on a private Ethereum network consisting of one node, namely the DropBox running on Truffle v4.5. We have used a single machine with Intel® Core™ CPU with 2.0 GHz, 8 GB of RAM, and 20 GB of persistent storage to meet the minimum hardware requirement for running Ethereum. Also, as a wallet for message transferring, we have used Geth v 1.7.0. An initial gas of 3,000,000 wei is used.

10.8.3 Simulation Metrics

Performance of the proposed BlockCent scheme is measured based on design metrics, network performance metrics, and blockchain metrics.

Table 10.1 Parameters used for simulation on ONE.

Parameter	Value
Simulation area	100 sq. km
No. of forwarder nodes	88
No. of shelters	11
Message generation rate	5 per hour
Message size	120 bytes
Message TTL	500 min
No. of zones	5
No. of observer nodes	5 (1 in each zone)
Forwarder node speed	0.5 m/s to 1.5 m/s
Observer node speed	1.5 m/s to 5.5 m/s
Transmission range	10 m
Movement model	Post-disaster mobility model
Routing protocol	PRoPHET
Simulation period	12 hours (43,200 sec)

10.8.3.1 Design Metrics

Selfishness ratio: It is the fraction of the number of selfish forwarder nodes to the total number of such nodes. It is defined as

$$SelfishnessRatio = \frac{No.of\ selfish\ forwarder-nodes}{Total\ no.of\ forwarder-nodes}$$

This metric is used to evaluate the ability of the BlockCent scheme to convert selfish nodes to cooperative nodes.

Average reward: It is the average of the rewards obtained by all forwarder nodes for forwarding all messages. It is defined as

$$AverageReward = \frac{\sum_{all\ forwarder-nodes} R_{FN}}{No.of\ forwarder-nodes}$$

where R_{FN} is the total reward obtained during a simulation period.

10.8.3.2 Network Performance Metrics

Delivery ratio [5]: It is the fraction of the messages delivered to destination nodes to those created by source nodes. It is defined as

$$DeliveryRatio = \frac{No.of\ msgs.delivered}{No.of\ msgs.created}$$

Average delay [5]: This indicates the average time taken by the messages from sources to destinations, including buffer delays, queuing delays, retransmission delays, and propagation time. It is defined as

$$AverageDelay = \frac{\sum_{forallM_D}(T_D - T_S)}{No.of\ msgs.delivered}$$

where T_D is the time when message reaches its destination and T_S is the time when the message was created at the source.

Overhead ratio [5]: Indicates the ratio of the number of control messages (including route request/reply/update/error packets) to the number of data messages. Defined as

$$OverheadRatio = \frac{No.of\ control\ packets\ relayed}{No.of\ data\ packets\ relayed}$$

10.8.3.3 Blockchain Metrics

Processing time (t_p) [28]: It is the amount of time a user has to wait, after pressing the "send" transaction button (t_1), to see their transaction appear on the blockchain (t_2) and is defined as

$$t_p = t_2 - t_1$$

Gas consumption (G) [29]: It is the gas consumed in processing transactions (tx = 1, 2, ..., n) for the current block and is defined as

$$G = \sum_{t_x=1}^{n} GasConsumption_{t_x}$$

10.8.4 Results and Discussion

In this section, we discuss the results of the experiments conducted for evaluating the BlockCent scheme in terms of achieving major design goals, network performance and block-chain overhead.

10.8.4.1 Achieving Major Design Goals

Two sets of experiments are conducted for evaluating the performance of the BlockCent scheme by measuring the extent of attaining design objectives.

The first set of experiments measure the selfishness ratio against time, shown in Figure 10.6(a), where time varies from 1 to 12 hours. We start with 50% selfish forwarder nodes and observe that the percentage of selfish forwarder nodes decrease with time. As more and more cooperative nodes are rewarded for forwarding messages, selfish nodes are encouraged to cooperate and participate in forwarding activities. Thus, the Block Cent scheme successfully

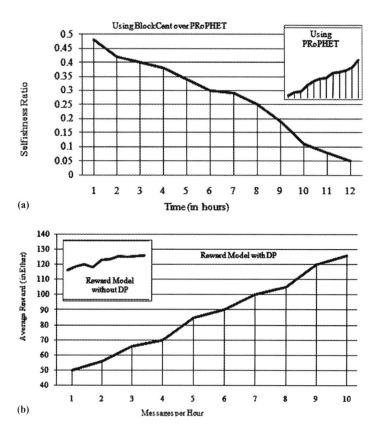

Figure 10.6 (a) Selfishness ratio with time. (b) Average reward with no. of messages per hour.

converts selfish nodes to cooperative nodes. The graph in the inset depicts the increase in selfishness ratio when the PRoPHET routing protocol is used without the BlockCent scheme.

The reward model used in the BlockCent scheme, described in Section 10.6.3, includes delivery predictability of forwarder nodes with the control node. Thus, messages are always forwarded through forwarder nodes having higher delivery predictability, which in turn reduces the number of hops and maximizes reward for cooperating nodes. The second set of experiments measure the average reward of the forwarder nodes against the number of messages per hour. Figure 10.6(b) shows that the average reward increases with increasing number of messages per hour using this reward model. The graph in the inset depicts that average reward does not increase substantially when delivery predictability is not included in the reward model, which is justified as most eligible forwarders are not chosen and the number of hops is increased.

10.8.4.2 Evaluation of Network Performance

We compare our BlockCent scheme with three other competing schemes, namely SMART [9], MobiCent [13], and MobiID [14], in terms of network performance metrics. To evaluate network performance, three sets of experiments are conducted.

In the first set of experiments, the performance of BlockCent is compared with other competing schemes in terms of delivery ratio with varying numbers of selfish forwarder nodes in the network. It is observed from Figure 10.7(a) that delivery ratio reduces with increasing number of selfish nodes for all schemes. However, BlockCent performs the best due to the appropriate reward model that incentivizes cooperative nodes for forwarding messages, thus boosting delivery performance. SMART performs second best, by having 12% less delivery ratio than BlockCent.

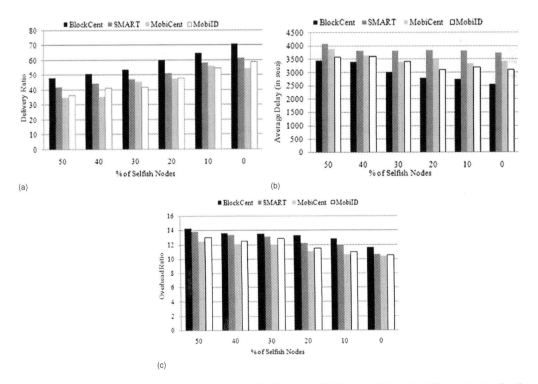

Figure 10.7 (a) Delivery ratio with different levels of selfishness. (b) Average delay with different levels of selfishness. (c) Overhead ratio with different levels of selfishness.

In the second set of experiments, the performance of BlockCent is compared with other schemes in terms of average delay (in seconds) with varying number of selfish nodes in the network. Figure 10.7(b) indicates that BlockCent performs best, followed by MobiID. Including delivery predictability into the reward model encourages forwarder nodes to forward messages to forwarders with highest DP to reduce the number of hops for maximizing their profit. This in turn results in fast message delivery. The gain in average delay over the SMART scheme is 22%.

The third set of experiments measures overhead ratio with varying number of misbehaving nodes for all three techniques. As shown in Figure 10.7(c), overhead ratio of BlockCent is high in comparison to the other schemes. This increase can be attributed to the control packets, consisting of reward TX, DPs, and incentive redemption values exchanged between the forwarder nodes and the observer nodes used for reward redemption. The MobiCent scheme performs best followed by the MobiID and SMART schemes. However, BlockCent has only 15% greater overhead than MobiCent, which is not alarming.

It is evident from the preceding discussion that BlockCent maximizes network performances in terms of delivery ratio and average delay while maintaining a tolerable overhead ratio.

10.8.4.3 Overheads Introduced for Blockchain Activities

Two sets of experiments are conducted to observe the overheads introduced by the BlockCent scheme for blockchain activities.

In the first set of experiments, the impact of varying number of shelters and zones on the processing time of BlockCent is measured, as depicted in Figure 10.8(a). We observe that t_p increases with the increase in the number of shelters due to the higher number of transactions to be verified. For example, with 4 zones, t_p with 8 shelters is 22.85% and 15.82% higher than the t_p with 4 and 6 shelters, respectively. It is also observed that t_p does not increase significantly with increasing number of zones, as a slight increase in computation is involve in this case. Only 2.5% is the average increase in t_p with increasing number of zones.

In the second set of experiments, the impact of varying numbers of shelters and zones on gas consumption of BlockCent is measured, as shown in Figure 10.8(b). We observe that gas consumption increases with increasing number of shelters, which incurs higher number of transactions to be verified. For example, with 4 zones, G with 8 shelters is 10.2% and 5.75% higher than the G with 4 and 6 shelters respectively. However, the rate of increase in the case of zones is not so significant. Only 2.95% is the average increase in G with increasing number of zones.

10.9 CONCLUSION

This chapter proposed BlockCent, a blockchain-based node incentivizing scheme for a DTN enabled smartphone leveraged communication network that uses Ethereum to incentivize nodes for cooperation, built on top of a DTN-blockchain integrated environment. This environment develops an alternative way of broadcasting and authorizing blockchain transactions without having to rely on Internet connectivity. It uses the store-carry-forward feature of DTN nodes in conjunction with stationary relay nodes (e.g., smartphone, laptop), called DropBoxes, to transmit and authorize transactions, hence enabling the usage of blockchain in intermittently connected network environments. The incentivizing scheme is fully decentralized and does not rely on any central trusted authority. A novel reward model is used to make the incentivizing process fair and rational. The proposed mechanism is applied to the disaster management use case, based on the Ethereum platform using smart contracts in

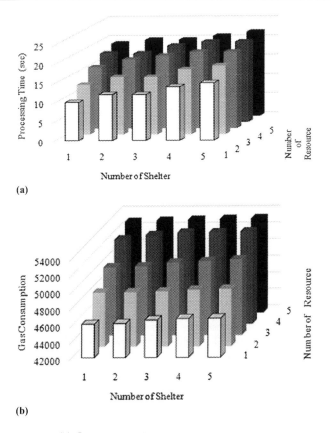

Figure 10.8 (a) Processing time. (b) Gas consumption.

Solidity. It successfully integrates blockchain with the DTN environment towards improving disaster management services in absence of end-to-end to the Internet.

As future work, we intend to make the proposed scheme resilient to collusion attacks by introducing anonymity of the forwarder nodes. One-time public key techniques, such as CryptoNote [30], can be used for this purpose.

REFERENCES

[1] A. Verma, M. Singh, K. K. Pattanaik, and B. K. Singh (2019) Future networks inspired by opportunistic network, *Opportunistic Networks: Mobility Models, Protocols, Security & Privacy*, CRC Press (Taylor & Francis Group), chapter 12, pp. 229–246, 2019. doi:10.1201/9780429453434

[2] S. Basu, A. Biswas, S. Roy, and S. DasBit (2018) Wise-PRoPHET: A watchdog supervised PRoPHET for reliable dissemination of post disaster situational information over smartphone based DTN. *J Netw Comput Appl* 109(2018), 11–23.

[3] S. Basu and S. Roy (2014) *A global reputation estimation and analysis technique for detection of malicious nodes in a post-disaster communication environment. In Proceedings of AIMoC 2014.*

[4] S. Basu and S. Roy (2014) *A group-based multilayer encryption scheme for secure dissemination of post-disaster situational data using peer-to-peer delay tolerant network. In Proceedings of ICACCI 2014*, pp. 1566–1572.

[5] C. Chakrabarti and S. Basu (2019) *A blockchain based incentive scheme for post disaster opportunistic communication over DTN. In Proceedings of the 20th International Conference on Distributed Computing and Networking*, January 2019, pp. 385–388, https://doi.org/10.1145/3288599.3295584.

[6] J. Ren, Y. Zhang, K. Zhang, and X. S. Shen (2015) Exploiting mobile crowdsourcing for perva-
     sive cloud services: Challenges and solutions. *IEEE Commun Mag* 53(3), 98–105.

[7] Y. Park, C. Sur and K. H. Rhee (2018) A secure incentive scheme for vehicular delay tolerant
     networks using cryptocurrency. *Secur Commun Netw* 2018, 1–13.

[8] Y. He et al. (2018) A blockchain based truthful incentive mechanism for distributed P2P applica-
     tions. *IEEE Access* 2018(6), 27324–27335.

[9] H. Zhu et al. (2009) SMART: A secure multilayer credit-based incentive scheme for delay-toler-
     ant networks. *IEEE Trans. Veh. Tech.* 58(8), 4628–4639.

[10] R. Lu, X. Lin, H. Zhu, X. Shen, and B. Preiss (2010) "Pi: A practical incentive protocol for delay
     tolerant networks." *IEEE TWC.*

[11] U. Shevade, H. H. Song, L. Qiu, and Y. Zhang (2008) *Incentive-aware routing in DTNs. In
     Proceedings of ICNP.*

[12] B. B. Chen and M. C. Chan (2010) *MobiCent: A credit-based incentive system for disruption
     tolerant network. In Proceedings of INFOCOM.*

[13] T. Ning, Z. Yang, X. Xie, and H. Wu (2011) *Incentive-aware data dissemination in delay-tolerant
     mobile networks. In Proceedings of SECON 2011.*

[14] L. Wei, H. Zhu, Z. Cao, and X. S. Shen (2011) *MobiId: A user centric and social-aware reputa-
     tion based incentive scheme for delay/disruption tolerant networks. In Proceedings of ADHOC-
     NOW 2011.*

[15] S. Nakamoto (2008) Bitcoin, A peer-to-peer electronic cash system. https://bitcoin.org/bitcoin.
     pdf.

[16] M. Crosby et al. (2015) Blockchain technology: Beyond Bitcoin. Sutardja Center for
     Entrepreneurship & Technology Technical Report, Berkeley University of California 2015, 1–35.

[17] A. M. Antonopoulous (2017) Mastering Bitcoin – programming the open blockchain. *O'Reilly
     Media* 2017.

[18] Understanding blockchain: A beginners guide to Ethereum Smart Contract Programming,
     https://www.codemag.com/Article/1805061/Understanding-Blockchain-A-Beginners-Guide-to-
     Ethereum-Smart-Contract-Programming (Accessed 9 July 2020).

[19] S. Basu, S. Bhattacharjee, S. Roy, and S. Bandyopadhyay (2015) *SAGE-PRoPHET: A security
     aided and group encounter based PRoPHET routing protocol for dissemination of post disaster
     situational data. In Proceedings of ICDCN 2015*, Article No. 20.

[20] C. Chakrabarti, S. Roy, and S. Basu (2018) Intention aware misbehavior detection for post-
     disaster opportunistic communication over peer-to-peer DTN. *Peer-to-Peer Networking and
     Applications* 2018, pp. 1–19.

[21] N. Das, S. Basu, and S. Das (2020) Efficient DropBox deployment towards improving post
     disaster information exchange in a smart city. In: *ACM Transactions on Spatial Algorithms and
     Systems*, 6(2), pp. 1–18.

[22] Map of water, food, shelter and medical resources. https://www.google.com/maps/d/viewer?mid
     =1Iv7GILViqyJAFn5o5hi1 F2Fg8mc&hl=en_US. 2015.

[23] A. Keranen, J. Ott, and T. Karkkainen (2009) *The ONE simulator for DTN protocol evaluation.
     In Proceedings of SIMUTools*, Rome, Italy, March 2–6, 2009.

[24] G. Wood (2014) Ethereum: A secure decentralized generalized transaction ledger. *Ethereum
     Project Yellow Paper*, 151, pp. 1–32.

[25] M. Y. S. Uddin, D. M. Nicol, T. F. Abdelzaher, and R. H. Kravets (2009) *A post-disaster mobil-
     ity model for delay tolerant networking. In Proceedings of WSC*, Texas, USA, December 13–16,
     2009.

[26] Doctors For You. http://doctorsforyou.org.

[27] A. Lindgren, A. Doria, E. Davies, and S. Grasic (2011) Probabilistic routing protocol for inter-
     mittently connected networks. draft-lindgren-dtnrg-prophet-09.txt, 2011.

[28] Regio A. Michelin et al. (2018). *SpeedyChain: A framework for decoupling data from blockchain
     for smart cities. In Proceedings of 15th EAI International Conference on Mobile and Ubiquitous
     Systems: Computing, Networking and Services*, ACM, 145–154.

[29] Gas usage, https://www.investopedia.com/terms/g/gas-ethereum.asp (Accessed 28 June 2020).

[30] N. van Saberhagen (2013) Cryptonote v2.0 2013. https://cryptonote.org/whitepaper.pdf.

Chapter 11

Evaluation of Energy Efficiency of Opportunistic Network Routing Protocols

Md. Khalid Mahbub Khan, Muhammad Sajjadur Rahim and Abu Zafor Md. Touhidul Islam

University of Rajshahi, Rajshahi, Bangladesh

CONTENTS

11.1 INTRODUCTION AND RELATED WORKS

The opportunistic network (OppNet) is an enhanced version of the multi-hop ad hoc network since the application fields of this network have been increased. This type of networking scenario was proposed for military purposes. To improve the success rate of data delivery on the battlefield among different military units was the primary issue. But in present days, OppNet's applications range has been broadened in civil aspects since the innovations of many smart communicating devices, and people also want connectedness during their daily activities [1]. The absence of a dedicated contact link from source to destination is the key feature of OppNet. In this case, the performance of the communication link is variable in either extreme. Conventional TCP/IP protocol will fail to perform efficiently as an end-to-end transmission path may exist for a short or uncertain period in OppNet. Moreover, transmission delays as well as the error rate are also higher. So, node mobility is exploited here to carry on legitimate data transmission in such cases, and mobile nodes are responsible for the delivery of data [2]. Here, nodes select the next relay for data forwarding based on the "store-carry-forward" manner. In this fashion, nodes can temporarily store data, and carry it for a time by adapting the advantages of both node mobility and opportunistic contacts. Besides, due to the mobility of node, OppNet becomes a heterogeneous network [3].

To identify the proper relay node for successful data delivery with reduced latency is the major issue in OppNet. Regarding this issue, several forwarding strategies have been proposed for OppNet from various points of view. The replication-based routing approach was the first-generation protocols proposed for OppNet. Epidemic [4] was the first replication-based protocol. Later, many improved and sophisticated protocols, viz. Spray and Wait [5], Spray and Focus [6], etc. were proposed within this generation. Then the probabilistic concept emerged in data forwarding. The PRoPHET [7] protocol is one of the examples of such a genre. After that, social-based forwarding strategies were developed for improved data forwarding in OppNet. Several social characteristics are considered to route data in such a generation of protocol. SCORP [8], dLife [9], dLifeComm [9], Bubble Rap [10], etc. belong to this category. Moreover, research has been going on in OppNet to find out the most efficient routing solutions.

However, nodes in OppNet are mostly lithium-ion-battery driven. So, they are energy-constrained. Nodes have to perform several activities such as transmitting and receiving messages, scanning neighboring nodes, etc., within their limited lifetime. A significant amount of energy is also invested due to the mobility of nodes. So, the energy consumption of nodes can be regarded as a vital aspect for OppNet. Furthermore, in practice, it is not easy, and sometimes not possible, to recharge the battery of a node. If the energy of nodes is consumed completely, then the temporal contact link between source and destination would be disconnected. Finally, data would not be delivered successfully to the destination [11]. Besides, the battery technology has not sufficiently improved to keep pace with the increasing requirement of energy. Early draining energy is one of the major deficiencies for mobile devices [12]. So, energy should be utilized efficiently for better networking performance as well as successful data delivery. Thus, the routing strategy should be energy-efficient.

Energy consumption with relevant performance evaluation of different routing protocols has been considered in many pieces of research in OppNet scenarios previously. In [13–17], the energy consumptions of different protocols, including some conventional OppNet protocols and a few social-based protocols, have been investigated with their comparative performance analysis. These investigations have been performed on behalf of several metrics such as average remaining energy, count of dead nodes, delivery ratio, average latency, overhead ratio, etc. In the aforementioned research works, networking scenarios have been varied after considering some networking aspects such as node density, buffer size, message generation rate, message TTL, networking area, etc. A performance evaluation of two social-based protocols, SCORP and dLife, has been presented against the replication-based and probabilistic protocols Epidemic, Spray and Wait, and PRoPHET in [18]. Moreover, in [11], the authors provided a detailed study of energy-aware, social-based protocols. But the experimental analysis has not exhibited there. A brief study about energy consumption and comparative performance evaluation of six distinct conventional protocols in the OppNet environment except considering any social-based protocol is provided in [19]. Here, also, the impact of node density and TTL has been reflected. A perfect routing protocol in the OppNet scenario must have a higher delivery ratio and minimum energy consumption, latency, and overhead ratio. To choose the best forwarding strategy among all that exist, it is necessary to know their comparative performance in the OppNet environment. This chapter provides an extended study relative to energy efficiency as well as modified performance evaluations of existing replication-based protocols (Epidemic, Spray and Wait, Spray and Focus), probabilistic (PRoPHET), and social-based protocols (Bubble Rap, SCORP, dLife, dLifeComm) in the OppNet. Besides, node density and message time-to-live (TTL) have important roles in message forwarding in the OppNet. With the increase of node density, the network size is also extended, and the increase of TTL means the message will be active for a longer time. These may affect the performance of the OppNet. In this research, we have investigated for a more energy-efficient protocol among the aforementioned protocols. A brief performance analysis of these protocols is also addressed in this study. The entire experimental evaluation has been performed from the impact of both node density and TTL after considering four performance criteria: average remaining energy, delivery ratio, average latency, and overhead ratio.

The structure of this chapter is as follows: a brief description of various OppNet routing protocols from various perspectives are provided in Section 11.2. Section 11.3 provides a general explanation of the simulation tool and settings. Performance metrics that are essential to our study are defined in Section 11.4. A comparative discussion on simulation outcomes is given in Section 11.5, and finally, Section 11.6 gives the conclusion of our study.

11.2 ROUTING PROTOCOLS FOR OppNet

The main objective of a routing protocol for OppNet is to deliver the message to the destination node successfully from a message carrier node. This forwarding operation should act to increase the message delivery rate and to reduce the latency, overhead, and number of message copies [20]. Furthermore, a routing protocol should be energy-efficient since nodes' energy is restricted. So, for better networking performance, less energy should be consumed by each node [19]. Several routing protocols have been developed for the OppNet. Stochastic-natured routing protocols are the main priority for data forwarding in most of the OppNet scenarios. Here, determining the future position of nodes is difficult, and the characteristics of the network are unpredictable [20]. However, routing protocols have been designed from various networking aspects to meet the requirements of such a challenging environment as the OppNet. In this chapter, we have considered replication-based (i.e., Epidemic, Spray and Wait, and Spray and Focus), probabilistic (i.e., PRoPHET), and social-based protocols (i.e., Bubble Rap, SCORP, dLife, and dLifeComm) for our studies, and a brief discussion of these protocols is provided in this section.

11.2.1 Replication-Based Protocols

Message replication-based protocols are the first-generation protocols for OppNet. The replication-based protocol can be referred to as a flooding-based protocol. In this case, the source node sends multiple copies of a message to the encountered nodes. Thus, flooding of the same message can occur in the network. The replication-based protocols have two major categories [21]:

- *Uncontrolled replication*: When a source node sends an unlimited replica of a message to all possible relays, this situation is referred to as uncontrolled replication. Epidemic is a replication-based protocol in which an uncontrolled replication technique is exploited.

- *Controlled replication*: In this case, a limited number of message replicas have been sent to a limited number of relays. Spray and Wait, and Spray and Focus protocols adopt this strategy to forward messages.

11.2.1.1 Epidemic

Epidemic is the first representative of the replication-based OppNet protocol. The major issue behind this forwarding strategy is copying messages. This strategy deploys blind replication of messages. The source node repeatedly copies its message and sends all the message copies to the encountered nodes who have not received any copy of the message yet. In this manner, the message is distributed into the full network, and finally, the target node will get the message. In Epidemic strategy, successful message delivery would be possible without regarding some issues such as congestion of network, buffer capacity, delivery delay, etc. Such unawareness behavior about the resources of Epidemic makes it a weak candidate for data routing in OppNet [21, 22].

11.2.1.2 Spray and Wait

Spray and Wait comes from the controlled replication genre. In this strategy, the blind replication of Epidemic has been replaced by controlled replication. So, the utilization of resources is a minimum here. In Spray and Wait, L number of message replicas are spread into the whole network, where L denotes the maximum number of copies of a message that are allowed to route from the source. The value of L can be fixed based on the number of nodes and the average delay that is required to deliver a message [18]. This protocol is composed of two phases [5]:

- Phase 1: This phase is known as the "Spray" phase, where the source node will forward L copies of a message towards L distinct encountered nodes with the source.
- Phase 2: If the destination node is not found in the "Spray" phase, then the message accepting L distinct nodes in the previous phase will wait until they can encounter the destination node to forward the message copy directly. This phase is known as "Wait".

Basically, the Spray and Wait forwarding protocol has two modes: Binary and Vanilla. The major differences between these two modes exist in the basic strategy adopted in the "Spray" phase to spray L replica of a message. In "Binary" mode, the message carrying source node begins with L copy of a message. The source node simply forwards L/2 copy of the message to that node who first encountered the source. Then each encountered message receiving node delivers half of the message copies carried by it toward those nodes they may meet in the future and those have no message copy. Finally, a node switches to the "Wait" phase while it delivers all of the message copies except one. In the "Wait" phase, it will wait for the direct opportunity to encounter with destination node for successful message transmission. In contrast, another mode named "Vanilla" where the source node would deliver a single copy of the message toward the L distinct relay nodes which encounter the source node after the generation of that message in the source. However, the "Binary" mode has the advantage over the "Vanilla" mode in Spray and Wait, and this advantage is that the message is disseminated much faster than in "Vanilla" mode [5]. In this study, only the "Binary" mode is considered for the Spray and Wait routing protocol.

11.2.1.3 Spray and Focus

Spray and Focus is the modified version of the Spray and Wait protocol. This forwarding technique has set up an upgraded single-copy routing strategy. Within this technique, once a relay node receives a single replica of a message, it would retransmit that copy instead of direct delivery. This routing strategy also consists of two phases: "Spray" and "Focus". Here, the first phase is similar to the Spray and Wait protocol. But the dissimilarity between the two takes place in the "Focus" phase. Unlike the "Wait" phase in Spray and Wait, relay nodes are not waiting for the destined node to encounter. They simply forward a single message copy toward a more consistent relay node. This forwarding is performed according to a utility-driven model. For a distinct message, if there is any relay with a token of single transmission, it would be turned into the "Focus" phase. So, a particular relay node may be able to receive a replica of the message in the "Focus" phase, instead of the destination node receiving it through direct transmission performed in the "Wait" phase of Spray and Wait. Finally, the functionalities of Spray and Focus protocols can be explained as follows [6]:

- *Message summary vector*: A vector is maintained by each node which contains the IDs of all messages stored by it and also those for which it acts as a relay. When two nodes encounter each other, they will swap their vectors. Then, they check the common message between them, and this message has a lifetime. The message would be discarded if this lifetime ends, and the entry of such discarded message would be erased in the message summary vector.
- *Last encounter timers*: A timer $\tau_i(j)$ for every other node j in the network is maintained by each node denoted by i. Let $t_m(d)$ denote the required time in which a node has to move a distance d under a given mobility model, m. If a node A encounters a node B at distance d_{AB}, then the following condition must be satisfied:

$$\tau_i(j) = \tau_j(i) = 0;$$

$$\forall j \neq B : \tau_B(j) < \tau_A(j) - t_m(d_{AB}),$$

$$\text{Where,} \tau_A(j) = \tau_B(j) + t_m(d_{AB}).$$

- *Spray and Focus forwarding*: At the source node, while a message is created, L "forwarding tokens" are also created in the source. If a node, either source or relay, carries a message copy, it would forward the message based on either one of two conditions:
  i. n > 1 forwarding tokens, in which binary spraying would be performed; or,
  ii. n = 1 forwarding token, which would perform utility-based forwarding according to the last encounter timers used as the utility function; here, n denotes the number of forwarding tokens.

11.2.2 Probabilistic Protocols

To address the efficient message forwarding problem in OppNet, researchers have introduced probabilistic concepts in the forwarding decision. Although a dedicated and fully connected end-to-end link is not guaranteed in OppNet, in reality, there are many cases where the probability of communication is nevertheless expected. So, this point is considered as the background behind the implementation of the probabilistic idea in data routing for the OppNet environment to improve the success rate of message delivery. The Probabilistic Routing Protocol using History of Encounters and Transitivity (PRoPHET) is an instance of this protocol class [7].

11.2.2.1 PRoPHET

The Probabilistic Routing Protocol using History of Encounters and Transitivity (PRoPHET) algorithm ensures the proper utilization in OppNet. To make sure that the perfect message delivery is performed on the proper destined node, a set of probabilities is conserved within this routing algorithm of PRoPHET. Unlike the blind replication property of the uncontrolled replication-based protocol, the likelihood of the nodes' real-world confrontation is adopted by this algorithm for efficient delivery of a message. After the encounter of other nodes with the message carrier source node, the message would be forwarded to the destined node following the higher probability of message transmission. The message is received first by that particular node which has a higher probability of transmission [7, 18].

11.2.3 Social-Based Protocols

In many OppNet environments, a large number of mobile nodes may be allocated in a scattering manner within a small and dense area. Thus, the topology is fully heterogeneous and forwarding protocols of the previous generation may not be suitable to route data effectively in such an application. Social-based protocols have presented a new generation of forwarding protocols to cope with this challenge by exploiting some social metrics such as community, centrality, etc., in forwarding decisions. The social attitudes and characteristics of node groups and the reflections of their inter-relationship are also recognized to select the latest eligible relay node. Thus, nodes interact in such divergent ways with each other where some nodes are used to more frequent encounters in a social-based network. However, nodes' social behaviors and attributes are the long-term properties, and this seems to be less volatile in comparison with nodes' mobility. Bubble Rap, SCORP, dLife, and dLifeComm protocols belong to this protocol type [23, 24].

11.2.3.1 Bubble Rap

This forwarding technique involves both communities as well as centrality. A collection of nodes can form a community. Within it, there are a few nodes that have greater acceptance than others on behalf of social interactions. Thus, these nodes may be regarded as having higher centrality. A particular node with the higher centrality may be selected as the eligible next relay by a message-containing source node to forward with a view to deliver towards the targeted final node. Nevertheless, if a Bubble Rap-implemented community in the OppNet scenario becomes extended by increasing the density of nodes with lower centrality, then the success rate of message delivery is assumed to be a minimum. So, the Bubble Rap forwarding technique is not applicable for pervasive OppNet scenarios, because to estimate the centrality for nodes, overall information relevant to the total network topology is needed. For this reason, the computation of centrality is not practicable for a spreading distributed network where a large number of nodes are employed [10, 25].

11.2.3.2 SCORP

The Social-Aware Content-Based Opportunistic Routing Protocol (SCORP) is another social-aware forwarding protocol where the interest of message content is the major concerning issue. In this strategy, knowledge about message content is required by each node including the type of message content, interested hosts, etc. A routing decision in SCORP is taken to identify the next relay depending on the content interest as well as nodes' social proximity. Here, a message owner source node forwards the message to that distinct encountered node which has a similar interest in the content of the message as the source node. Alternatively, the node which has a higher degree of relations with the message-containing node will have priority to receive the message first [8, 18].

11.2.3.3 dLife

dLife is a social-based forwarding approach which is based on the regular activities of users' everyday lives. To understand users' daily dynamic activity, a weighted social graph is required to be prepared. In this strategy, the source node routes a single message copy to that encountered node which has a strong relationship with the target node. So, there exists a probability that in the future it will encounter the target node, and finally, the message reaches the destined node. Furthermore, if this relationship between the source and target node is unexplored, then the message would forward in accordance with the importance of the node. The source node will forward the message to the most important node, and this importance is computed by source [9, 18, 25].

11.2.3.4 dLifeComm

A community variant of dLife protocol is dLifeComm. The main concept of forwarding data in this strategy is quite similar to the dLife approach. Here, the routing decision is made on the basis of social strength within a community, and it is calculated from the source and encountered nodes towards the destined node. Additionally, to forward the message outside a community, node importance is considered, and this protocol is focused mainly on the community rather than the centrality. The K-Clique algorithm is adopted in this forwarding strategy to detect such communities [9].

11.3 SIMULATION ENVIRONMENTS

This section provides a brief discussion about the environment needed to simulate the investigated forwarding protocols for OppNet.

11.3.1 Simulation Tool

The Opportunistic Network Environment (ONE) simulator is adopted as a simulation tool where all protocols are simulated. ONE is a Java-based simulator that is specially designed for the opportunistic environment. Various lucrative features of ONE simulator, such as visualizing node mobility, endorsing new maps and mobility models, and generating reports after simulation, make it an ideal tool for simulating OppNet scenarios [26]. Figure 11.1 provides a view of the ONE simulator. In this chapter, selected OppNet forwarding protocols from various genres have been simulated with a similar environment in the ONE simulator to investigate the energy efficiency along with their relevant performance comparison.

11.3.2 Simulation Scenario Settings

The entire experimental area is assumed as 10 km × 10 km on the Helsinki city map, which is the default in ONE simulator. The total duration of the simulation is considered to be 6 hours. Three groups of nodes are used, such as pedestrian, tram, and car. The average velocity (minimum and maximum) of pedestrian, tram, and car nodes in each group is set to 0.8–1.4, 2.7–13.9, and 7–10 m/s respectively. Here, nodes are communicating through the Bluetooth interface, and the maximum transmission range of each node is 100 m. Nodes' movement is performed according to the shortest path map-based movement. Eight different OppNet routing protocols have experimented into this study such as Epidemic, Spray and Wait, Spray and Focus, PRoPHET, Bubble Rap, SCORP, dLife, and dLifeComm. These

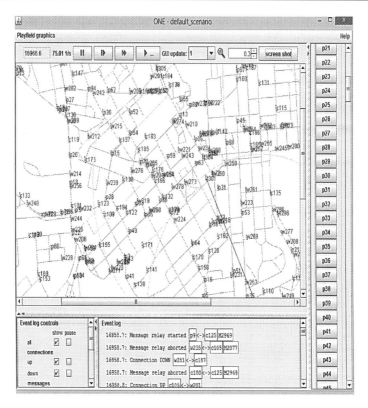

Figure 11.1 A view of the ONE simulator.

protocols are simulated in the ONE simulator by varying the node density (50, 80, 110, 140, 170, and 200) while TTL is kept fixed at 360 min (6 hours). Again, simulations are performed by varying time-to-live (TTL) from 120 to 360 min while keeping the node density per group fixed (50 nodes in each group). Table 11.1 exhibits the lists of necessary parameters that are required to perform simulations. Here, a maximum of 10 allowable copies is considered for Spray and Wait as well as Spray and Focus, and the number of seconds in the time unit is fixed at 30 for PRoPHET.

Table 11.1 Necessary settings for simulation arrangements.

Parameter	Value
Simulation duration	6 h
Node density	50, 80, 110, 140, 170, 200
Interface type	Bluetooth
Transmission rate	250 kbps
Transmission range	100 m
Routing Protocols	Epidemic, Spray and Wait, Spray and Focus, PRoPHET, Bubble Rap, SCORP, dLife, dLifeComm
Buffer capacity (MB)	5
Message TTL (in minutes)	120, 180, 240, 300, 360
Message size	500 KB to 1 MB
Mobility model	Shortest path map-based movement
Simulation area size	10 km × 10 km

Moreover, to examine the energy efficiency of experimented protocols, an energy module [27] is also implemented in the ONE simulator. Table 11.2 provides the necessary energy settings of nodes. Here, the initial battery energy of each node is given 5000 J before simulation. The energy expenditure which is spent per second on each scan for the nearest device discovery is known as "Scan energy". The amount of energy that is consumed per second in the time of message transmission is known as "Transmit energy". "Receive energy" can be denoted as the energy quantity expensed per second during the message reception in each receiving node. The time interval within which a node gets recharged is known as the "Interval of energy recharge". In this study, this time interval is regarded as 28,000 s, or 7.78 hours, which is greater than the entire simulation time (6 hours). This makes sure that the energy-constrained mobile nodes would not be able to get recharged during the overall simulation since they have been charged once before the start of the simulation.

11.4 PERFORMANCE METRICS

After performing simulations, the outcomes of simulated protocols are evaluated on behalf of four performance metrics: average remaining energy of each node, delivery ratio, average latency, and overhead ratio. This section briefly describes these metrics.

11.4.1 Average Remaining Energy

The average quantity of energy that remains in each node after fully completing the simulation is known as the average remaining energy. The value of this metric is expected as higher. A protocol that is implemented in an OppNet scenario can be considered as energy-efficient if the nodes of that network have higher remaining energy.

11.4.2 Delivery Ratio

The delivery ratio can be defined as the ratio of the message amount that successfully reached the destined node to the amount of message produced by the source node. The probability of successful message delivery is reflected in an OppNet by this metric. So, a higher success rate of message delivery depends on a higher delivery ratio.

11.4.3 Average Latency

This metric is defined as the average delay time required between a single message generation in the source node and its successful reception in the destination. For efficient OppNet operation, the value of this metric needs to be a minimum.

Table 11.2 Parameters related to energy of node.

Parameter	Settings
Capacity of battery	5000 J
Scan energy	0.92 mW/s
Transmit energy	0.08 mW/s
Receive energy	0.08 mW/s
Interval of energy recharge	28,000 s

11.4.4 Overhead Ratio

The overhead ratio measures how many redundant packets are relayed to deliver a single packet to the destination. Thus, it reflects the cost of transmission in OppNet, and the value of this metric is required to be a minimum.

11.5 RESULTS AND DISCUSSION

A comparative analysis of simulated OppNet protocols is given in this section regarding energy efficiency and performance evaluations. The discussions that follow are performed through the impact of both node density and TTL.

11.5.1 Impact of Node Density

Figure 11.2 exhibits the average energy of each node remaining after simulation with the increase of node density in each group. The increase of node density means the scanning, transmitting, and receiving operations are also increased, so energy would be consumed more. Here, SCORP and dLife both have higher remaining nodes' energy than other experimented OppNet protocols. The other two social-based protocols – Bubble Rap and dLifeComm protocols – have the average node energy (less than Spray and Wait, and less than Spray and Focus, but higher than Epidemic). The improved forwarding strategies of SCORP and dLife make them more energy-efficient, as both forwarding strategies consider interests of message content and daily behaviors of users, respectively. SCORP has the highest energy efficiency. So in SCORP, the least amount of energy is consumed for successful message delivery. On the other hand, Epidemic has the minimum energy efficiency than others because of the unawareness of resource (energy) during message forwarding. Thus, more energy is consumed by nodes for unnecessary as well as countless message replication and delivery to the next relays. Moreover, Spray and Wait, Spray and Focus, and dLifeComm protocols experienced less remaining node energy than SCORP and dLife. But these protocols have a higher value of remaining node energy than other simulated OppNet protocols (both PRoPHET and Bubble

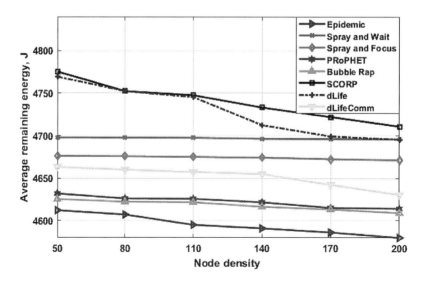

Figure 11.2 Average remaining energy vs. node density in each group.

Rap, respectively) with the variation of node density. Among these three protocols, the Spray and Wait protocol has much better performance for average remaining node energy, since a limited number of message copies is forwarded to the relay node. Additionally, Spray and Focus exhibits less remaining energy of each node than Spray and Wait, as more energy is invested by each node to further encounter with destination node. So, it can be decided that Spray and Wait, Spray and Focus, dLifeComm, PRoPHET, and Bubble Rap show the medium performance in this scenario in terms of average remaining energy, respectively, while the node density is increased.

From Figure 11.3, it can be clarified that dLife, Spray and Focus, and Spray and Wait have the highest delivery ratio with increasing node density in each group. But in comparison, both dLife and Spray and Focus have higher delivery ratios, and Spray and Wait experienced the second-highest performance among all simulated protocols. This is possible due to the selection of the next relay for dLife depending on either the relations with destination or node importance. So, the success ratio of message delivery to the destination node is higher. Like dLife, Spray and Focus has also the highest delivery ratio, because there are fewer message replicas here, and the relay nodes who receive message replicas would be intended for the final delivery of messages toward the destination node. So, the delivery ratio is the highest in this case. Again, the limited message replication process and selecting the proper relay nodes to forward messages in Spray and Wait protocol make delivery ratio higher in this case. But as compared to the best performers (dLife, Spray and Focus), the relay nodes who received message in the "Spray" stage would wait for the destination node for the final encounter. These activities of intermediate relay nodes may reduce the delivery ratio of Spray and Wait than Spray and Focus as well as dLife. The Spray and Wait protocol exhibits the second-best performance in terms of delivery ratio. Oppositely, Epidemic has the lowest performance with increasing node density because of its implementation of the flooding technique. So, many redundant message copies are forwarded from hop to hop in Epidemic, which results in the reduction of message delivery probability. The delivery ratio of another social-based protocol, dLifeComm, fluctuates with the increase of node density. The delivery ratio of SCORP remains constant with the extension of nodes per group. So, we can consider SCORP as a medium performer for delivery ratio. PRoPHET and Bubble Rap both have a minimum delivery ratio as compared to other protocols except the worst-performing protocol, Epidemic.

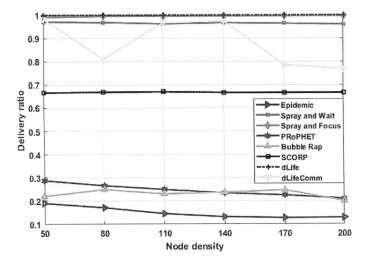

Figure 11.3 Delivery ratio vs. node density in each group.

The variation of average latency with varying node density is demonstrated in Figure 11.4. From this figure, it is realized that dLifeComm has the highest delivery delay if the network size is expanding. Because, in a community, it is possible that message would be delivered with delay to the known node. Moreover, the dLife protocol has the second-highest latency, since there may be a tendency of the nodes to deliver the message with delay towards those who are closely related to the destination node. Here, both dLifeComm and dLife protocols have comparatively higher latency than other simulated OppNet protocols. Spray and Focus has the lowest latency due to its sophisticated forwarding strategy, which makes it the best-performing protocol here. Simulated OppNet protocols like Bubble Rap, PRoPHET, Spray and Wait, Epidemic, and SCORP demonstrated medium performance in this case. The average delay time which is necessary to deliver a message in these protocols is not more than 30 minutes. Message forwarding in these protocols would take medium delay time as compared to dLifeComm, dLife (maximum delay time), and Spray and Focus (minimum delay time).

The overhead ratio is plotted with the variation of node density per group, as shown in Figure 11.5. Here, it is observed that Epidemic has the highest overhead ratio since the

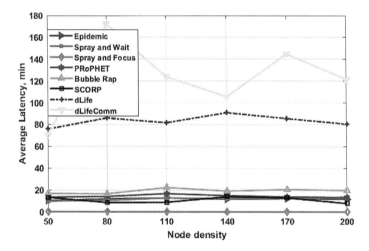

Figure 11.4 Average latency vs. node density in each group.

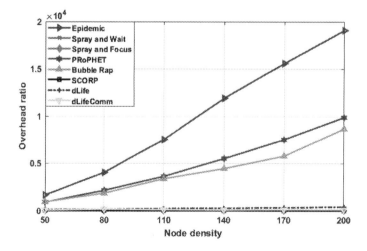

Figure 11.5 Overhead ratio vs. node density in each group.

unbound flooding of messages is deployed there. As a significant number of message copies have been wasted for the successful delivery of messages to the destinations, the overhead ratio is increased gradually with the increase of network size. Thus, Epidemic has the worst performance for overhead ratio. On the other hand, SCORP, Spray and Wait, Spray and Focus, dLifeComm, and dLife protocols (listed in order of performance) have the minimum overhead ratio. So, these protocols have the best performance for overhead ratio. Along with that, both PRoPHET and Bubble Rap have medium performance as compared to the best- and worst-performing protocols. Between these two protocols, the Bubble Rap protocol exhibits a higher overhead ratio than PRoPHET. Bubble Rap becomes ineffective when the size of the network is larger, due to the increase of node density.

11.5.2 Impact of TTL

In Figure 11.6, the average remaining energy is plotted with varying TTL (from 120 to 360 min). The increase in the value of message TTL means the longevity of a message would also increase so that the message can be active for a longer time. More energy would be required to deliver more active messages. As with the results of Figure 11.2, both SCORP and dLife have higher remaining energy than the others. Successful message delivery is possible for SCORP and dLife due to the effective forwarding techniques whether the message lifetime is increased. So, SCORP and dLife are the best performers on behalf of average remaining energy. Furthermore, SCORP exhibits better performance than dLife. In contrast, Epidemic exhibits the worst performance in this case due to resource unaware replications. Spray and Wait, Spray and Focus, and dLifeComm (listed in order of performance) are the medium performers in this case. PRoPHET and Bubble Rap exhibit slightly better performance than the worst performer, Epidemic, and between these two protocols, PRoPHET performs quite better than Bubble Rap.

The impact of TTL on the delivery ratio is visualized in Figure 11.7. In this figure, Spray and Focus and dLife have the highest ratio of message delivery, so we can consider both these forwarding protocols as the best performers among other examined OppNet protocols. Spray and Wait has a slightly lower value of delivery ratio than Spray and Focus and dLife, since the Spray and Wait indicating line is positioned below those of the other two protocols. It is also realized from this figure that the delivery ratio of dLifeComm is growing with the increase of TTL and when TTL is 240 min, the delivery ratio of dLifeComm becomes higher

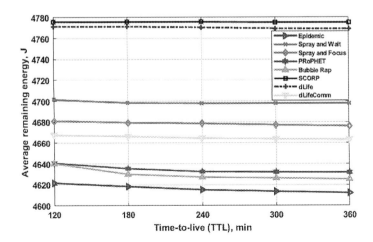

Figure 11.6 Average remaining energy vs. TTL.

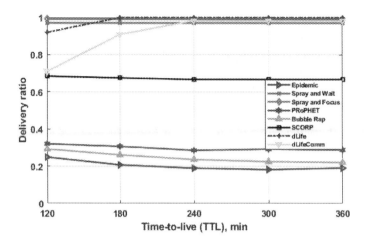

Figure 11.7 Delivery ratio vs. TTL.

as do Spray and Focus and dLife, since in a community the rate of successful message delivery would increase if message TTL is increased. Consequently, similar to the results of the node density impact, the delivery ratio is the lowest for Epidemic among the others. If the redundant message copies have a larger lifetime, it would not be effective for the successful transmission of those message copies.

Figure 11.8 represents the same result which is provided by Figure 11.4 for the best-performing protocol. For both cases, Spray and Focus experienced the lowest latency to deliver a successful message towards the destination. So, Spray and Focus performs the best among all simulated protocols. A limited number of message copies are successfully delivered to the destination with minimum delay. If the message lifetime is increased here, the delay time between the generation of a message in the source and its successful reception at the destination remains at a minimum. On the other hand, dLife and dLifeComm have a higher latency than others. But between these two protocols, dLife has quite higher latency than dLifeComm, So, it can be considered as the worst-performing protocol of all. Remaining protocols such as Bubble Rap, PRoPHET, SCORP, Epidemic, and Spray and Wait exhibit the medium performance, respectively.

From Figure 11.9, it is derived that Epidemic experienced the highest value for overhead ratio. This finding is similar to Figure 11.5. The impact of both increasing node density and

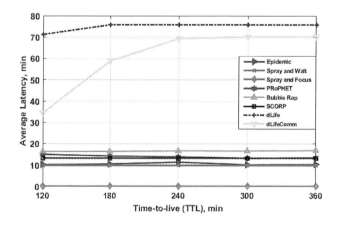

Figure 11.8 Average latency vs. TTL.

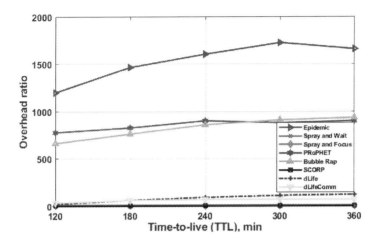

Figure 11.9 Overhead ratio vs. TTL.

TTL is similar for Epidemic. This protocol has the maximum overhead to deliver a single message to the destination. Many unnecessary message copies are generated by the source node in Epidemic. So, the ratio of overhead is increased and the network becomes congested for Epidemic protocol with the rise of message TTL. In opposition, three protocols – SCORP, Spray and Wait, and Spray and Focus – have minimum overhead than other OppNet protocols. So, these protocols can be recognized as the best-performing protocols for this metric, and their values of the overhead ratio are almost zero. So, very few messages would be wasted during message transmission for these protocols. This condition would not be violated if the message were to live for an additional time. Moreover, dLife and dLifeComm jointly exhibited a slightly higher value of overhead ratio than the best-performing protocols, and their indicating graph line is positioned slightly higher than the three best-performing protocols. Since within these two protocols, fewer messages would be wasted for a successful message delivery whenever the message lifetime grew. But this amount is slightly higher than the best-performing protocols SCORP, Spray and Wait, and Spray and Focus.

11.6 CONCLUSIONS

Within this chapter, we have evaluated the energy efficiency as well as the performance analysis of eight different OppNet forwarding protocols: replication-based (Epidemic, Spray and Wait, Spray and Focus), probabilistic (PRoPHET), and social-based (Bubble Rap, SCORP, dLife, dLifeComm). These evaluations were performed on the basis of varying both node density and TTL impact upon four performance metrics: average remaining energy, delivery ratio, average latency, and overhead ratio. Simulations of the mentioned OppNet protocols have been done on the ONE simulator. After analyzing the outcomes, we can conclude that with increasing node density and TTL, SCORP is the most energy-efficient protocol, while Epidemic is the least energy-efficient. Furthermore, dLife and Spray and Focus have the maximum delivery ratio. Moreover, Spray and Focus is the best-performing protocol for average latency. dLife (for TTL variation) and dLifeComm (for node density variation) have the least performance for average latency. For overhead ratio, SCORP, Spray and Focus, and Spray and Wait are the best-performing strategies, whereas Epidemic offers the worst performance for delivery probability and overhead ratio, respectively.

ACKNOWLEDGMENTS

This work has been supported by National Science and Technology (NST) MPhil fellowship, granted by Ministry of Science and Technology, Government of the People's Republic of Bangladesh.

REFERENCES

[1] Pelusi, L., Passarella, A., and Conti, M. (2006). Opportunistic networking: data forwarding in disconnected mobile ad hoc networks. *IEEE Communications Magazine*, 44(11), pp. 134–141.

[2] Huang, C. M., Lan, K. C., and Tsai, C. Z. (2008). *A survey of opportunistic networks. The 22nd International Conference on Advanced Information Networking and Applications-Workshops*, pp. 1672–1677.

[3] Trifunovic, S., Kouyoumdjieva, S. T., Distl, B., Pajevic, L., Karlsson, G., and Plattner, B. (2017). A decade of research in opportunistic networks: challenges, relevance, and future directions. *IEEE Communications Magazine*, 55(1), pp. 168–173.

[4] Vahdat, A., and Becker, D. (2000). Epidemic routing for partially connected ad hoc networks. Department of Computer Science, Duke University, Technical Report, CS-2000-06.

[5] Spyropoulos, T., Psounis, K., and Raghavendra, C. S. (2005). *Spray and Wait: an efficient routing scheme for intermittently connected mobile networks. Proceedings of the 2005 ACM SIGCOMM Workshop on Delay-Tolerant Networking*, pp. 252–259.

[6] Spyropoulos, T., Psounis, K., and Raghavendra, C. S. (2007). *Spray and Focus: efficient mobility-assisted routing for heterogeneous and correlated mobility. Fifth Annual IEEE International Conference on Pervasive Computing and Communications Workshops (PerComW'07)*, pp. 79–85.

[7] Lindgren, A., Doria, A., and Schelén, O. (2003). Probabilistic routing in intermittently connected networks. *ACM SIGMOBILE Mobile Computing and Communications Review*, 7(3), pp. 19–20.

[8] Moreira, W., Mendes, P., and Sargento, S. (2014). *Social-aware opportunistic routing protocol based on user's interactions and interests. International Conference on Ad Hoc Networks*, pp. 100–115. Springer, Cham.

[9] Moreira, W., Mendes, P., and Sargento, S. (2012). *Opportunistic routing based on daily routines. 2012 IEEE International Symposium on a World of Wireless, Mobile and Multimedia Networks (WoWMoM)*, pp. 1–6.

[10] Hui, P., Crowcroft, J., and Yoneki, E. (2010). Bubble Rap: social-based forwarding in delay-tolerant networks. *IEEE Transactions on Mobile Computing*, 10(11), pp. 1576–1589.

[11] Jagtap, P., and Kulkarni, L. (2019). Social energy-based techniques in delay-tolerant network. *Emerging Technologies in Data Mining and Information Security*, pp. 531–538. Springer, Singapore.

[12] Vardalis, D., and Tsaoussidis, V. (2014). Exploiting the potential of DTN for energy-efficient internetworking. *Journal of Systems and Software*, 90, pp. 91–103.

[13] Socievole, A., and Marano, S. (2012). *Evaluating the impact of energy consumption on routing performance in delay tolerant networks. 8th International Wireless Communications and Mobile Computing Conference (IWCMC)*, pp. 481–486.

[14] Cabacas, R. A., Nakamura, H., and Ra, I. H. (2014). Energy consumption analysis of delay tolerant network routing protocols. *International Journal of Software Engineering and Its Applications*, 8(2), pp. 1–10.

[15] Kaviani, M., Kusy, B., Jurdak, R., Bergmann, N., and Liu, V. (2016). Energy-aware forwarding strategies for delay tolerant network routing protocols. *Journal of Sensor and Actuator Networks*, 5(4), p. 18.

[16] Bista, B. B., and Rawat, D. B. (2016). *Energy consumption and performance of delay tolerant network routing protocols under different mobility models. 7th International Conference on Intelligent Systems, Modeling and Simulation (ISMS)*, pp. 325–330.

[17] Spaho, E. (2019). Energy consumption analysis of different routing protocols in a delay tolerant network. *Journal of Ambient Intelligence and Humanized Computing*, 11, pp. 3833–3839.

[18] Khan, M. K. M., and Rahim, M. S. (2018). *Performance analysis of social-aware routing protocols in delay tolerant networks*. 2018 International Conference on Computer, Communication, Chemical, Material and Electronic Engineering (IC4ME2), pp. 1–4.

[19] Khan, M. K. M., Roy, S. C., Rahim, M. S., and Islam, A. Z. M. T. (2020). On the energy efficiency and performance of delay-tolerant routing protocols. *Lecture Notes of the Institute for Computer Sciences, Social Informatics and Telecommunications Engineering*, 325, pp. 553–565. Springer, Cham.

[20] Mota, V. F., Cunha, F. D., Macedo, D. F., Nogueira, J. M., and Loureiro, A. A. (2014). Protocols, mobility models and tools in opportunistic networks: a survey. *Computer Communications*, 48, pp. 5–19.

[21] Aloui, E. A. A., Said, A., and Moha, H. (2015). The performance of DTN routing protocols: a comparative study. *WSEAS Transactions on Communications*, 14, pp. 121–130.

[22] Ababou, M., and Bellafkih, M. (2018). Energy efficient routing protocol for delay tolerant network based on fuzzy logic and ant colony. *International Journal of Intelligent Systems and Applications*, 10(1), pp. 69–77.

[23] Qirtas, M. M., Faheem, Y., and Rehmani, M. H. (2020). A cooperative mobile throwbox-based routing protocol for social-aware delay tolerant networks. *Wireless Networks*, 26, pp. 3997–4009.

[24] Zhang, X., Huang, P., Guo, L., and Fang, Y. (2019). Social-aware energy-efficient data offloading with strong stability. *IEEE/ACM Transactions on Networking*, 27(4), pp. 1515–1528.

[25] Haq, A., and Faheem, Y. (2020). A peer-to-peer communication based content distribution protocol for incentive-aware delay tolerant networks. *Wireless Networks*, 26(1), pp. 583–601.

[26] Keränen, A., Ott, J., and Kärkkäinen, T. (2009) *The one simulator for DTN protocol evaluation*. 2nd International Conference on Simulation Tools and Techniques (SIMUTools '09), pp. 1–10.

[27] Opportunistic Network Environment (ONE) simulator project page. https://www.netlab.tkk.fi/tutkimus/dtn/theone. Accessed 7 November 2020.

Chapter 12

Mobility Models in Opportunistic Networks

Jagdeep Singh
Sant Longowal Institute of Engineering and Technology, Longowal, India

Sanjay Kumar Dhurandher
Netaji Subhas University of Technology, New Delhi, India

Vinesh Kumar
University of Delhi, New Delhi, India

CONTENTS

12.1 INTRODUCTION

Due to the randomly disconnected topology, mobility plays a significant role in opportunistic networks (OppNets), as the mobility characteristics are used to formulate the mobility models in a realistic scenario. From the analysis and simulation, it is observed that the mobility models based on realistic topology have an important place in OppNets. The unrealistic scenario paves to unrealistic behavior or results when the mobility models based on unrealistic topology are used for a real-world scenario. Therefore, these unrealistic mobility models do not perform well in the real world. The evolution of a system in OppNets basically utilizes traffic patterns and mobility models [1, 2].

Mobility characteristics include the speed, predictability of movement patterns, and uniformity of mobile nodes in communication networking. Thus, the characteristics and features are the basis for the classification of mobility models. There exist three types of mobility models, namely trace-based models, synthetic models, and stochastic models. In conventional MANETs, networking protocols were generally evaluated using stochastic movement models as the Random Walk Mobility model (RWM), the Random Waypoint mobility model (RWP), the Random Direction Mobility model (RDM), etc. [3]. However, users' mobility is dependent upon the users' movement behavior, their social and personal characteristics, and environmental factors, because in reality users' movement is rarely random; thus, the conventional stochastic random-based mobility models fail to estimate the accuracy of the networking protocols in OppNets. In the literature, various studies demonstrate that users' social and personal behavior has a significant correlation with the users' movement patterns [4]. For instance, in recent studies and experimental analysis, it has been proven that users frequently visit a few places where they stay most of their time and with which they have strong social relationships. Furthermore, they rarely travel long distances and very often travel over short distances. Hence, based on the mobility characteristics discussed in the following section, an appropriate selection of mobility traces and models is necessary to emulate the movement behavior of humans and vehicles. Mobility models are categorized into four categories: a trace-based model of mobility, a stochastic model of mobility, a map-based model of mobility, and a synthetic model of mobility. Figure 12.1 represents the classifications for the mobility models.

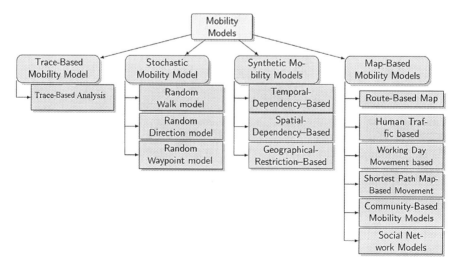

Figure 12.1 Mobility models classification.

The remainder of the chapter is discussed as follows. Section 12.2 covers the trace-based model. A detailed overview of models of stochastic mobility is given in Section 12.3. In Section 12.4, synthetic models are discussed. The geographical restriction dependent mobility model is introduced in Section 12.5. Section 12.6 focuses on the map-based mobility model. The findings and interpretation of the simulation are discussed in Section 12.7. Lastly, Section 12.8 concludes the chapter.

12.2 TRACE-BASED MODEL

12.2.1 Trace-Based Analysis

Trace-based models [5] can have an insight into actual mobility patterns by tracing mobile hosts in real-world scenarios. There are three methods for acquiring traces. The first method is to trace the location of devices by monitoring them with a particular tool. Currently, Global Positioning System (GPS) is used for monitoring the localization system. The second method is to use the communication system for monitoring communication devices. The accuracy of monitoring the communications is based upon the density of the access point to which communication devices communicate. This method may not be precise, but it can be used to validate the mobility models. The third method is to acquire contact traces using Bluetooth or WLAN by monitoring the contacts among mobile devices (Table 12.1).

Table 12.1 List of abbreviations used in the chapter.

Acronym	Description
GPS	Global Positioning System
RWM	Random Walk Model
RDM	Random Direction Model
RWP	Random Waypoint Model
SPMBM	Shortest Path Map-Based Movement Mobility Model
ONE	Opportunistic Network Environment
WLAN	Wireless Local Area Network
WDM	Working Day Movement Model
CMM	Community-Based Mobility Model
POI	Point of Interest
WKT	Well-Known Text
GIS	Geographic Information System
DTNs	Delay-Tolerant Networks
OppNets	Opportunistic Networks
RBMM	Route-Based Map Mobility Model
WSN	Wireless Sensor Networks
RPGM	Reference Point Group Mobility Model
RBM	Random Border Mobility Model
PMM	Pursue Mobility Model

12.3 STOCHASTIC MOBILITY MODEL

Mobility models are relatively easy to research under stochastic mobility [6] but show little or no resemblance to practical scenarios. Generally, they depend on random node movement.

12.3.1 Random-Based Mobility Models

In the random-based mobility models [7, 8], mobile nodes travel independently in the network without any constraints. The speed of nodes and destination are both randomly selected. The random-based mobility models are very easy to implement; hence, they are used in different types of simulation studies. These mobility models are further classified into three classes: RWM, RWP, and RDM.

12.3.1.1 Random Walk Model

This model [9] is the oldest mobility model. It is a quite simple model based on the Brownian Motion model. In this model, the nodes are selected randomly within a fixed range with the direction and degree. The nodes move around the first or initial position and cannot be out of the simulation area, and in the end the nodes return to initial position in one- and two-dimensional space. Figure 12.2 introduces the most general case of such a network: nodes travel around in space, and they may gain from the opportunity to forward a message whenever they are in random contact (given a transmission range) with another. The movement of nodes is not dependent on others. The simulation area should be sufficiently large for the efficient use of this mobility model.

12.3.1.2 Random Direction Model

This is a model that moves nodes to the border before changing direction and speed of the simulation area. The main aim of this model is to remove the limitation of the RWM [10]. The probability of a node passing through the simulation is greater, as compared to the RWM. It is simple, flexible, and easy to implement. Thus many simulations use this model. Although, there are certain limitations of this as it considers the only direction as the main parameter and not be a good response node in many states.

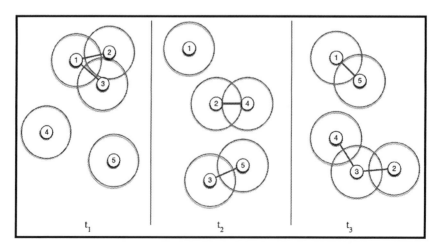

Figure 12.2 We can see the randomized opportunity ties between nodes in blue.

12.3.1.3 Random Waypoint Model

This model [11] is used in various simulations in mobile wireless ad hoc networks, and it includes the pause time between changes in direction or location. The implementation process is easy for this model. Several deficiencies exist in this model; for instance, as time increases, the simulation process is attenuated, and due to spatial distribution of nodes, the simulation process is changing from time to time. This model follows the memoryless property; that is, it does not contain information about past behavior. These deficiencies have been addressed by the modified version, namely the RBM and Restricted RWP. In Figure 12.3, the pseudo code of RWP model is presented [12].

12.4 SYNTHETIC MODELS

Synthetic models [13] are the mathematical models that capture the movement of nodes by imposing constraints like obstacles, pathways, etc. They do not rely on the random movement of nodes; rather, they move in a correlated manner to capture the movement of the nodes realistically.

12.4.1 Temporal Dependency-Based Mobility Models

The physical constraint of the mobile nodes affects the velocity of the mobile node which gradually or abruptly changes. The temporal dependency property describes the velocity of the nodes which are near each other at two-time instances, i.e., the current velocity of the node is dependent upon the previous velocity. These models of mobility based on temporal

```
public class RandomWaypoint extends MovementModel {private static final int PATH_LENGTH = 1;
        private Coord lastWaypoint;
        public RandomWaypoint(Settings settings) {super(settings);}

        protected RandomWaypoint(RandomWaypoint rwp) {super(rwp);}

        public Coord getInitialLocation() {
                assert rng != null : "MovementModel not initialized!";
                Coord c = randomCoord(); this.lastWaypoint = c; return c;
        }

        public Path getPath() {
                Path p;
                p = new Path(generateSpeed());
                p.addWaypoint(lastWaypoint.clone());
                Coord c = lastWaypoint;

                for (int i=0; i<PATH_LENGTH; i++) {
                        c = randomCoord();
                        p.addWaypoint(c);
                }
                this.lastWaypoint = c; return p;
        }
        public RandomWaypoint replicate() {return new RandomWaypoint(this);}

        protected Coord randomCoord() {
                return new Coord(rng.nextDouble() * getMaxX(), rng.nextDouble() * getMaxY());
        }
}
```

Figure 12.3 Pseudo code of RWP model.

dependence are further classified into two categories: the model of Gauss mark and the model of smooth random mobility. Each mobile node is initially assigned a current speed and direction in the Gauss mobility model, and the movement of each node occurs by updating its speed and direction at a fixed speed. The time intervals are also important. Every mobile node's speed and direction are associated over time and are calculated upon the previous speed and direction of the corresponding mobile node. In the smooth mobility model, the speed of the mobile node changes gradually rather than with sudden acceleration and sharp turns. Each node is defined by a motion vector which consists of speed and direction.

12.4.2 Spatial Dependency-Based Mobility Models

Under this group, the mobility model [14] is characterized by a spatial dependence property in which the movement of the mobile node is associated with that of the neighboring nodes. The Reference Point Group Mobility Model (RPGM), the Pursue Mobility Model (PMM), and the Community Mobility Model (CMM) are known as various spatial-dependency mobility models. In the RPGM, each mobile node in a group is randomly distributed around the predefined reference point, which allows individual random movement of nodes along with the group motion behavior. Each group of nodes has a logical center that determines the entire movement of the group through the group motion vector. As the reference point moves from the first point to the second point, the position of the actual reference point is modified according to the logical center of the group. The motion vector can be designed on the basis of predefined paths, or it can be selected at random. If the community motion vector has passed the checkpoint, the next checkpoint is again computed and transferred to that checkpoint. For example, a group of soldiers march toward their enemy by forming a line. A minor modification of the mobility model is where a set of mobile nodes are placed in a single-file line and are allowed to follow one another about their initial position. The *PMM* is another spatially dependent-based mobility model, in which mobile nodes attempt to track a particular target. It can be used for law enforcement and signal source tracking, as the nodes in the model try to capture the single target node [15].

12.5 GEOGRAPHICAL RESTRICTION-BASED MOBILITY MODEL

In random mobility models, mobile nodes are free to move anywhere within the simulation area without any geographical restrictions like environmental obstacles such as pathways, streets, buildings, etc. Geographical-restriction-based mobility models are models that are bounded by environmental restrictions [16].

12.5.1 Pathway Mobility Model

The Pathway mobility model [17] is a geographical-restriction-based mobility model where the model is bounded by pathways. In this model, the map is predefined for the simulation area, and it models the map as a random graph where vertices in the graph represent the buildings, and edges represent pathways.

12.5.2 Obstacle Mobility Model

The Obstacle mobility model [18] is another geographical-restriction-based mobility model that tries to model the movement of the node considering the environmental obstacles. Environmental obstacles affect the movement pattern of the nodes, as when the node encounters the obstacle, it has to change its path. Therefore, obstacles play a significant part in modeling the mobility of the node [19].

12.5.3 Manhattan Mobility Model

The Manhattan mobility model [20] is also a geographical-restriction-based mobility model which uses predefined grid topology for the movement of the nodes. It models the movement pattern of the nodes in well-organized streets of urban areas. It moves in a horizontal and vertical direction by using a probabilistic approach to move straight, left, and right on the street.

12.5.4 City Segment Model

The mobility model of the City Segment mimics the realistic movement of the node in a town segment. In the mobility model, the mobile nodes follow the predefined route and are limited, as in real scenarios, by obstacles and traffic regulation. Each mobile node begins its movement from a defined point on the street toward the randomly chosen destination through the shortest path. After reaching the destination, it stays at that location for a specific pause time, and again it chooses a random destination for the movement and repeats the process until the simulation ends. The City Section mobility model presents the realistic movement of the node by restricting the movement pattern, that is, following the predefined path like in real scenarios, as people move on a predefined path following traffic regulations.

12.6 MAP-BASED MOBILITY MODELS

In these models, using a subset of the Well-Known Text (WKT) format, map data is used. In reality, WKT is based on the ASCII format commonly used in the programs of the Geographic Information System (GIS) [4]. Map-based mobility models [21, 22] are based on the traffic and user mobility patterns to evaluate and deal with networking issues in the OppNets. Nodes in map-based mobility models move randomly on a predetermined path defined by the map data in a well-known text format file [23–25]. In map-based mobility models, a group of nodes can select a certain destination or point on the map, and after traveling to a specified distance, they pause for a certain pause time and then resume their journey.

12.6.1 Route-Based Map Mobility Model

The Route-Based Map Movement (RBMM) mobility model [26] mimics the movement of nodes based on predefined routes on the simulation maps. Nodes in this model have predetermined routes that they follow in the simulation map after getting the map data, but during the movement of nodes, the model does not select the destination randomly; instead, they select the next destination to which they are currently moving. The routes in the model have many stop points which the nodes follow while traveling; they stop for a specified pause time on the stop points to reach the destination with the shortest path for modeling the movement pattern.

12.6.2 Rush Hour (Human) Traffic Model

The Rush Hour Traffic model [27] deals with the traffic condition of rush hour. The Rush Hour Traffic model focuses on high traffic area and choice of destination, as people tend to travel to nearby places rather than going to further destinations. The algorithm is based on the exponential distribution.

12.6.3 Working Day Movement Model

The WDM [28] was proposed to capture movement patterns of real-life scenarios. To deal with inter-contact time and contact time distributions that are contained in the traces of real-world measurement experiments, this model was proposed for DTNs and OppNets. With four sub-models, the WDM is developed, representing the work routine that most individuals follow every day and repeatedly. Home activity, workplace activity, evening activity, and transport activity are these sub-models. The nodes move on a map containing all the positions of activities at home, office, and evening, and identifying spaces and routes for the nodes to move. The WDM tries to model each scenario that people follow in everyday life. In Figure 12.4, the pseudo code of WDM model is presented [12].

```
public class WorkingDayMovement extends ExtendedMovementModel {
public WorkingDayMovement(Settings settings) {
            super(settings);
            busTravellerMM = new BusTravellerMovement(settings);
            workerMM = new OfficeActivityMovement(settings);
            homeMM = new HomeActivityMovement(settings);
            eveningActivityMovement = new EveningActivityMovement(settings);
            carMM = new CarMovement(settings);
            ownCarProb = settings.getDouble(PROBABILITY_TO_OWN_CAR_SETTING);
            if (rng.nextDouble() < ownCarProb) {
                    movementUsedForTransfers = carMM;
            } else {
                    movementUsedForTransfers = busTravellerMM;
            }
            doEveningActivityProb = settings.getDouble(
                            PROBABILITY_TO_GO_SHOPPING_SETTING);

            setCurrentMovementModel(homeMM);
            mode = HOME_MODE;
}
public WorkingDayMovement(WorkingDayMovement proto) {
            super(proto);
            busTravellerMM = new BusTravellerMovement(proto.busTravellerMM);
            workerMM = new OfficeActivityMovement(proto.workerMM);
            homeMM = new HomeActivityMovement(proto.homeMM);
            eveningActivityMovement = new EveningActivityMovement(
                            proto.eveningActivityMovement);
            carMM = new CarMovement(proto.carMM);

            ownCarProb = proto.ownCarProb;
            if (rng.nextDouble() < ownCarProb) {
                    movementUsedForTransfers = carMM;
            } else {
                    movementUsedForTransfers = busTravellerMM;
            }
            doEveningActivityProb = proto.doEveningActivityProb;

            setCurrentMovementModel(homeMM);
            mode = proto.mode;
}
public Coord getInitialLocation() {
            Coord homeLoc = homeMM.getHomeLocation().clone();
            homeMM.setLocation(homeLoc); return homeLoc;
}
public MovementModel replicate() {return new WorkingDayMovement(this);}
public Coord getOfficeLocation() {return workerMM.getOfficeLocation().clone();}
public Coord getHomeLocation() {return homeMM.getHomeLocation().clone();}
public Coord getShoppingLocation() {return eveningActivityMovement.getShoppingLocation().clone();}}
```

Figure 12.4 Pseudo code of WDM model.

12.6.4 Shortest Path Map-Based Movement Model

The Shortest Path Map-Based Movement (SPMBM) mobility model [29] proposed for DTNs is based on the shortest path available within the simulation map scenario. In this model, rather than moving randomly around the simulation map, the shortest path is chosen over the available paths between two random nodes and points of interest (POIs) from the simulation map, as people mostly use the shortest path to travel. POIs are the places on the simulation map to which nodes within the group travel.

In the SPMBM, all the nodes are placed randomly on the simulation map, and then nodes travel to a certain destination in the simulation map following the Dijkstra algorithm to discover the shortest path from the available paths on the simulation map. After reaching a specified destination, the nodes pause for a certain period of time and again move to a newly selected destination through the shortest path. Map-based mobility models generally model movement patterns based on the map, the routes people follow, the traffic pattern, the working activities pattern, etc. These mobility models are useful in modeling various real-life scenarios, but the map-based mobility models have certain limitations in that the paths in the model are predefined, which, in some cases, is not possible in real life. In Figure 12.5, the pseudo-code of the SPMBM model is presented [12].

12.6.5 Social Network-Based Mobility Models

The social network-based mobility models [30–32] capture the behavior of movement based on human decisions and the social nature of humans, or the entities just like the social

```
public class ShortestPathMapBasedMovement extends MapBasedMovement implements
        SwitchableMovement {
    public ShortestPathMapBasedMovement(Settings settings) {
        super(settings);
        this.pathFinder = new DijkstraPathFinder(getOkMapNodeTypes());
        this.pois = new PointsOfInterest(getMap(), getOkMapNodeTypes(), settings, rng);
    }
    protected ShortestPathMapBasedMovement(ShortestPathMapBasedMovement mbm) {
        super(mbm);
        this.pathFinder = mbm.pathFinder;
        this.pois = mbm.pois;
    }
    public Path getPath() {
        Path p = new Path(generateSpeed());
        MapNode to = pois.selectDestination();

        List<MapNode> nodePath = pathFinder.getShortestPath(lastMapNode, to);

        // this assertion should never fire if the map is checked in read phase
        assert nodePath.size() > 0 : "No path from " + lastMapNode + " to " +
                to + ". The simulation map isn't fully connected";

        for (MapNode node : nodePath) { // create a Path from the shortest path
            p.addWaypoint(node.getLocation());
        }

        lastMapNode = to; return p;
    }
    public ShortestPathMapBasedMovement replicate() {
        return new ShortestPathMapBasedMovement(this);
    }
}
```

Figure 12.5 Pseudo-code of the SPMBM model.

behavior of humans in a battlefield, during disaster relief, etc. They rely upon the structure of the relationships among the individuals.

12.6.6 Community-Based Mobility Models

The CMM is a social network-based mobility model which mimics the movement pattern based on social network [33]. The *CMM* mobility model groups the node according to the community to which they belong. Nodes with the same community are grouped as friends, and nodes that belong to different communities are grouped as non-friends. Initially, a cell is allotted to each community where they share a social link between the friends' and non-friends' communities for the movement of the nodes in the network.

12.6.7 Social Network Models

In this model, artificial users periodically move between abstract locations. A set of abstract locations, which they have to visit at a fixed time, is given to the users. In [22], the author proposed a social network model is based on social network theory as it depends upon human relationships. The model allows several hosts to be grouped based on the social relationships among them and mapped to a topographical space. It models the social network through a weighted graph of the relationships. Social network models are based on the social relationship between the nodes, which affects the movement pattern of the node in the network; hence, it is necessary to model social behavior based on social network theory [34].

12.7 SIMULATION ANALYSIS

The Simulator for the Opportunistic Network Environment (ONE) [12] is a simulation tool commonly used by researchers studying DTNs and OppNets. Its salient features are listed as follows:

- The ONE simulator is a tool for researchers to carry out research-based simulations related to DTNs, WSN, and OppNets
- It develops and tests different movement models for nodes
- It adds message routing capabilities between nodes
- It uses *JUnit* and the *GUI* to debug the protocols being developed as well as visualize them
- It generates results and statistics for tests effortlessly.

The ONE simulator's default movement mobility model settings are shown in Figure 12.6. The map-based movement model requires that settings define the files being used to construct the movement models. It can be seen that the four map files used are essentially defining the paths and places of the real world that are being simulated. The reader can define their own map files in the WKT format. This will be covered later in more detail. One can also define the area of the world that they want for their simulations as well as the warmup time, which is the time duration for which hosts have moved around in the defined area until the onset of the real simulation. The router classes are connected. This is clearly shown in Figure 12.7. One class inherits another class, and subsequent inheritance takes place.

In Figure 12.8, the interaction of simulator files is shown. The simulation parameters are listed in Table 12.2.

```
1. MovementModel.rngSeed = 1
2. MovementModel.worldSize = 4500, 3400
3. MovementModel.warmup = 1000
4. MapBasedMovement.nrofMapFiles = 4
5. MapBasedMovement.mapFile1 = $data/roads.wkt$
6. MapBasedMovement.mapFile2 = $data/mainroads.wkt$
7. MapBasedMovement.mapFile3 = $data/pedestrianpaths.wkt$
8. MapBasedMovement.mapFile4 = $data/shops.wkt$
```

Figure 12.6 Default movement mobility model settings.

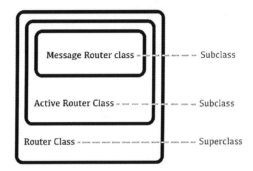

Figure 12.7 Inheritance between routers.

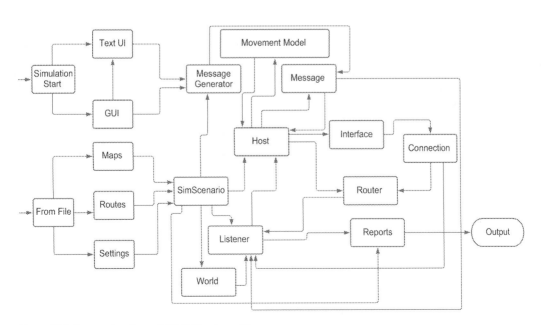

Figure 12.8 Interaction flow of ONE Simulator files.

Table 12.2 Simulation parameters.

Parameter	Value
Duration of simulation	43200s
Dimension	4500 × 3200 m
Nodes count in simulation	130–455
Velocity of nodes	0.5–1.5 m/s
Range	10 m
Speed	25 Kbytes/s
Size of message	1 Kbytes
Number of host groups	4
Group waiting time	0–120s
Group speed	0.5–1.5 m/s
Number of events	1
Events interval	25–35s
Map	Helsinki Map
GUI Under lay image offset	(64, 20)
GUI Under lay image scale	4.75
GUI Under lay image rotate	−0.015
GUI Event log panel	30
GUI Under lay Image	helsinki underlay
Report warmup time	0s
Movement model warmup	1000s
Movement model random number generator seed	1

12.7.1 Testing Tool

In this subsection, we show the flow of simulation files that are required for each simulation. Figure 12.9 represents *init* model flow. The DTNSim file is the main file of ONE simulation. This *initModel()* is directly linked to the main file and respective setting files. *EventQueue()* and *createHosts()* are called by *initModel()* for simulation. Figures 12.10 and 12.11 represent runtime flow and update flow, respectively. The *world()* is called when the simulation is running. The router *update()* is one of the main update functions for routing. Figure 12.12 represents the execution window of the *ONE* simulator at an instance. The Helsinki map model is used in the simulation.

12.7.2 Simulation Results

In this subsection, the PRoPHET protocol [35] is selected for simulation. The reason behind the PRoPHET routing scheme selection is better delivery probability.

PRoPHET's performance is simulated and compared against that of benchmark mobility models SPMBM, RBMM, and RWP mobility models using the ONE simulator. The performance metrics for simulating the mobility models for the PRoPHET routing scheme are delivery probability and average latency. Figure 12.13 shows that RBMM offers better distribution than the RWP and SPMBM with the PRoPHET routing protocol. This means that the delivered packets are high in the RBMM mobility model, while the number of nodes in the network is smaller. This is good for OppNets since, with a limited number of nodes, the network is first generated and then extended. If the network is sparse, better delivery probability

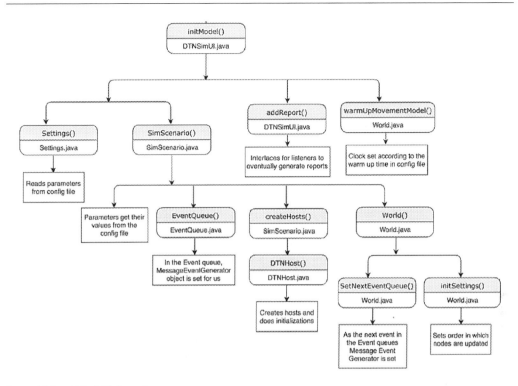

Figure 12.9 initModel() flow chart.

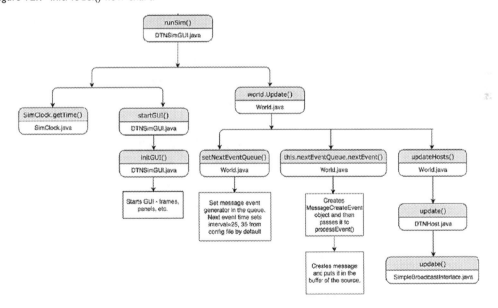

Figure 12.10 runSim flow() chart.

can also be done. Figure 12.14 shows PRoPHET's performance when the delay is considered as the performance metric.

The buffer capacity is varied and the effect of this adjustment is investigated on the likelihood of delivery. In Figure 12.15, the output of buffer policies is captured. It is found that with increasing buffer power, the likelihood of delivery is also improved. Because of a node's

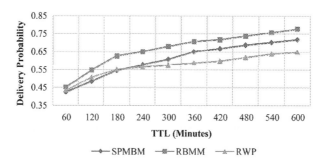

Figure 12.11 Delivery Probability vs TTL.

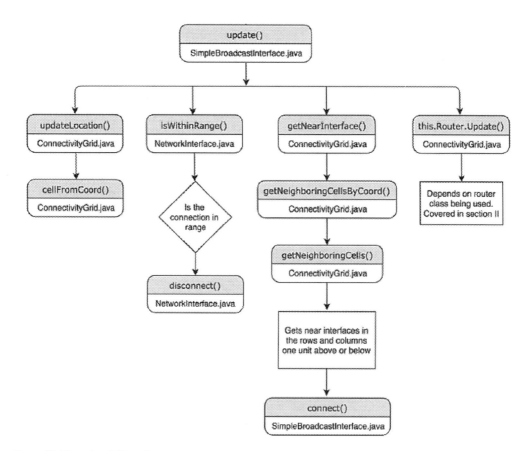

Figure 12.12 update() flow chart.

buffer capacity, the more messages stored in that buffer are increased, the more messages that are sent to the receiver node. The buffer capacity is varied, and the impact of this adjustment is examined on the average latency. In Figure 12.16, the effects of the buffer policies are captured. The TTL is diverse and the effect on the latency of this transition is observed. Figure 12.17 shows the results of delivery probability versus time to live. Figure 12.18 displays the results of average latency versus time to live. It is found that the average latency often increases when the TTL is boosted. This is because a substantial TTL value often increases the message's stay in the buffer of the node.

Figure 12.13 ONE simulator: execution screenshot at an instance.

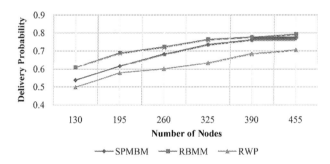

Figure 12.14 Delivery probability vs. number of nodes.

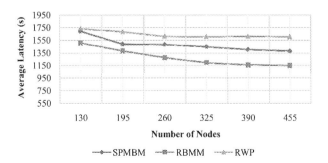

Figure 12.15 Average latency vs. number of nodes.

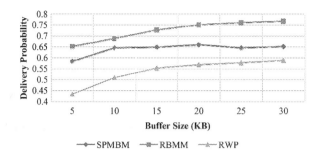

Figure 12.16 Delivery probability vs. buffer size.

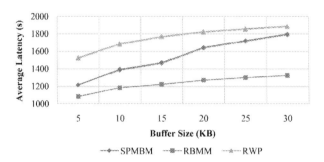

Figure 12.17 Average latency vs. buffer size.

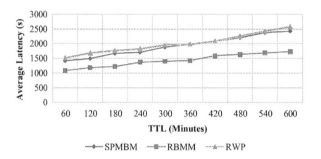

Figure 12.18 Average latency vs. TTL.

12.8 CONCLUSION

In this chapter, the mobility models were discussed for opportunistic networks in detail and summarized according to simulation outcomes. The simulation study showed that the best mobility model is dependent on regular routes. We conducted a performance simulation of PRoPHET using the RWP, SPMBM, and RBMM models. The results of the simulation indicate that PRoPHET produces better performance in the case of the RBMM model.

Humans are found in our research as fundamental components of many opportunistic applications. We should therefore bring greater emphasis on balancing distinct human behavior with other problems of modeling. Since contact times are a microscopic human mobility property, the research built around them has tremendous significance in realistic modeling of mobility, especially when one departs from the exponential assumption or introduces a small amount of heterogeneity. In mobility modeling, it is believed that human mobility and subsequent contacts occur by purpose and position. Secondly, social links between nodes direct a node to choose the destination and the timing of a mobility journey, while the location provides guidance on the direction to be followed. This produces a very complex system of touch that is not easily visible at the inter-contact level.

Therefore, to capture the array of node interrelations in a more tractable way, a more abstract and macroscopic view of mobility is needed. Greater emphasis should be placed on integrating various human behaviors together with modeling problems in future researchers.

REFERENCES

[1]   M. Conti et al., "From Opportunistic Networks to Opportunistic Computing," *IEEE Commun. Mag.*, vol. 48, 2010, pp. 126–139.

[2]   C. Boldrini, M. Conti, and A. Passarella, "Exploiting User's Social Relations to Forward Data in Opportunistic Networks: The HiBOp Solution," *Pervasive Mob. Comput.*, vol. 4, 2008, pp. 633–657.

[3]   M. C. Gonzalez, C. A. Hidalgo, and A.-L. Barabasi, "Understanding Individual Human Mobility Patterns," *Nature*, vol. 453, 2008, pp. 779–782.

[4]   C. Song et al., "Modeling the Scaling Properties of Human Mobility," *Nature Phys.*, vol. 6, 2010, pp. 818–823.

[5]   P. Hui et al., "Pocket Switched Networks and Human Mobility in Conference Environments," *Proceedings of the 2005 ACM SIGCOMM Workshop on Delay-Tolerant Networking*, 2005, pp. 244–251.

[6]   D. Brockmann, L. Hufnagel, and T. Geisel, "The Scaling Laws of Human Travel," *Nature*, vol. 439, 2006, pp. 462–465.

[7]   A. Chaintreau et al., "Impact of Human Mobility on Opportunistic Forwarding Algorithms," *IEEE Trans. Mobile Comp.*, vol. 6, 2007, pp. 606–620.

[8]   B. D. Walker, T. C. Clancy, and J. K. Glenn, "Using Localized Random Walks to Model Delay-Tolerant Networks," in *Military Communications Conference, 2008. MILCOM 2008*, 2008, pp. 1–7. IEEE.

[9]   T. Karagiannis et al., "Power Law and Exponential Decay of Intercontact Times between Mobile Devices,"*IEEE Trans. Mobile Comp.*, vol. 9, 2010, pp. 1377–1390.

[10] C. Song et al., "Limits of Predictability in Human Mobility," *Science*, vol. 327, 2010, pp. 1018–1021.

[11] J. Ghosh, S. J. Philip, and C. Qiao, "Sociological Orbit Aware Location Approximation and Routing (SOLAR) in MANET," *Ad-Hoc Net.*, vol. 5, 2007, pp. 189–209.

[12] A. Kernen, J. Ott, and T. Krkkinen, "The ONE Simulator for DTN Protocol Evaluation," in *Proceedings of the 2nd International Conference on Simulation Tools and Techniques*, 2009 (March), pp. 1–10.

[13] W. J. Hsu et al., "Modeling Time-Variant User Mobility in Wireless Mobile Networks," *IEEE INFOCOM 2007*, 2007, pp. 758–766.

[14] K. Lee et al., "SLAW: A New Mobility Model for Human Walks," *IEEE INFOCOM 2009*, 2009, pp. 855–863.

[15] A. Mei and J. Stefa, "SWIM: A Simple Model to Generate Small Mobile Worlds," *IEEE INFOCOM 2009*, 2009, pp. 2106–2113.

[16] I. Rhee, K. Lee, S. Hong, S. J. Kim, and S. Chong, "Demystifying the Levy-Walk Nature of Human Walks," Technical Report, NCSU, http://netsrv.csc.ncsu.edu/export/DemystifyingLevyWalkPatterns.pdf, 2008.

[17] Q. Zheng et al., "Agenda Driven Mobility Modeling," *Int'l. J. Ad Hoc Ubiquitous Comput.*, 5, 2010, pp. 22–36.

[18] F. Ekman et al., "Working Day Movement Model," *Proceedings of the 1st ACM SIGMOBILE Workshop Mobility Models*, 2008, pp. 33–40.

[19] M. Musolesi and C. Mascolo, "Designing Mobility Models Based on Social Network Theory," *SIGMOBILE Mob. Comp. Commun. Rev.*, 11, 2007, pp. 59–70.

[20] C. Boldrini and A. Passarella, "HCMM: Modeling Spatial and Temporal Properties of Human Mobility Driven by User's Social Relationships," *Computer Commun.*, 33, 2010, pp. 1056–1074.

[21] V. Borrel et al., "SIMPS: Using Sociology for Personal Mobility," *IEEE/ACM Trans. Net.*, 17, 2009, pp. 831–842.

[22] S. Yang et al., "Using Social Network Theory for Modeling Human Mobility," *IEEE Netw.*, 24, 2010, pp. 6–13.

[23] D. Thakore and S. Biswas, "Routing with Persistent Link Modeling in Intermittently Connected Wireless Networks," in *Military Communications Conference, 2005. MILCOM 2005*, 2005 (October), pp. 461–467. IEEE.

[24] F. Warthman, "Delay Tolerant Network – A Tutorial," version 1.1, May 2003.

[25] R. S. Mangrulkar and M. Atique, "Routing Protocol for Delay Tolerant Network: A Survey and Comparison," in *2010 IEEE International Conference on Communication Control and Computing Technologies (ICCCCT)*, 2010 (October), pp. 210–215. IEEE.

[26] D. Fischer, K. Herrmann, and K. Rothermel, "GeSoMo: A General Social Mobility Model for Delay Tolerant Networks," *2010 IEEE 7th International Conference on Mobile Ad Hoc Sensor System*, 2010, pp. 99–108.

[27] A. Bar-Noy, I. Kessler, and M. Sidi, "Mobile Users: To Update or Not to Update?" *Wirel. Netw.*, 1(2), 1995, pp. 175–185.

[28] M. Grossglauser and D. Tse, "Mobility Increases the Capacity of Ad-hoc Wireless Networks," in *INFOCOM 2001, Twentieth Annual Joint Conference of the IEEE Computer and Communications Societies. Proceedings. IEEE*, Anchorage, AK, vol. 3, 2001, pp. 1360–1369. IEEE.

[29] W. J. Hsu, T. Spyropoulos, K. Psounis, and A. Helmy, "Modeling Time-Variant User Mobility in Wireless Mobile Networks," in *INFOCOM 2007, 26th IEEE International Conference on Computer Communications. IEEE*, Barcelona, Spain, 2007 (May), pp. 758–766.

[30] W. J. Hsu, T. Spyropoulos, K. Psounis, and A. Helmy, "Modeling Spatial and Temporal Dependencies of User Mobility in Wireless Mobile Networks," *IEEE/ACM Trans.Network (ToN)*, 17(5), 2009, 1564–1577.

[31] W. Gao, Q. Li, B. Zhao, and G. Cao, "Multicasting in Delay Tolerant Networks: A Social Network Perspective," in *Proceedings of the Tenth ACM International Symposium on Mobile Ad Hoc Networking and Computing*, 2009, pp. 299–308. ACM.

[32] T. Hossmann, T. Spyropoulos, and F. Legendre, "Know Thy Neighbor: Towards Optimal Mapping of Contacts to Social Graphs for DTN Routing," in *INFOCOM, 2010 Proceedings IEEE*, 2010, pp. 1–9. IEEE.

[33] J. Lakkakorpi, M. Pitknen, and J. Ott, "Adaptive Routing in Mobile Opportunistic Networks," in *Proceedings of the 13th ACM International Conference on Modeling, Analysis, and Simulation of Wireless and Mobile Systems*, 2010 (October, pp. 101–109. ACM.

[34] B. D. Walker, J. K. Glenn, and T. C. Clancy, "Analysis of Simple Counting Protocols for Delay-Tolerant Networks," in *Proceedings of the Second ACM Workshop on Challenged Networks*, 2007, pp. 19–26. ACM.

[35] A. Lindgren, A. Doria, E. Davies, and S. Grasic, (2012, August), "Probabilistic Routing Protocol for Intermittently Connected Networks," *IETF 6693*.

Chapter 13

Opportunistic Routing in Mobile Networks

Zehua Wang
University of British Columbia, Vancouver, Canada

Yuanzhu Chen, and Cheng Li
Memorial University of Newfoundland, St. John's, Canada

CONTENTS

13.1 INTRODUCTION

The multi-hop wireless network has drawn a great deal of attention in the research community. Within the long period after the network was proposed, the routing and forwarding operations in the network have remained quite similar to those in the multi-hop wired network or the Internet. However, all the data transmission over the wireless medium in the wireless networks is by broadcasting in nature, which is different from the Internet. Because of the broadcast nature, many opportunities based on overhearing can be used to enhance the data transmission ability in the wireless network. ExOR is the first practical data forwarding scheme which tries to promote data transmission ability by utilizing the broadcast nature in wireless mesh networks, and opportunistic data forwarding becomes a well-known term given by ExOR to name this kind of new data forwarding scheme. The basic idea in ExOR has triggered a great deal of derivations. However, almost all these derivations are proposed for wireless mesh networks or require the positioning service to support opportunistic data forwarding in mobile ad hoc networks (MANETs). In this chapter, we propose a series of solutions to implement opportunistic data forwarding in more general MANETs, which is called Cooperative Opportunistic Routing in Mobile Ad-hoc Networks (CORMAN). CORMAN includes three following important components. First, a new lightweight proactive source routing (PSR) scheme is proposed to provide source routing information in MANETs for both opportunistic data forwarding and traditional IP forwarding. Second, we analyze and evaluate the topology change with mathematical model and propose large-scale live update to update routing information more quickly with no extra communication overhead.

Third, we propose that small-scale retransmission utilizes the broadcast nature one step further than ExOR, and furthermore, it helps us to enhance the efficiency and robustness of the opportunistic data forwarding in MANETs.

13.1.1 Mobile Ad Hoc Networks (MANETs)

A mobile ad hoc network (MANET) is a wireless communication network, where nodes that are not within direct transmission range of each other will require other nodes to forward data. It can operate without existing infrastructure, supports mobile users, and falls under the general scope of multi-hop wireless networking. Such a networking paradigm was originated from the needs in battlefield communications, emergency operations, search and rescue, and disaster relief operations. Later, it found civilian applications such as community networks. A great deal of research results have been published since its early days in the 1980s [1]. The most salient research challenges in this area include end-to-end data transfer, link access control, security, and providing support for real-time multimedia streaming.

The network layer has received the most attention when working on MANETs. As a result, abundant routing protocols in such a network with differing objectives and for various specific needs have been proposed [2]. In fact, the two most important operations at the network layer, i.e., data forwarding and routing, are distinct concepts. Data forwarding regulates how packets are taken from one link and put on another. Routing determines what path a data packet should follow from the source node to the destination. The latter essentially provides the former with control input.

Routing protocols for MANETs can be classified based on a variety of factors. The most important distinction is the way in which routing information is maintained or updated. On the one hand, a protocol can be proactive, requiring network nodes to always maintain valid routes to all destinations. In this case, the protocol is also known as table driven. Destination-Sequenced Distance Distance-Vector (DSDV) [3] and Optimized Link State Routing (OLSR) [4] are two proactive routing protocols. A reactive protocol, on the other hand, is used when a node receives data from the upper layer for a given destination and must first figure out how to get there if it does not always possess the up-to-date routing information. Also known as on-demand, the reactive protocol is used in Dynamic Source Routing (DSR) [5] and Ad Hoc On-Demand Distance Vector (AODV) [6].

13.1.2 Extremely Opportunistic Routing

Despite the efforts in routing in ad hoc networks, data forwarding, in contrast, follows pretty much the same paradigm as in IP forwarding in the Internet. IP forwarding was originally designed for multi-hop wired networks, where one packet transmission can be received only by nodes attached to the same cable. For the case of modern Ethernet, an IP packet is transmitted at one end of the Ethernet cable and received at the other. However, in wireless networks, when a packet is transmitted over a physical channel, it can be detected by all other nodes within the transmission range on that channel. For the most part of the research history, overhearing a packet not intended for the receiving node had been considered as completely negative, i.e., interference. Thus, the goal of research in wireless networking was to make wireless links as good as wired ones. Unfortunately, this ignores the inherent nature of broadcasting of wireless communication links. For mobile ad hoc networks to truly succeed beyond labs and testbeds, we must tame and utilize its broadcasting nature rather than fight it. Cooperative communication is an effective approach to achieve such a goal.

The concept of cooperative communication was initially put forward by Cover and El Gamal [7], who studied the information theoretic properties of relay channels. More recent progress on this subject started to proliferate in the early 2000s [8]. In cooperative communication at the physical layer, multiple nodes overhearing the same packet may transmit it together as a virtual multiple-antenna transmitter. With enhanced digital signal-processing capabilities on the receiver side, the packet is more likely to be decoded. Yet, little research had been done on cooperative commutation at the link layer and above until ExOR (Extreme Opportunistic Routing) [9]. ExOR is a milestone piece of work in this area, and it is an elegant way to utilize the broadcasting nature of wireless links to achieve cooperative communication at the link and network layers of static multi-hop wireless networks. Therefore, in our research, we further extend the scenarios that the idea behind ExOR can be used, dubbed as *Cooperative Opportunistic Routing in Mobile Ad-hoc Networks (CORMAN)*.

13.2 RELATED WORK AND MOTIVATION

In this part of the chapter, we review recent work in two related fields. We first review opportunistic data forwarding, including its ancestors and derivations, the related math models built for it, and its effectiveness studies. We then review the importance of routing protocols, and hence we will review some routing protocols in the second part of this chapter. After the review of related work, we will highlight the motivation and introduce the framework in our research.

13.2.1 Recent Work Related to Opportunistic Data Forwarding

This part provides an overview of opportunistic data forwarding in multi-hop wireless networks by explaining and comparing existed ones. They may include many trails for different coordinate protocols, metrics, network types and so on. The discussion in this part is organized as follows. Section 13.2.1.1 will present and compare the mechanisms used to realize opportunistic data forwarding. Section 13.2.1.2 will list other derivations in opportunistic data forwarding. Section 13.2.1.3 will review the question of when the opportunistic data forwarding can achieve better performance than traditional IP forwarding in a multi-hop network. Section 13.2.1.4 contains the mathematical models that can help us to analyze the performance of opportunistic data forwarding and understand how to maximize such performances.

13.2.1.1 Protocols Based on Opportunistic Data Forwarding

In recent years, dozens of opportunistic data forwarding protocols have been proposed. Many of them are significant in this area. In this section, we present the challenges in opportunistic data forwarding and how to address them. First, we need to know where the advantages of opportunistic data forwarding comes from. The two most common scenarios that exist in wireless data transmission are shown in Figures 13.1 and 13.2 [9].

In Figure 13.1, the source node *src* transmits a packet to the destination node *dst*. By the topology and link delivery probabilities shown in the figure, we know that the probability of at least one intermediate node receiving the data from *src* is $1 - (1 - 0.1)^{100}$, which is greater than 10%, the probability of a particular node receiving it. Furthermore, any intermediate node can forward the packet. It is one reason why a better performance may be achieved in opportunistic data forwarding.

Figure 13.2 presents another possibility that one transmission may reach a node which is closer to the destination than the particular next hop in traditional IP forwarding. Assume in

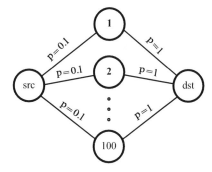

Figure 13.1 Topology and deliver probabilities as an example network.

Figure 13.2 Best node can be used in opportunistic data forwarding.

traditional IP forwarding, the next hop for a packet sent from *A* is *B*, and node *C* overhears the packet by opportunity. In the traditional approach, this is taken as the interference on node *C*, and the overheard packet will be discarded. Meanwhile, node *B* will transmit it to *C* again. What makes it worse is that if node *B* cannot decode the packet successfully, even though node *C* could properly decode the packet, the benefit cannot be utilized in the IP forwarding scheme, and node *A* should transmit the packet again until *B* receives it. Opportunistic data forwarding can postpone the decision to choose the best forwarder which may be far away from the transmitter but receives the packet by a certain opportunity. Note that some situations may diminish the benefits of opportunistic data forwarding because the next hop is not determined and any node receiving the data could be an actual relay. Hence the duplicated transmission from many intermediate nodes should be avoid. If the cooperation protocol cannot guarantee that only one forwarder candidate will forward the data, duplicated transmission would decrease the performance. Moreover, if the coordination mechanism is quite resource consuming, this protocol will be unusable. Therefore, we believe that the coordination protocol, which serves opportunistic data forwarding, is the key point. How to design a coordination protocol for forward candidates will be a dominant issue to make opportunistic data forwarding more efficient and more practical. As far as we know, two protocols for coordination, introduced next, are significant for opportunistic data forwarding.

13.2.1.1.1 Multiple Handshake

Larsson [10] proposes a four-way handshake approach as the coordination protocol in his Selection Diversity Forwarding (SDF). In SDF, if a node has a packet to transmit, it just broadcasts the packet to every neighbor. Then, every neighbor that received the packet successfully will send back an acknowledgement (ACK) with their local information to the transmitter. The transmitter makes a decision based on the ACKs and sends a forwarding order (FO) to the best forwarder candidate. Once the selected relay node receives the FO, it will send the forwarding order ACK (FOA) back to the transmitter and then proceed to data forwarding. This process continues until the final destination is reached. As a piece of pioneering work on opportunistic data forwarding, Larsson has made a significant contribution. However, people realize that two problems exist in Larsson's work. One is that the ACKs and FOA may be lost in the wireless environment, and the loss of either one will lead to unnecessary retransmissions. The other is that such a gossiping mechanism wastes a great deal of resources and introduces further delay. By this consideration, Rozner et al. [11] explored the approaches to make multiple- handshake coordination tolerate ACK loss and reduce the delay. Rozner et al. used selective ACKs to address ACK losses and minimize useless retransmissions, and used piggybacked ACKs and ACK compression to reduce ACK overhead. With selective ACKs, a single ACK loss will not trigger retransmissions because subsequent ACKs for later packets will guarantee that all packets have been received recently. In particular here, piggybacked ACKs are constructed by including acknowledgment information to a data frame, and when the packet is transmitted. The downstream nodes and upstream transmitter should decode the data and ACK information from such a packet separately. Another solution based on multiple-handshake is presented in Kurth et al. [12]. It suggests that a lower data rate can be used to transmit ACKs to enhance their reliability, and it also explores the transmitter diversity, which follows the same idea as receiver diversity. Naghshvar and Javidi [13] also use multiple-handshake to coordinate all forwarder candidates. They propose a new metric, which is the backlogs from all downstream links, to achieve better performance. However, Naghshvar and Javidi assume that the ACKs are loss free.

13.2.1.1.2 Route-Prioritized Contention

Route-prioritized contention is the other type of coordination protocols. In route-prioritized contention, all forwarding candidates will follow a given order to content the wireless medium, and the *best* node that receives the packet should grab the medium first. Therefore, the broadcast nature can be explored.

This approach is proposed by Biswas and Morris [9], called ExOR. ExOR is a milestone for opportunistic data forwarding for three reasons: (1) ExOR uses piggybacked ACK to tell the lower-priority nodes that the packet has been forwarded by higher-priority nodes. (2) The overhead in ExOR is small, containing only the ExOR header and the metric updating information. (3) Although it cannot completely avoid duplications, it can do so to a high degree.

Now, we try to explain this coordinate protocol in detail. The header of ExOR is shown in Figure 13.3. In the figure, *Ver*, *HdrLen*, *PatloadLen*, *Checksum*, and *Payload* are self-explanatory. In particular, to enhance the efficiency, ExOR delivers packets by *batch*, and each batch contains a number of packets. The quantity of batches is recorded by the *BatchSz*

Figure 13.3 ExOR header.

in the header, so the *BatchSz* of every packet in the same batch should be identical. The data transmission is based on batch, which means the source will flush all packets in the same batch into the network. Once all packets in the same batch are received by the destination (in ExOR, when the destination received over 90% of the total packets in the same batch), the next batch should be transmitted. The *Batch ID* in the header identifies which batch the current packet belongs to, and the *PktNum* is the index of the packet in the current batch. *Forwarder List* contains the forwarder candidates which are selected and sorted by source node when a batch is constructed. As a result, the *Forwarder List* should remain the same for all packets in one batch. If a node contained in the *Forwarder List* overhears a packet and the node's priority is higher than the transmitter's, the packet should be buffered in the node. In the forwarding step, to make a higher-priority node send buffered packets first, waiting time is calculated by every node based on its position in the *Forwarder List*. The time may be updated by the *Transmission Tracker,* which records the transmit rate and the quantity of packets that need transmit from current transmitter. To support the calculation, the *FragNum* and *FragSz* are added to the header. Here, a *Fragment* is a series of packets heard by the node with ID *ForwarderNum*. *FragSz* is the quantity of packets that the current node should transmit when its turn comes, and *FragNum* is the index of this packet in the current *Fragment*. When a node receives a packet for the first time, the *Batch Map* maintained by the node should be updated. For every packet, the highest priority forwarder that has received it should be recorded in the *Batch Map*. A copy of this *Batch Map*, which is maintained on the node, will be added to the header of the packet to be transmitted. Hence, when lower-priority nodes overhear a packet from a higher-priority node, they will continue to merge the *Batch Map* maintained by themselves. At the same time, they will look up the *Batch ID* and *PktNum* in their buffers, and once a same packet is found, the transmission of this packet would be canceled. By looking up the *Batch Map,* many duplications can be avoided. All necessary information for coordinating forwarder candidates is integrated in the ExOR header, thus no time is wasted for gossiping between nodes.

Many papers were published after ExOR, such as Zeng et al. [14], Yang et al. [15], and Zhong et al. [16], and they use nearly the same coordinate protocol to make trails on other directions. However, ExOR has its disadvantage. Because the forwarding timer is always initialized according to the node's priority in the *Forwarder List,* the nodes far from destination will always wait for a long time. It quite constrains us from exploring the spatial reuse in the multi-hop network. Furthermore, ExOR is quite suitable for unicast, but in multicast, it may not perform well, Chachulski et al. [17]. At last, piggybacked ACKs may lose, so duplicated transmission may happen.

It is worth mentioning that Chachulski et al. [17] proposed MORE (Mac-independent opportunistic routing) which is an opportunistic forwarding approach with spatial reuse, and it also belongs to the Route-Prioritized Contention scope. As far as we know, it is the first paper that uses network coding to realize opportunistic data forwarding. Its motivation for using network coding is that it wants to explore the spatial reuse in ExOR, without duplicated transmission. Recall that in ExOR, the *Forwarder List* is generated by the source node, and the nodes which are far from the destination will wait for a long time, even though some of these nodes can transmit (or receive) some packets to (or from) some other nodes without much interference on the forwarding taking place far away. The reason why ExOR cannot operate like a pipeline is that for each packet we cannot tell how far it has already been transmitted in the last hop. To enable pipeline opportunistic data forwarding without unnecessary data transmission, the network coding is used by Chachulski et al.

In MORE, some predefined number of packets compose a batch as that in ExOR. All packets in the same batch will be encoded by linear network coding before sending out from the source node. The source node will calculate a sorted *Forwarder List* by the routing metric of expected transmission count (ETX) [18], which is also quite similar to ExOR. When a node in the *Forwarder List* receives a packet of a batch, and if this packet is linear independent from all received packets in the same batch, the packet should be forwarded. The exact time to forward the packet depends on two factors: one factor is the position of the forwarder in the *Forwarder List*, and the other factor depends on the 802.11 MAC. The difference from ExOR is that the time used to wait for transmission can be shortened, whereas if the timer expires and the medium is free, the linear encoded packet can be transmitted.

13.2.1.2 Other Variants of Opportunistic Forwarding

Besides the contributions on coordination protocol for opportunistic data forwarding, several works make many other trials by changing the network types or metrics, or combining with other technologies to help opportunistic data forwarding make decisions.

13.2.1.2.1 Network Types

A multi-hop wireless network is like a family which is composed by mesh networks, mobile ad hoc networks, sensor networks, and vehicular networks. Each network can borrow the idea of opportunistic data forwarding from mesh networks to promote theirs performance. By following this way, Ma et al. [19] uses opportunistic forwarding in sensor networks to enhance the probability that a data packet is forwarded successfully. Hence, more sensor nodes can remain in the sleep mode for a longer time to save energy without influencing performance. Vehicular networks can also explore the benefits of opportunistic data forwarding, and usually they use the positioning service to get the topology of the vehicular networks and select the *best* forwarder by the perspective of distance, like Leontiadis and Mascolo [20].

13.2.1.2.2 Metrics

In ExOR, ETX is used as the metric to evaluate which candidate is better. However, numbers of metrics have been proposed recently, such as distance, expected any-path transmissions, and backlog. Distance is quite suitable for position-based opportunistic forwarding because it is quite easily calculated by positioning service and has relatively good performance in Leontiadis and Mascolo [20] and in Yang et al. [15]. Expected any-path transmissions (EAX) is proposed in Zhong et al. [16], which are equal to the expected number of transmissions required to deliver a packet to its destination under the perspective of opportunistic forwarding.

13.2.1.2.3 Network Coding-Based Opportunistic Forwarding

Network coding-based opportunistic data forwarding was mentioned in the previous section, which usually uses the linear network coding method to encode the received or original packets in a series of random linear combinations. The exploration on this includes Chachulski et al. [17], Radunović et al. [21], and Koutsonikolas et al. [22]. The coordination protocols will not work as aggressively as those in ExOR. Usually, ACKs should be sent back when the destination receives enough information to decode all packets in the batch, and the source node will stop sending packets or go on sending the packets in the next batch when the ACK for the current batch is received.

13.2.1.2.4 Position-Based Opportunistic Forwarding

Position based opportunistic forwarding is explored in Yang et al. [15]. The metric is the distance from any forward candidates to the final destination, and the coordinate protocol is quite similar to that in ExOR. The node which is closer to the destination will access the medium earlier to forward buffered packets.

13.2.1.3 Scenarios Suitable for Opportunistic Forwarding

In the previous sections, we have presented the advantages of opportunistic data forwarding and the challenges in it. Both Shah et al. [23] and Kim et al. [24] make a comparison between traditional IP forwarding and opportunistic data forwarding. They believe that the opportunistic data forwarding is suitable for the networks whose mobility is not high, and the higher the node density, the better performance that can be achieved. The reason to explain this is that, if the mobility is high, the forwarder candidates cannot always keep in communication range with each other, so a packet may be transmitted by two or more forwarders several times. The duplication becomes the most important reason for the decreases in the benefits we get from opportunistic data forwarding, and the performance gain by opportunistic forwarding will be marginal or even negative because the coordination overhead always exists in coordination protocols.

13.2.1.4 Performance Modeling for Opportunistic Forwarding

Many performance analyses and optimization works also contributed to the study of opportunistic data forwarding. These works run in two directions. One direction is the building of models to analyze opportunistic data forwarding, and the other is using existing models to enhance the performance in opportunistic data forwarding.

13.2.1.4.1 Performance Optimization

No matter what kind of data transmission protocols people choose to use, they want to optimize the network's performance. Moreover, if a boundary of performance can be proved, it will be quite useful. Two performance optimization works have been done by Radunović et al. [21] and Zeng et al. [25]. In opportunistic data forwarding, many forwarder candidates can overhear a same packet, but a different forwarding decision will affect the performance significantly. Hence, Radunović et al. uses an optimal flow-decision to maximize the links' utility of the whole network. A comprehensive model built by Radunović et al. gives us the relationship between the links' utility and flow-decision set. To maximize the utility, Radunović et al. handle the optimization problem with two steps. First, they propose transport credits, which are used to denote the quantity of packets that have been sent out. Second, for each possible flow on each node, Radunović et al. studied the relationship between the flow's transport credits and a three-tuple composed by *scheduling strategy*, *power control*, and *transmit rate control*. Hence, the transport credits for a particular node can be presented by the three-tuple, so the summation of all nodes' transport credits in the network can be related with the three-tuple as well. When we take the three-tuple as the varying variable, the optimized network utility is achieved by selecting the tuples that give us the maximum value of transport credits.

Zeng et al. use *"Transmitter Conflict Graph"* rather than *"Link Conflict Graph"* to find what the lower and upper bounds are of throughput in the opportunistic forwarding-based multi-hop wireless networks. Three important concepts are introduced in Zeng et al. [25]. They are Concurrent Transmitter Sets (CTS), Conservative CTS (CCTS), and Greedy CTS (GCTS). CTS is self-explained, and CCTS is one extreme case in CTS that all links associated with any node in the CCTS can be used simultaneously. GCTS is another case in CTS that at least one link of any node in the node set can be used concurrently. Hence, the lower bound and the upper bound can be calculated in GCTS and CCTS separately. The lower bound can be calculated in the protocol model with linear programming, and the upper bound can be figured out in the physical model because of the fact that SNR will decrease when more concurrent links are used.

13.2.1.4.2 Modeling from Markov Process and Game Theory

Cerdà-Alabern et al. [26] and Lu et al. [27] are two typical works based on the Markov Process. The idea lies on the packet-received probability between every node pair in a network that can be estimated, either by radio's propagation model or by exchanging the probe messages periodically. Moreover, in opportunistic forwarding, the node with the smallest metric should forward the data first, so we can predict the performances statistically, such as hop count, transmission count, and even energy consumption. In Lu et al. [27], the Markov Process is constructed with the states which are the combinations of different events, and the probability distribution of the ETX for a node to deliver a packet is proposed. However, Cerda-Alabern et al. [26] use just the forwarder candidates to be the states in Markov Process. The receive probability between every pair of nodes is estimated by the radio propagation model. Therefore, the expectations of hop count and throughput could be found.

Game Theory is used in Wu et al. [28], and the purpose of their work is to try to build a mechanism by which the rewards and punishments are given to each node to make it honestly announce its link quality and help other nodes forward packets. The reason to do such work is that nodes in opportunistic forwarding may be selfish, and such a behavior is unfair to other nodes and degrades the total performance. Wu et al. [28] use the Strict Dominant Strategy Equilibrium to design a mechanism in which nodes will get the most benefit if and only if they honestly report link quality and do their best to forward packets for others.

13.2.2 Routing Algorithms Review for Opportunistic Data Forwarding in MANETs

In this section, we will briefly review the existing routing protocols. The routing protocol is important for our CORMAN because we want to utilize the Route-Prioritized Contention (Section 13.2.1.1.2) approach to equip mobile ad hoc networks (MANETs) with the ability of opportunistic data forwarding. This thesis aims to present CORMAN, a series of solutions of opportunistic data forwarding in MANETs, so we will look at whether the existing routing protocols could support opportunistic data forwarding and how they would perform if they could.

13.2.2.1 Timing Strategy in Routing Protocols

Basically, routing protocols in MANETs can be categorized using an array of criteria. The most fundamental among these is the timing of routing information exchange. On one hand, a protocol may require that nodes in the network should maintain valid routes to all destinations all the time. In this case, the protocol is considered to be *proactive*, a.k.a. *table driven*. Examples of proactive routing protocols include Destination-Sequenced Distance-Vector (DSDV) [3] and Optimized Link State Routing (OLSR) [4]. On the other hand, if nodes in the network do not always maintain routing information, when a node receives data from the upper layer for a given destination, it must first find out how to reach the destination. This approach is called *reactive*, a.k.a. *on demand*. Dynamic Source Routing (DSR) [5] and Ad Hoc On-Demand Distance Vector (AODV) [6] fall in this category.

13.2.2.2 Fundamental Algorithms – Link State and Distance Vector

These well-known routing schemes can also be categorized by their fundamental algorithms. The most important algorithms in routing protocols are Distance Vector (DV) and Link State (LS) algorithms. In LS, every node will share its best knowledge of links with other nodes in the network, so nodes can reconstruct the topology of the entire network locally, e.g., OLSR. In DV, a node only provides its neighbors with the cost to every given destination, so nodes know the costs to a given destination via different neighbors as the next hop, e.g., DSDV and AODV. DSR is an early source routing protocol in MANETs. In DSR, when a node has data for a destination node, a route request is flooded to all nodes in the network. An intermediate node that has received the request adds itself in the request and rebroadcasts it. When the destination node receives the request, it transmits a route reply packet to send the discovered route back to the source node via the route in request packet reversely, and the source node will use the discovered route in data transmission.

13.2.2.3 Tree-Based Routing Protocols Derived from the Internet

As the scalability of the Internet also suffers from the overhead problem, many lightweight routing protocols for the Internet have been proposed. The pioneer work by Garcia-Luna-Aceves and Murthy [29] proposed a new routing algorithm called Loop-Free Path-Finding Algorithm (LPA), based on the path-finding algorithm (PFA). To avoid loops, LPA proposed a Feasibility Condition (FC). The FC can effectively avoid the temporary loop but may overkill some possible valid paths in the network. Another piece of work, by Behrens and Garcia-Luna-Aceves [30], proposed a new routing algorithm called Link Vector (LV). In LV, the basic element in the update message contains the destination, the precursor to destination, the cost between them, and a sequence number to avoid loops. LV is different from LS, because in LS, an updater sends all the link information it knows by flooding; but in LSA, an updater only sends the information of links which are preferred. So, the node in LV can construct only a partial topology of the entire network. An algorithm proposed by Levchenko et al. [31] called *approximate link state* (XL) conclusively summarizes the LV and PFA in a nutshell. It uses *soundness* and *completeness* to define the correctness of the routing protocols, and it uses *stretch* to evaluate the optimality. XL also uses the *lazy* update concept to further decrease the overhead. Hence, in XL, a node sends an update message that contains only the changed links which are on the only way that must be passed to a given destination, or the updated link can improve the cost to a destination node by a given parameter.

Similar trails to reduce wireless routing overhead have been made during the development listed previously. Murthy and Garcia-Luna-Aceves proposed the Wireless Routing Protocol (WRP) [32], which utilizes the basic idea of PFA within wireless networks. Every node in WRP has a tree structure for the network, and every time the node sends out only the routing update message by differential update. However, WRP requires the receiver of an update message to transmit the ACK. Such a requirement introduces more overhead and consumes more channel resources, but it may not improve the performance dramatically. Furthermore, WRP uses *RouterDeadInterval* to detect the loss of neighbors which may adversely affect the response time of topology change, considering only one route is maintained in the tree structure. An extensional work based on WRP is Source Tree Adaptive Routing (STAR) [33], which is proposed by Garcia-Luna-Aceves and Spohn. In STAR, every node maintains a tree structure for the network and adopts a tree update strategy that is neither proactive nor reactive. Instead, it uses a lazy approach, where an update message will be transmitted only when the local tree structure is considered sufficiently inferior to the original optimum.

13.2.2.4 Suitability of Existing Routing Protocols for Opportunistic Data Forwarding in MANETs

In fact, none of the previous listed protocols can ideally support opportunistic data forwarding in MANETs. In particular, AODV [6], DSDV [3], and other DV-based routing algorithms were not designed for source routing and hence are not suitable for opportunistic data forwarding. The reason is that every node in these protocols knows only the next hop to reach a given destination node, not the complete path. OLSR [4] and other LS-based routing protocols could support source routing, but their overhead is still fairly high for the load-sensitive MANETs. DSR together with its derivations, such as [34], [35], and [36], are not suitable because they have a long bootstrap delay and are therefore not efficacious for frequent data exchange. Furthermore, the reactive routing protocols will inject too much route request packets in the mobile networks, especially when there are a large number of data sources. Moreover, the route reply message may be lost since it is sent based on IP forwarding via recorded route reversely, so reactive routing schemes suffer from even more unpredictable delay in data transmission. The WRP [32] and STAR [33] – the early attempts to port the routing capabilities in linking state routing protocols to MANETs – are built on the framework of PFA for each node to use a tree for loop-free routing. Although WRP is an innovative exploration in the research on MANETs and STAR uses a "lazy" routing strategy to reduce the topology update burden, they have a very high operation overhead due to the routing procedures being triggered by the event of topology changes, which introduce a great deal of information exchanged and stored by the nodes. Our intention was to include WRP in our experimental comparison later in this chapter, and we have implemented WRP in NS-2. Unfortunately, our preliminary tests indicate that its communication overhead is at least an order of magnitude higher than the other main-stream protocols.

In a nutshell, designing a new lightweight proactive source routing protocol is a very important component to develop opportunistic data forwarding in MANETs.

13.2.3 Motivation and Framework

When we review the related work, we can see that a systematic solution to equip MANETs requires the ability of opportunistic data forwarding. Hence, in our research, we propose CORMAN as a network layer solution to implement opportunistic data forwarding in MANETs. Its node coordination mechanism is largely inline with that of ExOR, and it is an extension to ExOR in order to accommodate node mobility. Here, we first highlight our objectives and challenges in order to achieve them. Later in this section, we provide a general description of CORMAN. The details of the major components of CORMAN are presented in Sections 13.3 (proactive source routing), 13.4 (large-scale live update), and 13.5 (small-scale retransmission).

13.2.3.1 Objectives and Challenges

CORMAN has two objectives: (1) it broadens the applicability of ExOR to mobile multi-hop wireless networks without relying on external information sources, such as node positions; and (2) it incurs a smaller overhead than ExOR by including shorter forwarder lists in data packets.

The following challenges are thus immanent.

1. *Overhead in route calculation*: CORMAN relies on the assumption that every source node has complete knowledge of how to forward data packets to any node in the network at any time. This calls for a proactive source routing protocol. Link state routing,

such as OLSR, would meet our needs, but it is fairly expensive in terms of communication costs. Therefore, we need a lightweight solution to reduce the overhead in route calculation.

2. *Forwarder list adaptation*: When the forward list is constructed and installed in a data packet, the source node has updated knowledge of the network structure within its proximity, but its knowledge about further areas of the network can be obsolete due to node mobility. This becomes worse as the data packet is forwarded towards the destination node. To address this issue, intermediate nodes should be able to update the forwarder list adaptively with their new knowledge when forwarding data packets.

3. *Robustness against link quality variation*: When used in a dynamic environment, a mobile ad hoc network must inevitably face the drastic link quality fluctuation. A short forwarder list carried by data packets implies that they tend to take long and possibly weak links. This could be problematic for opportunistic data transfer, since the list may not contain enough redundancy in selecting intermediate nodes. This should be overcome with little additional overhead.

13.2.3.2 CORMAN Fundamentals

CORMAN forwards data in a similar batch-operated fashion as ExOR. A flow of data packets are divided into batches. All packets in the same batch carry the same forwarder list when they leave the source node. To support CORMAN, we have an underlying routing protocol, *Proactive Source Routing* (PSR), which provides each node with the complete routing information to all other nodes in the network. Thus, the forwarder list contains the identities of the nodes on the path from the source node to the destination. As packets progress in the network, the nodes listed as forwarders can modify the forwarder list if the network topology has observed any changes. This is referred to as *large-scale live update* in our work. In addition, we also allow some other nodes that are not listed as forwarders retransmit data if this turns out to be helpful, referred to as *small-scale retransmission*. Note that CORMAN is a complete network layer solution and can be built upon off-the-shelf IEEE 802.11 networking commodities without any modification.

Therefore, the design of CORMAN has the following three modules. Each module answers to one of the challenges stated previously.

1. *Proactive source routing*: PSR runs in the background so that nodes periodically exchange network structure information. It converges after the number of iterations equal to the network diameter. At this point, each node has a spanning tree of the network, indicating the shortest paths to all other nodes. The amount of information broadcast by each node in an iteration is $O(n)$, where n is the number of nodes in the network. Such an overhead is the same as distance vector algorithms and is much smaller than link state. Technically, PSR can be used without CORMAN to support conventional IP forwarding.

2. *Large-scale live update*: When data packets are received by and stored at a forwarding node, the node may have a different view of how to forward them to the destination from the forwarder list carried by the packets. Since this node is closer to the destination than the source node, such discrepancy usually means that the forwarding node has more updated routing information. In this case, the forwarding node updates the part of the forwarder list in the packets from this point on towards the destination according to its own knowledge. When the packets with this updated forwarder list are broadcast by the forwarder, the update of network topology change propagates back to the upstream neighbor. The neighbor incorporates the change to the packets in its

cache. When these cached packets are broadcast later, the update is further propagated towards the source node. Such an update procedure is significantly faster than the rate at which a proactive routing protocol disseminates routing information.

3. *Small-scale retransmission*: A short forwarder list forces packets to be forwarded over long and possibly weak links. To increase the reliability of data forwarding between two listed forwarders, CORMAN allows nodes that are not on the forwarder list but are situated between these two listed forwarders to retransmit data packets if the downstream forwarder has not received these packets successfully. Since there may be multiple such nodes between a given pair of listed forwarders, CORMAN coordinates retransmission attempts among them extremely efficiently.

13.3 PROACTIVE SOURCE ROUTING – PSR

Before describing the details of PSR, we will first review some graph theoretic terms used here. Let us model the network as an undirected graph $G = (V, E)$, where V is the set of nodes (or vertices) in the network and E is the set of wireless links (or edges). Two nodes u and v are connected by an edge $e = (u, v) \in E$ if they are close to each other and can communicate directly with a given reliability. Given a node v, we use $N(v)$ to denote its open neighborhood, i.e., $\{u \in V | (u, v) \in E\}$. Similarly, we use $N[v]$ to denote its closed neighborhood, i.e., $N(v) \cup \{v\}$. In general, for a node v, we use $N_l(v)$ $(l \geq 1)$ to denote the distance-l open neighborhood of v, i.e., the set of nodes that are exactly l hops away from v. Similarly, $N_l[v]$ $(l \geq 1)$ denotes the distance-l closed neighborhood of v, i.e., the set of nodes that are within l hops of v. As special cases, $N_1[v] = N[v]$, $N_1(v) = N(v)$, $N_0[v]$ is v itself, and $N_0(v) = \emptyset$. Also as a convention in graph theory, for any $S \subseteq V$, we use $\langle S \rangle$ to denote the subgraph induced by S, i.e., $\langle S \rangle = (S, E')$, where E' is the set of edges where each element has both endpoints in S. The readers should refer to West [37] for other graph theoretic notions and other details.

The research findings from CORMAN entirely have been published to the *IEEE Journal on Selected Areas in Communications* [38]. The research findings from PSR and its early refinement have been published to *IEEE GLOBECOM '11* [39] and *IEEE Communications Letters* [40]. PSR's last improvement was submitted to *IEEE INFOCOM '12* [41]

13.3.1 Design of PSR

Essentially, PSR provides every node with a Breadth First Spanning Tree (BFST) of the entire network rooted at itself. To do that, nodes periodically broadcast the tree structure to its best knowledge in each iteration. Based on the information collected from neighbors during the most recent iteration, a node can expand and refresh its knowledge about the network topology by constructing a deeper and more recent BFST. This knowledge will be distributed to its neighbors in the next round of operation (Section 13.3.1.1). On the other hand, when a neighbor is deemed lost, a procedure is triggered to remove its relevant information from the topology repository maintained by the detecting node (Section 13.3.1.2). Intuitively, PSR has about the same communication overhead as distance-vector-based protocols. We go the extra mile to reduce the communication overhead incurred by PSR's routing agent. Details about such overhead reduction will be discussed in Section 13.3.1.3.

13.3.1.1 Route Update

Due to its proactive nature, the update operation of PSR is iterative and distributed among all nodes in the network. At the beginning, a node v is aware only of the existence of itself, so there is only a single node in its BFST, which is the root node v. By exchanging the BFSTs with the neighbors, it is able to construct a BFST within $N[v]$, i.e., the star graph centered at v, denoted by S_v.

In each subsequent iteration, nodes exchange their spanning trees with their neighbors. From the perspective of node v, towards the end of each operation interval, it has received a set of routing messages from its neighbors packaging the BFSTs. Note that, in fact, more nodes may be situated within the transmission range of v, but their periodic updates were not received by v due to, say, bad channel conditions. After all, the definition of a neighbor in MANETs is a fickle one. (We have more details on how we handle lost neighbors subsequently.) Node v incorporates the most recent information from each neighbor to update its own BFST. It then broadcasts this tree to its neighbors at the end of the period. Formally, v has received the BFSTs from some of its neighbors. Including those from whom v has received updates in recent previous iterations, node v has a BFST, denoted T_u, cached for each neighbor $u \in N(v)$. Node v constructs a union graph

$$G_v = S_v \cup \bigcup_{u \in N(v)} (T_u - v). \tag{13.1}$$

Here, we use $T - x$ to denote the operation of removing the subtree of T rooted at node x. As special cases, $T - x = T$ if x is not in T, and $T - x = \emptyset$ if x is the root of T. Then, node v calculates a BFST of G_v, denoted T_v, and places T_v in a routing packet to broadcast to its neighbors.

In fact, in our implementation, the aforementioned update of the BFST happens multiple times within a single update interval so that a node can incorporate new route information to its knowledge base more quickly. To the extreme, T_v is modified every time a new tree is received from a neighbor. Apparently, this is a trade-off between the routing agent's adaptivity to network changes and computational cost. Here, we choose routing adaptivity as a higher priority assuming that the nodes are becoming increasingly powerful in packet processing. Nevertheless, this does not increase the communication overhead at all because one routing message is always sent per update interval.

Assume that the network diameter, i.e., the maximum pairwise distance, is D hops. After D iterations of operation, each node in the network has constructed a BFST of the entire network rooted at itself. This information can be used for any source routing protocol. The amount of information that each node broadcasts in an iteration is bounded by $O(|V|)$ and the algorithm converges in at most D iterations.

13.3.1.2 Neighborhood Trimming

The periodically broadcast routing messages in PSR also double as "Hello" messages for a node to identify which other nodes are its neighbors. When a neighbor is deemed lost, its contribution to the network connectivity should be removed, called "neighbor trimming". Considering node v, the neighbor trimming procedure is triggered at v about neighbor u when

- no routing update or data packet has been received from this neighbor for a given period of time, or
- a data transmission to node u has failed as reported by the link layer.

Node v responds by

1. first, updating $N(v)$ with $N(v) - \{u\}$;
2. next, constructing the union graph with the information of u removed

$$G_v = S_v \cup \bigcup_{w \in N(v)} (T_w - v), \tag{13.2}$$

3. then, computing the BFST T_v.

Notice that T_v thus calculated is not broadcast immediately to avoid excessive messaging. With this updated BFST at v, it is able to avoid sending data packets via lost neighbors. Thus, multiple such neighbor trimming procedures may be triggered within one period.

13.3.1.3 Streamlined Differential Update

In addition to dubbing route updates as hello messages in PSR, we interleave the "full dump" routing messages as stated previously with "differential updates". The basic idea is to send the full update messages less frequently than shorter messages, containing the difference between the current and previous knowledge of a node's routing module. Both the benefit of such an approach and how to balance between these two types of messages have been studied extensively in earlier proactive routing protocols. In this work, we further streamline the routing update in two new avenues. First, we use a compact tree representation in full dump and differential update messages to halve the size of these messages. Second, every node attempts to maintain a "stable" BFST as the network changes, so that the differential update messages are even shorter.

- *Compact tree representation*: For the full dump messages, our goal is to broadcast the BFST information stored at a node to its neighbors in a short packet. To do that, we first convert the general rooted tree into a binary tree of the same size, say s nodes. Then we serialize the binary tree using a bit sequence of $34 \times s$ bits, where the IPv4 (Internet Protocol version 4) is assumed. Specifically, we scan the binary tree layer by layer. When processing a node, we first include its IP address in the sequence. In addition, we append two more bits to indicate if it has the left and/or right child. For example, the binary tree in Figure 13.4 is represented as A10B11C11D10E00F00G11H00I00. As such, the size of the update message is a bit over half compared to the traditional approach, where the message contains a discrete set of edges.

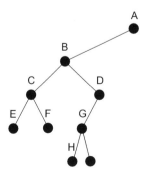

Figure 13.4 Binary tree.

The difference between two BFSTs can be represented by the set of nodes who have changed parents, which are essentially a set of edges connecting to the new parents. We observe that these edges are often clustered in groups. That is, many of them form a sizable tree subgraph of the network. Similar to the case of full dump, rather than using a set of loose edges, we use a tree to package the edges connected to each other. As a result, a differential update message usually contains a few small trees, and its size is noticeably shorter.

- *Stable BFST*: The size of a differential update is determined by how many edges it includes. Since there can be a large number of BFSTs rooted at a given node of the same graph, we need to alter the BFST maintained by a node as little as possible when changes are detected. To do that, we modify the computation described earlier in this section, such that a small portion of the tree needs to change either when a neighbor is lost or when it reports a new tree.

Consider node v and its BFST T_v. When it receives an update from neighbor u, denoted by T_u, it first removes the subtree of T_v rooted at u. Then it incorporates the edges of T_u for a new BFST. Note that the BFST of $(T_v - u) \cup T_u$ may not contain all necessary edges for v to reach every other node. Therefore, we still need to construct the union graph

$$\left(T_v - u\right) \cup \bigcup_{w \in N(v)} \left(T_w - v\right), \tag{13.3}$$

before calculating its BFST. To minimize the alteration of the tree, we add one edge of $(T_w - v)$ to $(T_v - u)$ at a time. During this process, when there is a tie, we always try to add edges that were originally removed from T_v.

When node v thinks that a neighbor u is lost, it deletes the edge (u,v) but still utilizes the network structure information contributed by u earlier. That is, even if it has moved away from v, node u may still be within the range of one of v's neighbors. As such, T_v should be updated to a BFST of

$$\left(T_v - u\right) \cup \left(T_u - v\right) \cup \bigcup_{w \in N(v)} \left(T_w - v\right). \tag{13.4}$$

Note that, since $N(v)$ no longer contains u, we need to explicitly put it back into the equation. Similarly in this case, edges of $(T_u - v) \cup \bigcup_{w \in N(v)} (T_w - v)$ are added to $(T_v - u)$ one at a time, with those just removed because of u taking priority.

13.3.2 Implementation

As the implementation contains many algorithms, here we can only introduce some important ones. Before looking at the algorithms, we summarize the notations we used for these algorithms in Table 13.1.

13.3.2.1 Routing and Neighborhood Update Algorithm

As the operations in routing update and neighborhood update are similar, we take the routing update as an example to introduce their implantation. Equation 13.3 theoretically enables us to implement the routing update algorithm. However, direct implementation needs a high algorithm complexity, and, according to the procedure in Equation 13.3, every time a node v receives an update from neighbor u_k, we have to check every node in subtrees $T_w - v$ ($w \in N(v)$). That is a time-consuming job in both reality and simulation. In fact,

Table 13.1 Notations in algorithms.

Notation	Explanation
stack ⇑ [v]	Pop the top of *stack* and store to v, v is optional
stack ⇓ v	Push v on the top of *stack*
queue ⇑ [v]	Get an element from FIFO *queue* and keep in optional v
queue ⇓ v	Put v into *queue* in FIFO manner
R(T)	Get the root node of tree T
$B_{l/r}(n)$	The left/right brother node of n
$L_T(n)$	Look up node with same ID of n from tree T
$G_{s/b/l}(n)$	Create a spanning/binary tree node or linear element of n
$A_T(n^p, n^c)$	Add a new node n^c as a child of n^p in T
$U_T(n^p, n^c)$	Update n^p as the new parent of n^c in tree T
$H_T(n)$	Hops from node n to its root node in tree T
$C_x(n)$	The xth child of node n in spanning tree
$C_{l/r}(n)$	The left/right child of node n in binary tree
P(n)	The parent of node n
$c_l(e)$	The left tag of a linear element e
$c_r(e)$	The right tag of a linear element e
$a \cup e$	append an element e at the end of an array a

from Section 13.3.1.1 we can see that the shorter path is always selected, and if two paths with same hops are found in the union graph, we always keep the original one. Hence, in our implementation, we look up every node in trees $T_w - v$ ($w \in N(v)$) from the BFST being constructed. If a duplicated node is found and the new discovered path is shorter, the parent of the node should be changed. Therefore, the incorporating work for every subtree $T_w - v$ ($w \in N(v)$) can be finished by Algorithm 13.1.

The routing update for neighbor u_k would be finished if we run Algorithm 13.1 for all subtrees in the last parameter of Equation 13.3. Hence the implementation of neighborhood update would be finished by the same way, and the only difference is the initial T_v.

13.3.2.2 Algorithms for Transformation of Tree Structures

In general, each node maintains the topology by BFST as we introduced before, which is quite suitable to provide a route to a destination node. When a routing update packet is needed, either the entire spanning tree in a full dump update or the forest in a differential update should be presented in linear form. In fact, a single tree can be taken as a forest with one member. As we talked before, we first convert a spanning forest to a binary forest, and then convert the binary forest into its linear form. When a node receives the update, the receiver should restore the binary form of a linear expression and then transform the binary forest to a spanning forest. Depending on which kind of update is used, we can choose how to deal with the spanning forest. If the update is a full dump one, there must be only one tree in the spanning forest, and we direct replace the original BFST cached for such a neighbor by the new one. Otherwise, we should update the original BFST according to the spanning forest as introduced in Section 13.3.2.3. In this part we talk only about the algorithms related with the transformation of tree structures. In a nutshell, we have five algorithms to help us finish all these work:

ALGORITHM 13.1   INCORPORATE $T_{u_i} - v\left(u_i \in N(v)\right)$ **TO** T_v

$stack \Downarrow R(T_{u_i} - v)$
  loop
    $stack \Uparrow n_{u_i}^c$
    if $n_{u_i}^c$ is null **then**
      if $stack$ is not null **then**
        $stack \Uparrow n_{u_i}^c;\ stack \Downarrow B_r\!\left(n_{u_i}^c\right)$
      else
        return;
      end if
    else
      $n_v^c \Leftarrow L_{T_v}\!\left(n_{u_i}^c\right)$
      if $stack$ is null **then**
        $n_v^p \Leftarrow R(T_v)$
      else
        $stack \Uparrow n_{u_i}^p;\ n_v^p \Leftarrow L_{T_v}\!\left(n_{u_i}^p\right);\ stack \Downarrow n_{u_i}^p$
      end if
      if n_v^c is null **then**
        $n_v^c \Leftarrow G_s\!\left(n_{u_i}^c\right);\ A_{T_v}\!\left(n_v^p, n_v^c\right)$
      else if $H_{T_v}\!\left(n_v^c\right) > H_{T_{u_i} - v}\!\left(n_{u_i}^c\right) + 1$ **then**
        $U_{T_v}\!\left(n_v^p, n_v^c\right)$
      end if
      $stack \Downarrow n_{u_i}^c$
      if $n_{u_i}^c$ is leaf node **then**
        $stack \Downarrow$ null
      else
        $stack \Downarrow C_0\!\left(n_{u_i}^c\right)$
      end if
    end if
  end loop

- Algorithm to convert spanning tree to binary tree – Algorithm 13.2.
- Algorithm to convert binary tree to spanning tree – Algorithm 13.3.
- Algorithm to convert binary tree to linear tree – Algorithm 13.4.
- Algorithm to convert linear tree to binary tree – Algorithm 13.5.
- The last one is the ancillary algorithm to separate the linear-forest to a set of linear-trees – Algorithm 13.6.

13.3.2.3 Implementation of Reconstruction in Differential Update

In this part, we will cover the algorithm used on the receiver side to reconstruct the neighbor's BFST in a differential update. The implementation of setting up a differential update can be more easily introduced after talking about Algorithms 13.1, 13.4, 13.5, and 13.6.

By operating all four of these algorithms, the linear-forest in a differential update can be separated into a set of linear-trees, and each linear-tree can be converted to a binary tree and further to a spanning tree. As we specified before, every edge in a differential spanning tree indicates a new parent in a neighbor's BFST; hence, the differential update can be easily handled by incorporating all spanning trees in the local cached one. The incorporation process is similar with Algorithm 13.1; however, there are two differences.

1. We do not check the hop count of path anymore. Instead, we replace all edges with the new ones indicated by the differential spanning tree.
2. A spanning tree with root ID 255.255.255.255 is a spatial case in the differential update. All nodes in such spanning tree should be trimmed from the BFST of the neighbor.

ALGORITHM 13.2   CONVERT SPANNING TREE T_v TO BINARY TREE $T_{v'}$

Initialize $T_{v'}$ with node v as root
  $stack \Downarrow R(T_v)$
  $w_v^c \Leftarrow R(T_v)$
  loop
    $stack \Uparrow n_v^c$
    if n_v^c is null **then**
      if $stack$ is not null **then**
        $stack \Uparrow n_v^c$
        for $i = 1$ to $N_c(n_v^c)$ **do**
          $w_{v'}^c \Leftarrow P(w_{v'}^c)$
        end for
        $n_v^c \Leftarrow B_r(n_v^c)$
        if n_v^c is not null **then**
          $n_{v'}^c \Leftarrow G_b(n_v^c); P(n_{v'}^c) \Leftarrow w_{v'}^c; C_r(w_{v'}^c) \Leftarrow n_{v'}^c; w_{v'}^c \Leftarrow n_{v'}^c$
        end if
        $stack \Downarrow n_v^c$
      else
        return;
      end if
    else if n_v^c is leaf node **then**
      $stack \Downarrow$ null
    else
      $n_v^c \Leftarrow C_0(n_v^c); n_{v'}^c \Leftarrow G_b(n_v^c); P(n_{v'}^c) \Leftarrow w_{v'}^c; C_l(w_{v'}^c) \Leftarrow n_{v'}^c; w_{v'}^c \Leftarrow n_{v'}^c;$
      $stack \Downarrow n_v^c$
    end if
  end loop

ALGORITHM 13.3   CONVERT BINARY TREE T_v TO SPANNING TREE $T_{v'}$

Initialize $T_{v'}$ with node v as root

Initialize $tag \Leftarrow 0$ $(tag \in \{0, 1, 2\})$

$stack \Downarrow R(T_v)$

$w_{v'}^{c} \Leftarrow R\left(T_{v'}\right)$

if $w_{v'}^{c}$ is leaf node **then**

  return;

else

  loop

      $stack \Uparrow n_{v}^{c}$

      if n_{v}^{c} is null **then**

        if $tag = 0$ **then**

          $stack \Uparrow$; $tag \Leftarrow 1$

        else if $tag = 1$ **then**

          $stack \Uparrow$; $tag \Leftarrow 2$

        else if $tag = 2$ **then**

          return;

        end if

      else

        if $tag = 0$ **then**

          if $C_{l}\left(n_{v}^{c}\right)$ is not null **then**

            $stack \Downarrow n_{v}^{c}$; $stack \Downarrow C_{l}\left(n_{v}^{c}\right)$; $n_{v'}^{c} \Leftarrow G_{s}\left(C_{l}\left(n_{v}^{c}\right)\right) P\left(n_{v'}^{c}\right) \Leftarrow w_{v'}^{c}$;

            $C_{N_{c}\left(w_{v'}^{c}\right)}\left(w_{v'}^{c}\right) \Leftarrow n_{v'}^{c}$; $w_{v'}^{c} \Leftarrow n_{v'}^{c}$; $tag \Leftarrow 0$

          else

            $stack \Downarrow$ null; $tag \Leftarrow 0$

          end if

        else if $tag = 1$ **then**

          if $C_{r}\left(n_{v}^{c}\right)$ is not null **then**

            $stack \Downarrow n_{v}^{c}$; $stack \Downarrow C_{r}\left(n_{v}^{c}\right)$; $n_{v'}^{c} \Leftarrow G_{s}\left(C_{r}\left(n_{v}^{c}\right)\right)$; $P\left(n_{v'}^{c}\right) \Leftarrow P\left(w_{v'}^{c}\right)$;

            $C_{N_{c}\left(P\left(w_{v'}^{c}\right)\right)}\left(P\left(w_{v'}^{c}\right)\right) \Leftarrow n_{v'}^{c}$; $w_{v'}^{c} \Leftarrow n_{v'}^{c}$; $tag \Leftarrow 0$

          else

            $stack \Downarrow$ null; $tag \Leftarrow 1$

          end if

        else if $tag = 2$ **then**

          if $n_{v}^{c} = C_{l}\left(P\left(n_{v}^{c}\right)\right)$ **then**

            $tag \Leftarrow 1$; $w_{v'}^{c} \Leftarrow B_{l}\left(w_{v'}^{c}\right)$

          else

            $tag \Leftarrow 2$; $w_{v'}^{c} \Leftarrow P\left(w_{v'}^{c}\right)$

          end if

          $stack \Uparrow$;

          if $w_{v'}^{c}$ is null **then**

            return;

(Continued)

                    end if
                  end if
                end if
              end loop
            end if

ALGORITHM 13.4   CONVERT BINARY TREE T_v TO A LINEAR-ELEMENT ARRAY a_e

Initialize a_e with null
   $queue \Downarrow R(T_v)$
   while $queue$ is not null do
      $queue \Uparrow n_v$
      $e_{ae} \Leftarrow G_1(n_v)$
      if $C_l(n_v)$ is null then
         $c_l(e_{ae}) \Leftarrow `0'$
      else
         $c_l(e_{ae}) \Leftarrow `1'$; $queue \Downarrow C_l(n_v)$
      end if
      if $C_r(n_v)$ is null then
         $c_r(e_{ae}) \Leftarrow `0'$
      else
         $c_r(e_{ae}) \Leftarrow `1'$; $queue \Downarrow C_r(n_v)$
      end if
      $a_e \Leftarrow (a_e \cup e_{ae})$
   end while

ALGORITHM 13.5   CONVERT LINEAR-ELEMENT ARRAY a_e TO BINARY TREE T_v

Require: a_e is not null
   Initialize $index = 0$; $m_1 = -1$; $m_2 = 0$
   Initialize binary-node array a_n with null
   $n_v \Leftarrow G_s(a_e[index + +])$
   $R(T_v) \Leftarrow n_v$
   $a_n[0] \Leftarrow R(T_v)$
   while $m_1 \neq m_2$ do
      $cnt \Leftarrow 0$
      for $i = 1$ to $m2 - m1$ do
         $n_v^p \Leftarrow a_n[m_1 + i]$
         if $C_l(n_v^p)$ is `1' then

$$n_v^c \Leftarrow G_s(a_e[index + +])$$
$$C_1\left(n_v^p\right) \Leftarrow n_v^c; P\left(n_v^c\right) \Leftarrow n_v^p; a_n \Leftarrow a_n \cup n_v^c; cnt + +$$
end if
if $C_r\left(n_v^p\right)$ is '1' then
$$n_v^c \Leftarrow G_s(a_e[index + +])$$
$$C_r\left(n_v^p\right) \Leftarrow n_v^c; P\left(n_v^c\right) \Leftarrow n_v^p; a_n \Leftarrow a_n \cup n_v^c; cnt + +$$
end if
$$m_1 = m_2$$
$$m_2 = m_2 + cnt$$
end for
end while

ALGORITHM 13.6 CALCULATE THE ARRAY OF INDEXES TO SEPARATE a_e INTO LINEAR-TREES

Require: a_e is not null
    Initialize *milestone_array* with null; *milestone_cnt* = 0; *storage* = 0
    milestone_array[*milestone_cnt* + +] = 0
    for $i = 0$ to $|a_e| - 1$ do
        if $c_1(a_e[i])$ then
            storage + +;
        end if
        if $c_r(a_e[i])$ then
            storage + +;
        end if
        if *storage* $= i$ then
            milestone_array[*milestone_cnt* + +] = $i + 1$; *storage* + +
        end if
    end for

13.4 LARGE-SCALE ROUTING UPDATE AND SMALL-SCALE RETRANSMISSION

13.4.1 Large-Scale Live Update

When a batch of packets are forwarded along the route towards the destination node, if an intermediate node is aware of a new route to the destination, it is able to use this new route to forward the packets that it has already received. There are a few implications of this. First, this new route will also be used to forward the subsequent packets of the same batch. Second, when packets are forwarded along the new route, such an updated forwarder list replaces the old list in the packets. As a result, the upstream nodes can be notified of the new route, and this information can propagate back to the source node quickly. Details of data forwarding and list update are described in this section with the help of Figure 13.5. In the figure, the source node v_1 has a flow of data packets for destination node v_{10}. According to

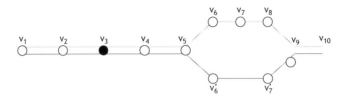

Figure 13.5 Route update.

its own routing module, v_1 decides that the best route to v_{10} is $v_1 v_2 v_3 v_4 v_5 v_6 v_7 v_8 v_9 v_{10}$; hence, the forwarder list.

At a given time point during the data transfer of a batch, there is a node on the forwarder list that has the highest priority and has received any packet of the batch. We call such a node the *frontier* of the batch. At the beginning, the frontier is the source node. When the destination has received at least one packet of a batch, it has become the frontier of the batch. Recall that a fragment (Section 13.2.1.1.2) is a subset of packets in the current batch which are sent together from a given forwarder. Here, the frontier has cached its first fragment of packets. Suppose at this point, the frontier in Figure 13.5 is v_3. When it is about to forward this fragment, if its routing module indicates there is a new route to the destination, e.g., $v_3 v_4 v_5 v_6' v_7' v_9 v_{10}$, it replaces the segment of the original forwarder list from itself to the destination (i.e., $v_3 v_4 v_5 v_6 v_7 v_8 v_9 v_{10}$) with this new route. That is, the forwarder list carried by these data packets is now $v_1 v_2 v_3 v_4 v_5 v_6' v_7' v_9 v_{10}$. When the packets of the fragment are forwarded, they will follow the new route. In addition, upstream nodes can overhear these packets, and thus their new forwarder list. These nodes can update their own routing information and will incorporate such information when forwarding their fragments. This backtrack continues until the source is aware of the latest route information.

We would like to bring the following notes to the readers' attention:

1. When the network diameter is large and nodes are moving fast, the routing information can be obsolete by the time it has propagated to a remote node. That is, a node's knowledge about the network topology becomes less accurate when the destination node is located farther away. Thus, the forwarder list composed by the source node needs to be adjusted as packets are forwarded towards the destination, where intermediate nodes closer to the destination could have better routing information. This is achieved effectively by allowing the frontier node to modify the forwarder list carried by the fragment of packets. As a result, CORMAN has a fairly good tolerance of route inaccuracy for any source node to start with.

2. When a frontier node updates an forwarder list, only the segment of the list between the frontier and destination is replaced, while the rest of the list (i.e., nodes that the fragment has gone through) remains intact. The reason for this design decision is that these upstream nodes should not be disturbed by the new route, so that the scheduling coordination among them is consistent.

3. We allow a frontier node to update a packet's forwarder list only according to its routing module. A node that is no longer a frontier should incorporate only the forwarder list that it overhears from downstream nodes. The purpose is to avoid unnecessary updates of route information, as the time needed to transfer a batch of data packets is very short. During this time, usually little has changed about the network topology, and nodes may not even have exchanged the routing information for the next periodic interval.

4. Consider a particular intermediate node on the forwarder list. As the frontier moves from the source to the destination, the forwarder list may be refreshed multiple times by different frontiers. Thus, this node may experience one update about its route to the destination every time a frontier decides to modify the list.

All of these are achieved rapidly and with no extra communication overhead compared to ExOR.

13.4.2 Small-Scale Retransmission

As we mentioned before, implementing opportunistic date transfer over the least hops path can thoroughly explore the broadcast nature in wireless transmission; however, the penalty of using such a path to transmit packets is that the connectivity between every pair of forwarders may be vulnerable. We propose a solution to this problem, and the basic idea is that the nodes not contained in the forwarder list [9] can run a distributed *small-scale retransmitter selection* algorithm to verify whether it should participate in the packet forwarding. Such a small-scale retransmitter should have the *best* position between two *listed forwarders* (the forwarders which are contained in the forwarder list) compared with other nodes. In general, our small-scale retransmitter selection algorithm has three important features. First, it is a distributed algorithm. In particular, if there are many nodes between two listed forwarders, after every node running the small-scale retransmitter selection algorithm, the best one will know it is the best and participate in the data forwarding without sending any announcement. Meanwhile, other nodes will directly refuse to be the small-scale retransmitter without any notification, either. Second, the small-scale retransmitter selection algorithm is used when the packets in a batch are received by the nodes which are not the forwarders in the forwarder list, so the small-scale retransmitter can be selected on time without the problem of inaccuracy. Third, all information required by the node to run, such as the algorithm, has already been collected through exchange of routing packets, so no additional information needs to be injected into the network.

In this section, we will first deeply analyze the reason why the small-scale retransmitter is required in CORMAN and then specify the details in our solution. The research findings from small-scale retransmission have been submitted to *IEEE ICC '12* [42]

13.4.2.1 Considerations of Small-Scale Retransmission

Previously, we proposed updating the forwarder list in a piggyback way when a data packet is being forwarded towards the destination. However, it is not helpful to enforce the link connectivity, because updating the forwarder list is helpful only when another node replaces the one moving away; but if no node replaces the position, the forwarder list will still be broken, as shown in Figure 13.6. Node Y moves from the position Y' to the current position, with the direction shown by the dashed arrow. Even though the wireless link is quite vulnerable, a transient link from Y to X may exist and over such a link, node X sends downstream nodes

Figure 13.6 Situation which cannot be solved by updating the forwarder list.

the information that the precursor to go to node *Y* is *X*. However, the vulnerable link is unreliable or broken when the source node wants to use the forwarder list "*Dest, Y, X, C, B, A, Src*", even by opportunistic data forwarding. To deal with this problem, we propose that node *Z*, which is a neighbor for both nodes *X* and *Y*, can be the small-scale retransmitter to offer a help when it receives packets from upstream nodes. That is because if node *Z* receives a packet with the forwarder list "*Dest, Y, X, C, B, A, SrC*", *Z* can figure out that *Y* and *X* are the designated forwarders in the forwarder list, and meanwhile is itself a neighbor for both of them. Hence, node *Z* could participate in forwarding the data packet. In particularly, node *Z* also waits for a period of time before its forwarding. Such a time period should be longer than the waiting time of node *Y* but shorter than the waiting time of node *X*. Or, if *Z* can overhear a fragment of packets being transmitted from node *Y*, it can predict the exact time when *Y* will finish its transmission and begin directly forwarding the rest of the useful packet. However, before we discuss the details about the small-scale retransmitter selection algorithm, it is necessary to introduce the changes of data structure kept on each node.

13.4.2.2 Design of Small-Scale Retransmission

In general, a *Neighbor Table* in CORMAN maintains the records of two hop topology of the current node and the link quality from its neighbors to its neighbors' neighbors. In CORMAN, we use the *Received Signal Strength Indicator* (RSSI) to evaluate the quality of links, because according to the work done by Charles Reis et al. [43] and Mei-Hsuan Lu et al. [44], RSSI can be reported by almost all commodity wireless cards, and a greater RSSI value usually indicates a better wireless channel quality.

In particular, *RSSI Table* is a sub-table maintained in every *Neighbor Table* entry which records the RSSI values from that neighbor to that neighbor's neighbors, and the entries in the *RSSI Table* have the content shown in Figure 13.7. The *Neighbor's Neighbor ID* keeps the address of the nodes which are in fact two hop neighbors via current neighbor in *Neighbor Table*; the *RSSI* records the RSSI value from the current neighbor to such two hop neighbors; the *Expire Time* holds the time when the entry should be removed; and the *next* helps us find the next entry in the same *RSSI Table*.

The operations on the *Neighbor Table* are presented as follows. When a routing update packet needs to be transmitted out from one node periodically, it not only contains the routing information but also collects the RSSI values from all its local *Neighbor Table* entries. All tuples (*Neighbor ID, RSSI*) are added into the routing packet as the *Neighbor RSSI List*. After the node sends out such a packet, all its neighbors update the *RSSI Table*s kept by the corresponding *Neighbor Table* entries.

To make the following discussion easier, we give a brief topology in Figure 13.8. In this topology, we assume the wireless link from node *X* to node *Y* is a transient link, and such a vulnerable link may successfully deliver a routing packet from *X* to *Y* or from *Y* to *X*. Hence, the link between *X* and *Y* may be used as a part of the forwarder list by a node far away. In this topology we can see the neighbors of node *X* are *A, B, C*, and *Y*, and node *Y*'s neighbors are *A, B, D* and *X*. Therefore, nodes *X* and *Y* have the same two neighbors *A* and *B*. We use *R(M,N)* to denote the RSSI value detected by node *N* from node *M*'s transmission, and to make a intuitional discussion, the *Neighbor Tables* and their *RSSI Tables* on both node *A* and *B* are briefly presented in Figure 13.9.

Neighbor's Neighbor ID	RSSI	Expire Time	next→

Figure 13.7 Entry of *RSSI Table* in *Neighbor Table* entry.

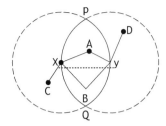

Figure 13.8 Topology example.

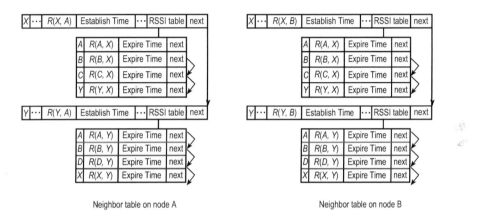

Neighbor table on node A   Neighbor table on node B

Figure 13.9 Neighbor table on node A and node B.

Without loss of generality, in Figure 13.8 we assume node *A* and node *B* overhear one or more packets in a same batch, and node *X* and node *Y* are contained in the forwarder list in these packets. In traditional opportunistic data forwarding, only the nodes in the forwarder list can participate in the packet forwarding process, and the probability that the packet was transmitted between *X* and *Y* may be quite low. That is because the least hop path may contain unreliable and transient links in the forwarder list, and such a situation becomes even worse in mobile cases. Such a problem can be solved by choosing one node from *A* and *B* as the small-scale retransmitter, which is selected by small-scale retransmitter selection algorithm as we present next.

13.4.2.3 Algorithm and Scoring Function

In particular, nodes *A* and *B* will decode the forwarder list from the overheard packet, if they are not contained in the forwarder list (which is true in our assumption; otherwise, the node will participate to forward packets certainly), it will check whether it is a neighbor node of two adjacent listed forwarder pair in the forwarder list. This work can be done easily by a searching algorithm with complexity $O(n \times m)$, where *n* is the number of neighbors of *A* or *B*, and *m* is the length of the forwarder list.

In our example shown in Figure 13.8, both *A* and *B* are the neighbors of *X* and *Y*, and we should evaluate further like follows. Take node *A* as an example: it will check whether $R(A, X)$ is greater than $R(Y, X)$ and whether $R(A, Y)$ is greater than $R(X, Y)$. If we assume that links are symmetric, $R(X, Y)$ should equal $R(Y, X)$, and four such RSSI values are maintained in the *Neighbor Table*. If both of the previous two comparisons are true, node *A* knows it is a *valid* small-scale retransmitter but cannot guarantee it is the *best* one. To check whether it is the *best* small-scale retransmitter, it will look up whether other valid small-scale retransmitters exist in the network. This can be done by running a simple searching algorithm on the listed forwarders' *RSSI Tables*. In our example, we should search the same *Neighbor's Neighbor ID* from *X*'s and *Y*'s RSSI Tables in *A*'s *Neighbor Table*; and the complexity is $O(n_X \times n_Y)$, where n_X and n_Y are the number of neighbors of nodes *X* and *Y*. In our example, another valid small-scale retransmitter would be *B* if both $R(B,X)$ is greater than $R(Y,X)$ and $R(B,Y)$ is greater than $R(X,Y)$ are true.

If *B* is also a valid small-scale retransmitter judged by node *A*, then node *A* has to make a decision regarding which one (*A* itself or *B*) is better. We propose a scoring function given in the following to evaluate the score of every valid small-scale retransmitter *i* between *M* and *N*.

$$
W(i) = \left(\sqrt[4]{\frac{1}{R(i,N)}} + \sqrt[4]{\frac{1}{R(i,M)}} \right)
$$
$$
\times \left(\frac{1}{4} \times \ln \frac{\max\left(\frac{1}{R(i,N)}, \frac{1}{R(i,M)} \right)}{\min\left(\frac{1}{R(i,N)}, \frac{1}{R(i,M)} \right)} + \ln\left(\sqrt[4]{\frac{1}{R(i,N)}} + \sqrt[4]{\frac{1}{R(i,M)}} \right) \right). \tag{13.5}
$$

Hence, A has to compare the score of itself, given by

$$
W(A) = \left(\sqrt[4]{\frac{1}{R(A,Y)}} + \sqrt[4]{\frac{1}{R(A,X)}} \right)
$$
$$
\times \left(\frac{1}{4} \times \ln \frac{\max\left(\frac{1}{R(A,Y)}, \frac{1}{R(A,X)} \right)}{\min\left(\frac{1}{R(A,Y)}, \frac{1}{R(A,X)} \right)} + \ln\left(\sqrt[4]{\frac{1}{R(A,Y)}} + \sqrt[4]{\frac{1}{R(A,X)}} \right) \right) \tag{13.6}
$$

to the score of node B by given by

$$
W(B) = \left(\sqrt[4]{\frac{1}{R(B,Y)}} + \sqrt[4]{\frac{1}{R(B,X)}} \right)
$$
$$
\times \left(\frac{1}{4} \times \ln \frac{\max\left(\frac{1}{R(B,Y)}, \frac{1}{R(B,X)} \right)}{\min\left(\frac{1}{R(B,Y)}, \frac{1}{R(B,X)} \right)} + \ln\left(\sqrt[4]{\frac{1}{R(B,Y)}} + \sqrt[4]{\frac{1}{R(B,X)}} \right) \right). \tag{13.7}
$$

and the less one should be chosen as the result of small-scale retransmitter selection algorithm. The same comparison will also be operated on node *B*, because node *B* knows node *A* is a valid small-scale retransmitter too.

The algorithm does not need an additional packet to coordinate the final decision on different nodes because the *RSSI Tables* kept for node *X* and *Y* on node *A* and the *RSSI Tables* kept for node *X* and *Y* on node *B* must be the same. That is because if *A* and *B* are neighbors for both X and Y, the RSSI information broadcasted from X and Y will be received simultaneous by nodes A and B, and both of them will refresh their corresponding *RSSI Tables* together. Therefore, every valid small-scale retransmitter will finally *agree* on a particular one as the *best* small-scale retransmitter out of themselves without gossip.

An important explanation must be clarified here, which is why we propose Equation 13.5 to evaluate the valid small-scale retransmitters. Even though the RSSI calculation method is not specified in the 802.11 standard, 802.11 device vendors should calculate the RSSI according to the received signal strength, and the higher the received signal strength, the greater RSSI value should be returned from hardware [45]. Ideally, the path loss model of wireless communication follows the rule that the received signal power is inversely proportional to d^n, where d is the distance between the transmitter and the receiver and n is usually an integer within 2, 3, and 4. If we take other parameters as constants, the ideal relationship between distance from M to N and RSSI can be presented as

$$R(M,N) = C \times d_{M,N}^{-n} \tag{13.8}$$

where C is a constant.

When we want to get the reciprocal of RSSI, we have

$$\frac{1}{R(M,N)} = \frac{d_{M,N}^{n}}{C} \tag{13.9}$$

The rationality of comparing scores calculated by Equation 13.5 is explained as follows. Considering Equation 13.9, the Equation 13.5 can be simplified to:

$$W(i) = \left(\frac{d_{i,M}^{\frac{n}{4}}}{\sqrt[4]{C}} + \frac{d_{i,N}^{\frac{n}{4}}}{\sqrt[4]{C}} \right) \times \left(\ln \frac{\max\left(\frac{d_{i,M}^{\frac{n}{4}}}{\sqrt[4]{C}}, \frac{d_{i,N}^{\frac{n}{4}}}{\sqrt[4]{C}} \right)}{\min\left(\frac{d_{i,M}^{\frac{n}{4}}}{\sqrt[4]{C}}, \frac{d_{i,N}^{\frac{n}{4}}}{\sqrt[4]{C}} \right)} + \ln\left(\frac{d_{i,M}^{\frac{n}{4}}}{\sqrt[4]{C}} + \frac{d_{i,N}^{\frac{n}{4}}}{\sqrt[4]{C}} \right) \right) \tag{13.10}$$

$$= \frac{\left(d_{i,M}^{\frac{n}{4}} + d_{i,N}^{\frac{n}{4}} \right)}{4 \times \sqrt[4]{C}} \times \left(n \times \ln \frac{\max(d_{i,M}, d_{i,N})}{\min(d_{i,M}, d_{i,N})} + 4 \times \ln\left(d_{i,M}^{\frac{n}{4}} + d_{i,N}^{\frac{n}{4}} \right) - \ln C \right)$$

We can see that a node which leads the minimum value of Equation 13.10 should be the node which has nearly the same distances to both node M and node N, and it should have the least summation of two such distances. The parameters in Equation 13.5 are selected by a great deal of trials in MATLAB. Finally, the figures we created in MATLAB when n equals 2, 3, and 4 are shown in Figures 13.10 through 13.12.

In Figures 13.10 through 13.12, the rectangle with the Width/Height ratio $\sqrt{3}$ is the $\pi/2$ rotated circumscribed quadrilateral of the olivary shape in Figure 13.8, which is surrounded by $\widehat{PXQ}$ and $\widehat{PYQ}$. The nodes X and Y are located on the position $\left(50 \times \sqrt{3}, 0\right)$ and $\left(50 \times \sqrt{3}, 100\right)$, and the nodes P and Q are located on the position $(0,50)$ and $\left(100 \times \sqrt{3}, 50\right)$. The contour line is the curve on which all nodes have the same score given by Equation 13.10. We can see

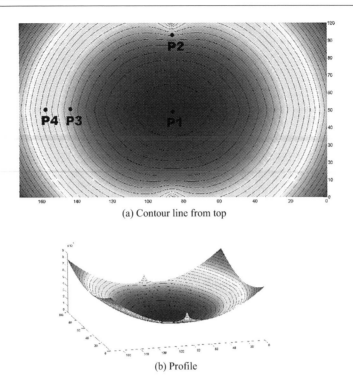

(a) Contour line from top

(b) Profile

Figure 13.10 Scoring function with *n* = 2 as the path loss parameter.

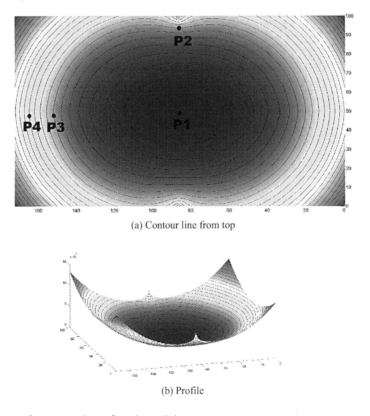

(a) Contour line from top

(b) Profile

Figure 13.11 Scoring function with *n* = 3 as the path loss parameter.

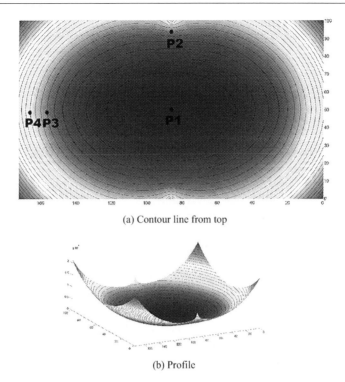

(a) Contour line from top

(b) Profile

Figure 13.12 Scoring function with *n* = 4 as the path loss parameter.

that, for all possible *n* from 2 to 4, the contour lines follow the same rule that when a node is nearer to one of two connected neighbors, the node's score is greater, so it has less opportunity to be selected. That is reasonable because the small-scale retransmitter which is too near with one of the two nodes that are connected by a vulnerable link contributes less than that in the middle of such two nodes. Furthermore, the center point on every figure always has the minimum value, which indicates to us where the best small-scale retransmitter is supposed to be.

For example, P_2 in all the aforementioned figures probably has too little of a contribution than P_1 to be a small-scale retransmitter. One more thing: P_2 should have a better position than P_4 because P_4 is near the position of P or Q in Figure 13.8 and has a longer distance to both nodes X and Y than to P_2, and from our scoring function, it does have a smaller score than P_4. Hence, the contribution made by sender diversity and receiver diversity from P_2 is more than the contribution given by P_4. Furthermore, considering P_2 is better than P_4 and worse than P_1, and P_4 and P_1 are on the same center line, we predict there must be a point on the center line which has the same score with P_2. In fact, we do find such a point: P_3, which is on the same contour line of P_2.

Based on the analysis before, we can see that the scoring function we proposed matches our predications very well, so we proved the function is suitable to be used to evaluate valid small-scale retransmitters.

13.5 PERFORMANCE EVALUATION

In this section, we will study the performance of the proposed solution that enables the opportunistic data forwarding in MANETs. In particular, we study not only the overall performance where all the solutions work together, but also the particular effectiveness of PSR

(Section 13.3) and the small-scale retransmission (Section 13.5). Hence, this part consists of the following three subsections: the performance study of PSR, the effectiveness study of small-scale retransmission, and the overall performance of CORMAN.

13.5.1 Performance Study of PSR

We study the performance of PSR using computer simulation with Network Simulator 2 (version 2.34). We compare PSR against OLSR [4], DSDV [3], and DSR [5], three fundamentally different routing protocols in MANETs, with varying network densities and node mobility rates. We measure the data transportation capacity of these protocols supporting TCP (Transmission Control Protocol) and UDP (User Datagram Protocol) with different data flow deployment characteristics. Our tests show that the overhead of PSR is indeed only a fraction of that of the baseline protocols. Nevertheless, as it provides global routing information at such a small cost, PSR offers similar or even better data delivery performance. In this section, we first describe how the experiment scenarios are configured and what measurements are collected. Then we present and interpret the data collected from networks with heavy TCP flows and from those with light UDP streams.

13.5.1.1 Experiment Settings

We use the Two-Ray Ground Reflection channel propagation model in our simulation. Without loss of generality, we select a 1 Mbps nominal data rate at the 802.11 links to study the relative performance among the selected protocols. With the default Physical Layer parameters of the simulator, the transmission range is approximately 250 m and the carrier sensing range is about 550 m.

We compare the performance of PSR with that of OLSR, DSDV, and DSR. We selected these baseline protocols that are different in nature for the following reasons. On one hand, OLSR and DSDV are both proactive routing protocols, and PSR is also in this category. On the other hand, OLSR makes a complete topological structure available at each node, whereas in DSDV, nodes have only distance estimates to other nodes via a neighbor. PSR sits in the middle ground, where each node maintains a spanning tree of the network. Furthermore, DSR is a well-accepted reactive source-routing scheme and, as with PSR, it supports source routing, which does not require other nodes to maintain forwarding lookup tables. For a more leveled comparison to the peer proactive protocols, we make PSR transport regular IP packets rather than source-routed packets, although it can and should be used as a source-routing scheme. All three baseline protocols are configured and tested out of the box of NS-2.

In modeling node mobility of the simulated MANETs, we use the Random Waypoint Model to generate node trajectories. In this model, each node moves towards a series of target positions. The rate of velocity for each move is uniformly selected from $[0, v_{max}]$. Once it has reached a target position, it may pause for a specific amount of time before moving towards the next position. All networks have 50 nodes in our tests. We have two series of scenarios based on the mobility model. The first series of scenarios has a fixed v_{max} but different network densities by varying the network dimensions. The second series has the same network density but varying v_{max}.

We study the data transportation capabilities of these routing schemes and their overhead in doing so by loading the networks with TCP data flows and UDP voice streams.

- To test how TCP is supported, in each scenario, we randomly select 40 nodes out of the 50 and pair them up. For each pair, we set up a permanent one-way FTP (File Transfer Protocol) data transfer. We repeat the selection of the 40 nodes five times and study their collective behavior. This essentially mimics heavily loaded mobile networks. For all four protocols, we measure their TCP throughput, end-to-end delay, and routing overhead in byte per node per second in each scenario.

- To study their performance in supporting UDP, we use two-way Constant Bit Rate (CBR) streams for compressed voice communications. Specifically, we select three pairs of nodes and feed each node with a CBR flow of 160 byte/pkt and 10 pkt/s, which simulates mobile networks with a light voice communication load. We measure the Packet Delivery Ratio (PDR), end-to-end delay, and delay jitter in each scenario.Results about TCP (Sections 13.5.1.2 and 13.5.1.3) and UDP (Sections 13.5.1.4 and 13.5.1.5) with regard to varying node densities and velocity rates are in the subsequent subsections.

13.5.1.2 TCP with Node Density

We first study the performance of PSR, OLSR, DSDV, and DSR in supporting 20 TCP flows in networks with different node densities. Specifically, with the default 250 m transmission range in ns-2, we deploy our 50-node network in a square space of varying side lengths that yield node densities of approximately $5, 6, 7, \ldots, 12$. These nodes move following the random waypoint model with $v_{\max} = 30$ m/s.

We plot in Figure 13.13 the per-node per-second routing overhead, i.e., the amount of routing information transmitted by the routing agents measured in byte/node/second, of the four protocols when they transport a large number of TCP flows. This figure shows that the overhead of PSR (20 to 30) is just a fraction of that of OLSR and DSDV (140 to 260), and more than an order of magnitude smaller than DSR (420 to 830). The routing overhead of PSR, OLSR, and DSDV goes up gradually as the node density increases. This is a typical behavior of proactive routing protocols in MANETs. Such protocols usually use a fixed time interval to schedule route exchanges. While the number of routing messages transmitted in the network is always constant for a given network, the size of such a message is determined

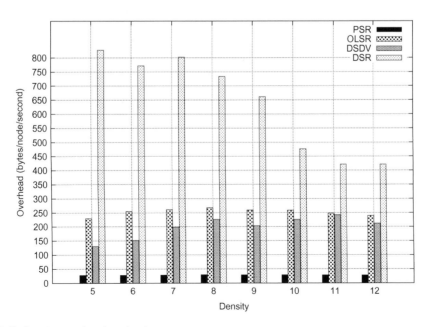

Figure 13.13 Routing overhead vs. density.

by the node density. That is, a node periodically transmits a message to summarize changes as nodes have come into or gone out of its range. As a result, when the node density is higher, a longer update message is transmitted even if the rate of node motion velocity is the same. Note that when the node density is really high, say around 10 and 12, the overhead of OLSR flattens out or even slightly decreases. This is a feature of OLSR when its Multi-Point Relay mechanism becomes more effective in removing duplicate broadcasts, which is the most important improvement of OLSR over conventional link-state routing protocols. PSR uses a highly concise design of messaging, allowing it to have a much smaller overhead than the baseline protocols. In contrast, DSR as a reactive routing protocol incurs a significantly higher overhead when transporting a large number of TCP flows because every source node needs to conduct its own route search. This is not surprising, as reactive routing protocols were not meant to be used in such scenarios. Later in our experiments (Sections 13.5.1.4 and 13.5.1.5), we test all four protocols supporting just a few UDP streams for a different perspective. Here, the routing overhead of DSR decreases with the node density going up and network diameter going down. This is because the number of hops to a destination is smaller in a denser network, so the shorter, more robust routes break less frequently and do not need as many route searches. Furthermore, compared to IP forwarding, the fact that DSR is source routing and that intermediate nodes cannot modify the routes embedded in data packets works against its performance in a mobile network, both in terms of the increase of search operations and the loss of data transportation capacity. The reason is that, because a source node can be quite a few hops away from the destination, its knowledge about the path as embedded in the packets can quickly become obsolete in a highly mobile network. As a packet progresses en route, if an intermediate node cannot reach the next hop as indicated in the embedded path, it will be dropped. This is very different from IP forwarding, where intermediate nodes can have more updated routing information than the source and can utilize that information in forwarding decisions.

Figure 13.14 plots the TCP throughput of the four protocols for the same node density levels as before. The total throughput of the 20 TCP flows of PSR, OLSR, and DSDV is noticeably higher than that of DSR. In addition, while the TCP throughput of DSR decreases with node density, that for the other three are somewhat unaffected, hovering at around 500 kbps. Apparently, the large routing overhead of DSR, especially in dense networks, consumes a fair amount of channel bandwidth, leaving less room for data transportation. In most cases, PSR has the highest throughput because it needs to give up the least network resources for routing.

Next, we focus on the end-to-end delay of TCP flows to investigate how well these protocols support time-sensitive applications. Figure 13.15 presents the delay measured for different node densities. As the density increases from 5 to 12 neighbors, the delay of DSR goes up from 0.58 to about 1.5 s, which is significantly higher than the typical value of 0.15 to 0.35 s for the other three protocols. Such a difference is caused by the initial route search when a TCP flow starts, and by the subsequent searches triggered by route errors. As the network becomes denser, all protocols show an increasing trend in end-to-end delay. This may seem counter-intuitive, as in denser networks, the average hop distance between source-destination pairs is smaller, which should lead to shorter round-trip time. However, this benefit is completely offset by more intense channel contention. Recall that the node density is inversely proportional to the square of network diameter. As such, in the interplay between route length and channel contention, the latter dominates the overall effect.

13.5.1.3 TCP with Velocity

We also study the performance of PSR and compare it to OLSR, DSDV, and DSR with different rates of node velocity. In particular, we conduct another series of tests in networks

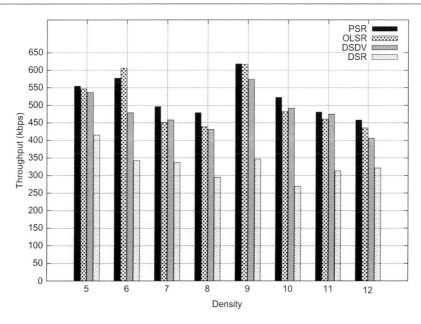

Figure 13.14 TCP throughput vs. density.

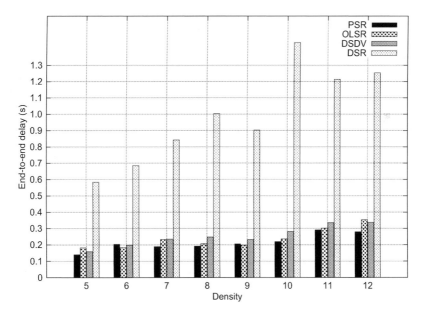

Figure 13.15 End-to-end delay in TCP vs. density.

of 50 nodes deployed in a 1100×1100 (m^2) square area with v_{max} set to 0, 4, 8, 12, ... , 32 (m/s). The network thus has an effective node density of around 7 neighbors per node, i.e., a medium density among those configured in the previous subsection. As before, 20 TCP one-way flows are deployed between 40 nodes, and we measure the routing overhead, TCP throughput, and end-to-end delay (Figures 13.16 through 13.18).

The routing overhead of all four protocols with varying rates of node velocity is plotted as in Figure 13.16. Note that the velocity to the right of the x-axis corresponds to the middle bars in Figure 13.13. We observe in the plot here, as v_{max} decreases, the overhead of all protocols

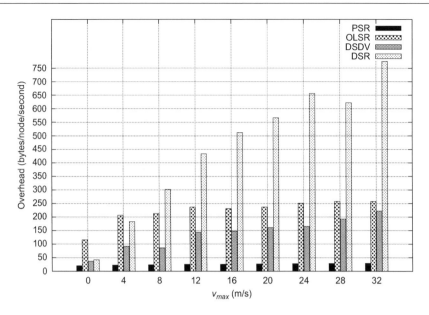

Figure 13.16 Routing overhead vs. velocity.

comes down. The reason for DSR is that, as the network structure becomes more stable, fewer route repair attempts are necessary. For the case of the proactive protocols, it is the reduction in the size of routing messages (i.e., fewer neighbors have changed positions) that cuts down the overhead. Still relative among these four protocols, when the network is not stationary ($v_{max} \neq 0$), the overhead of PSR (20 to 30 byte/node/second) is a fraction of that of OLSR and DSDV (90 to 300), and more than an order of magnitude lower than DSR (180 to 770).

The TCP throughput and end-to-end delay are plotted in Figures 13.17 and 13.18 respectively. From these figures, we observe that the performance of PSR, OLSR, and DSDV are

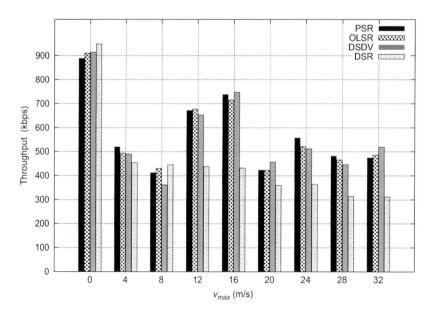

Figure 13.17 TCP throughput vs. velocity.

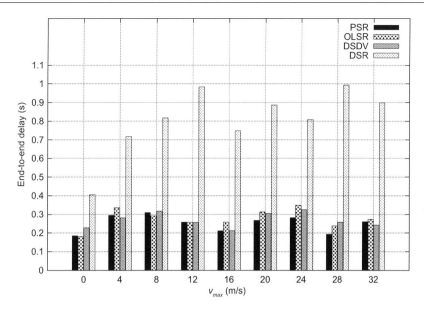

Figure 13.18 End-to-end delay in TCP vs. velocity.

similar, with PSR leading the pack in most cases. In addition, neither throughput or delay is affected by the different rates of velocity. The only exception is that when $v_{max} = 0$, all protocols yield a high throughput of 900 kbps. With a greater portion of the channel bandwidth devoured by routing messages in highly mobile networks, DSR suffers a noticeable performance penalty in TCP throughput and end-to-end delay.

13.5.1.4 UDP with Density

We also tested the four protocols for their performance in transporting a small number of UDP streams. This is a typical assumption for ideal scenarios of reactive routing protocols. Here, we deploy three two-way UDP streams in order to simulate compressed voice communications. To find out about how node density affects these protocols, we use the same network and mobility configurations as in Section 13.5.1.2. We measure and plot the PDR (Figure 13.19), delay (Figure 13.20), and delay jitter (Figure 13.21) against varying node densities.

In Figure 13.19, the PDRs of all four protocols are in the same ballpark across different node densities, with DSR slightly in the lead and OLSR trailing behind. This verifies that the traffic configuration is favorable for DSR. The relatively high loss rate of OLSR among the proactive routing protocols is caused by the higher routing overhead compared to PSR and DSDV. When the nodes are neither too sparse, so that the network connectivity is good, nor too dense, so that the channel can be spatially reused, these protocols have a fairly high PDR of over 70% for PSR, DSDV, and DSR, and 60% to 70% for OLSR.

When we turn to end-to-end delay (Figure 13.20), there is a noticeable difference between DSR and the proactive protocols. In particular, DSR as reactive protocol has a rather large delay in sparse networks. This is because the long, vulnerable routes discovered during the search procedure break frequently, forcing nodes to hold packets back for an extended period before new routes are identified. Oppositely, the network sparsity does not affect proactive protocols as much because their periodic routing information exchange makes them more prepared for network structure alteration. While the delay of DSR is off the chart, that of PSR is always less than 0.05 s, which is also much less than that for DSDV and OLSR (0.1–0.43 s).

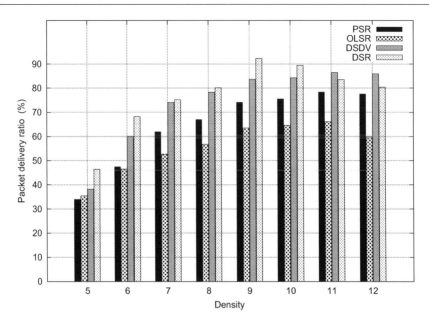

Figure 13.19 PDR in UDP vs. density.

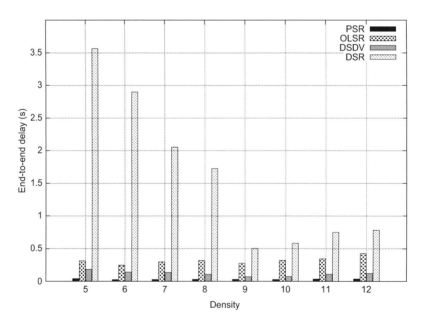

Figure 13.20 End-to-end delay in UDP vs. density.

On a related note, the delay jitter (Figure 13.21) of PSR is significantly lower than the other three. Note that Voice-Over-IP (VoIP) applications usually discard packets that arrive too late. Therefore, the jitter among the packets actually used by the VoIP receiving agent is much smaller. Nevertheless, our metric still reflects how consistent these protocols are in delivering best-effort packets.

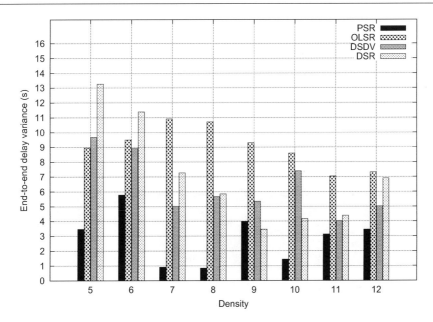

Figure 13.21 End-to-end delay jitter in UDP vs. density.

13.5.1.5 UDP with Velocity

The same measurements are taken to test these protocols in response to different rates of node velocity. As with the case of the previous subsection, we pick three node pairs out of the 50 nodes and give them two-way CBR streams.

For the entire series of different velocity caps $v_{max} = 0, 4, 8, 12,..., 32$ m/s, the node density is again set to around 7 neighbors per node.

From the plot of PDR (Figure 13.22), we observe that DSR is able to support three voice streams with little packet loss. Specifically, the PDR of DSR, PSR, and DSDV is always over 70% even when $v_{max} = 32$. The reliability of OLSR is relatively lower, which can go below 60% at high speed ($v_{max} = 28$ or 32 m/s). Note that all four protocols are very reliable in data delivery when $v_{max} = 0$ or 4 m/s, where the loss rates are well below 10%. Their performance in terms of PDR degrades gracefully as the rate of node velocity increases.

The end-to-end delay (Figure 13.23) presents a rather distinct landscape. In particular, the number for DSR is significantly higher than the other protocols except in low-mobility networks with $v_{max} = 0$ or 4 m/s. In all cases, the delay for PSR is much smaller compared to OLSR and DSDV. On the other hand, the measured delay jitter (Figure 13.24) indicates that all protocols become less consistent when nodes move faster. Relatively speaking, however, the variance of PSR is much smaller than the other three.

13.5.2 Effectiveness Study of Small-Scale Retransmission

In this section, we study the effectiveness of small-scale retransmission in particular by running computer simulation using NS-2. To investigate the effectiveness of small-scale retransmission, we test CORMAN with small-scale retransmission enabled and CORMAN with small-scale retransmission disabled separately. The performance improvement and explanations of the simulation results are explained in the rest of this section.

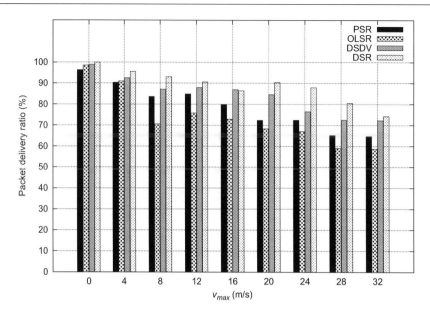

Figure 13.22 PDR in UDP vs. velocity.

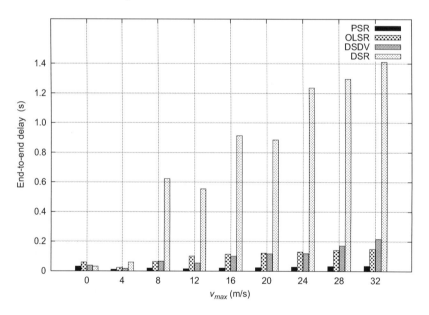

Figure 13.23 End-to-end delay in UDP vs. velocity.

13.5.2.1 Experiment Settings

The channel propagation model used in NS-2 had been predominantly the Two-Ray Ground Reflection model early on. However, this model is realized to be a simplified path loss model without considering fading. In our work, we choose the Nakagami propagation model to test CORMAN in a more realistic fading environment. The probability density function of Nakagami distribution of the received signal's amplitude $X = r(r \geq 0)$ is defined as:

$$f(r, \mu, \omega) = \frac{2\mu^{\mu}}{\Gamma(\mu)\omega^{\mu}} r^{2\mu-1} \exp\left(-\frac{\mu}{\omega} r^2\right), \quad (13.11)$$

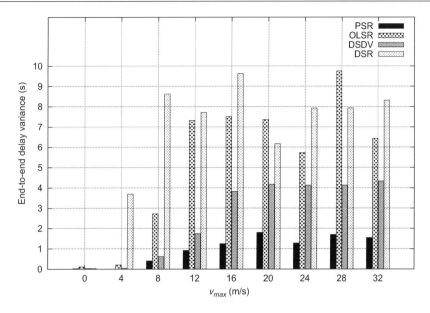

Figure 13.24 End-to-end delay jitter in UDP vs. velocity.

where $\mu = \dfrac{E^2\left[X^2\right]}{Var\left[X^2\right]}$ and $\omega = E[X^2]$. In NS-2, when a node has received a packet, it first calculates the received power using path loss based on the Frii Free-Space model. This value is compensated with Nakagami's fluctuation before further processing. We configure the nominal data rate at the 802.11 links to 1 Mbps. In modeling node motion, we also adopt the Random Waypoint model (Section 13.5.1.1) to generate the simulation scenarios.

We inject CBR (constant bit rate) data flows in the network, which are carried by UDP. Specifically, a source node generates 50 packets every second, each has a payload of 1000 bytes. This translates to a traffic rate of 400 kbps injected by a node. When comparing different CORMAN versions which has and does not have the small-scale retransmission functionality, we record the packet delivery ratio (PDR), i.e. the fraction of packets received by the destination out of all the packets injected, and end-to-end delay average and variance. We observe that the CORMAN with small-scale retransmission enabled outperforms the CORMAN with small-scale retransmission disabled in terms of all of these metrics.

13.5.2.2 Performance versus Network Dimension

We make performance comparison of CORMAN under different configurations. One is with the small-scale retransmission enabled and the other is with such a module disabled, and our tests are against the varying of the network density. In particular, we have network tomographies of $l \times l$ (m²), where $l = 450, 500, 550, ..., 950$. We deploy 50 nodes in each of these network dimensions to test the protocols with differing node densities, and every node moves randomly with the Random Waypoint model at $v_{max} = 20$ m/s. For each dimension scenario, we test performances of CORMAN with small-scale retransmission enabled and CORMAN with small-scale retransmission disabled in transporting CBR data flows between a randomly selected source-destination pair. We repeat this process 20 times for a given scenario. We measure the PDR, end-to-end delay, and delay jitter for both protocols and average them over the 20 repetitions of each scenario, as plotted in Figures 13.25 through 13.27, respectively.

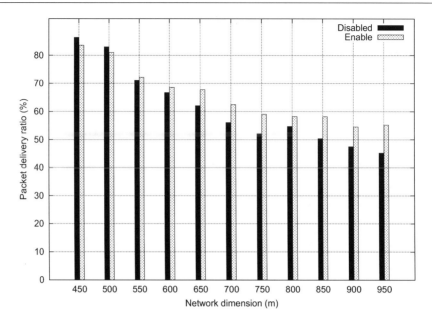

Figure 13.25 PDR vs. network dimension.

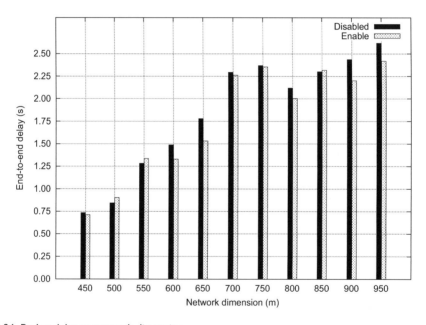

Figure 13.26 Packet delay vs. network dimension.

We observe that when the network density is relatively low, CORMAN with small-scale retransmission disabled has better PDR. In particular, when the side length of the square network boundary grows from 450 to 500 m, the CORMAN with disabled small-scale retransmission outperform the CORMAN with the proposed scheme enabled. When the side length of the square boundary kept increasing from 500 m, the CORMAN with small-scale retransmission enabled outperform its opposite, and the greater the side length, the more obvious PDR gain could be achieved.

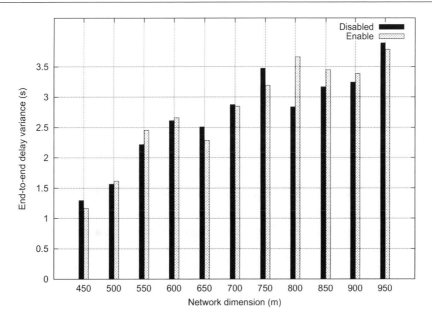

Figure 13.27 Delay jitter vs. network dimension.

To investigate the reason behind such a phenomenon, we studied the simulation trace files deeply and plotted out the number of packets forwarded by local relay nodes and the collisions on listed forwarders caused by local relay nodes in Figure 13.28. The left scale of Figure 13.28 gives us the number of packets forwarded by small-scale retransmitters and the collisions on listed forwarders caused by small-scale retransmitters, and the right scale gives us the ratio of the two values. It is obvious that both the small-scale forwarding and the collisions they caused increase when the network dimension goes up. Furthermore, the

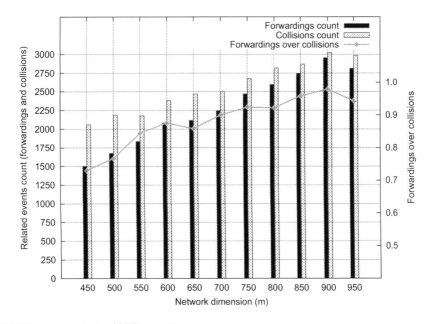

Figure 13.28 Reasons analysis of PDR reversion.

forwarding count increases faster than the collisions they caused. In our further investigation, we found the forwarding increases because the hops from source to destination node increase with the network dimensions. However, we found the increases of collisions are not due to the growth of hop count because we have granted nodes the priority to transmit the packets in the same batch [9]. The real reason is that a route with more hops from source to destination reduces the size of fragment transmitted by intermediate node, so the entire fragment has a higher probability of being lost on both listed forwarders and local relay nodes. As the result, nodes cannot predict the exact time to start their transmission, and that is the real reason for more collisions in the network. Therefore, the collision will not grow directly against the increases of network dimensions and with a lower growth speed than that the increasing of forwarding times.

In a summary, because wireless links in a dense network have relatively high reliability with relatively less hop count – note that the nodes have to try to deliver a same packet several times before discarding them – so the contribution of small-scale retransmission in dense network is in fact limited. When the density of the network decreases, the benefit achieved by using small-scale retransmission shows its advantage.

From Figure 13.26 we can see that the packet end-to-end delay in both tested schemes with and without small-scale retransmission are nearly the same in a dense network. The scheme with local cooperative relay enabled has shorter packet end-to-end delay in a sparse network, because the local cooperative relay will save packets and prevent them from being delivered from original source again. The Figure 13.27 indicate that both the CORMANs with and without small-scale retransmission have nearly the same packet end-to-end delay jitter.

13.5.2.3 Performance versus Velocity

We also study CORMAN's performance with the small-scale retransmission enabled or not by varying the node velocity. We conduct another set of tests in a network of 50 nodes deployed in a 700×700 (m²) space with a varying v_{max}, where $v_{max} = 0, 3, 6, \ldots, 30$ (m/s). For each velocity scenario, we test CORMAN with small-scale retransmission enabled and disabled in transporting CBR data flows between a randomly selected source-destination pair as well. We repeat this process 20 times for a given scenario. We collect the PDR, end-to-end delay, and delay jitter for both protocols averaged over each scenario, and plot them in Figures 13.29 through 13.31, respectively.

From Figure 13.29, we can see that for all velocity scenarios, the small-scale retransmission enabled scheme outperformed the small-scale retransmission disabled scheme for the same reasons as we analyzed in Section 13.5.2.2, so we will not explain that again here. What makes these results interesting is that, even though both the PDRs in two test conditions decrease together when nodes move faster, the PDR gain by using small-scale retransmission in CORMAN grows when the velocity goes up. That is due to the scoring function in the small-scale retransmitter being a real-time and distributed algorithm, and it can provide the best retransmitter on time to reduce the effect of nodes' mobility. As a result, the small-scale retransmission has more effectiveness to fix up the break links when the node velocities go up.

Also, CORMAN with small-scale retransmission enabled has shorter end-to-end delay and almost same end-to-end delay jitter compared to that with small-scale retransmission disabled, due to the same reasons we analyzed in Section 13.5.2.2.

13.5.3 Overall Performance of CORMAN

In this section, we study the performance of CORMAN by running computer simulation using NS-2 (version 2.34). We compare it against AODV with varying network densities and

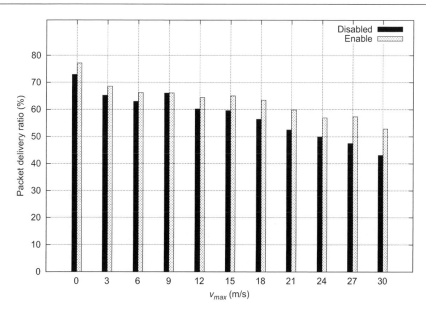

Figure 13.29 PDR vs. node velocity.

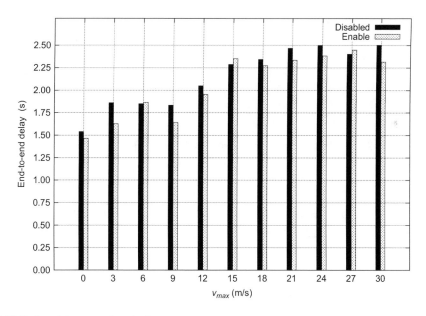

Figure 13.30 Packet delay vs. node velocity.

node mobility rates. The performance improvement and explanations of these results are explained in the rest of this section.

13.5.3.1 Experiment Settings

The basic configurations of the simulations for the overall performance in CORMAN are nearly the same as that in the simulations for the effectiveness of small-scale retransmission in Section 13.5.2, where we choose the Nakagami propagation model to test CORMAN's overall performance, configure the nominal data rate at the 802.11 links to 1 Mbps, and

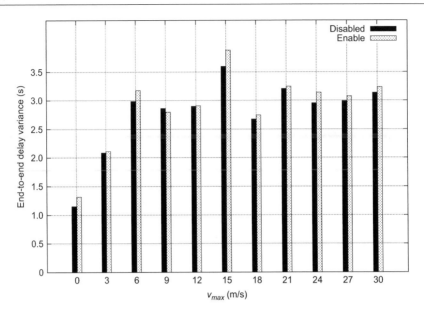

Figure 13.31 Delay jitter vs. node velocity.

adopt the Random Waypoint mobility model to generate the simulation scenarios. As well, we inject CBR data flows in the network carried by UDP with the same data rate as we specified in Section 13.5.2. We compare the overall performance of CORMAN with that of AODV [6]. We select AODV as the baseline because AODV is a widely adopted routing protocol in MANETs, and its behavior in both NS-2 and real network operations is well understood by the research community.

13.5.3.2 *Performance versus Network Dimension*

We first compare the performance of CORMAN and AODV with different network dimensions. Specifically, we have network tomographies of $l \times l(\mathrm{m}^2)$, where $l = 250, 300, 350, ...,$ 1000. We deploy 50 nodes in each of these network dimensions to test the protocols with differing node densities. These nodes move following the Random Waypoint model with $v_{\max} = 10$ m/s. For each dimension scenario, we test CORMAN and AODV's capabilities in transporting CBR data flows between a randomly selected source-destination pair. We repeat this process five times for a given scenario. We measure the PDR, end-to-end delay, and delay jitter for both protocols and average them over the five repetitions of each scenario, as plotted in Figures 13.32 through 13.34, respectively.

We observe that CORMAN has a PDR (Figure 13.32) of about 95% for dense networks (i.e., $250 \leq l \leq 500$ m). As the node density decreases, this rate gradually goes down to about 60%. In contrast, AODV's PDR ranges between 60% and 80% for dense networks and quickly drops to around 20% for sparse networks. (We use a red plotting series to indicate the relative performance of CORMAN over PDR in all of our figures.) There are two reasons for the PDR penalty for AODV to operate in sparse networks. First, data packets are forwarded using traditional IP forwarding in AODV. When channel quality varies (as emulated by the Nakagami model), a packet may be lost at the link layer. After a few failed retransmits, it will be dropped by the network layer. CORMAN, however, is designed to utilize such link effects so that at least one downstream node would be available despite the link variation. CORMAN facilitates opportunistic data forwarding using the link quality

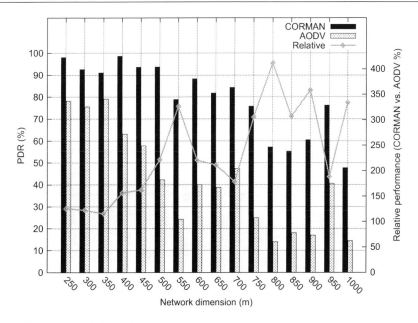

Figure 13.32 PDR vs. network dimension.

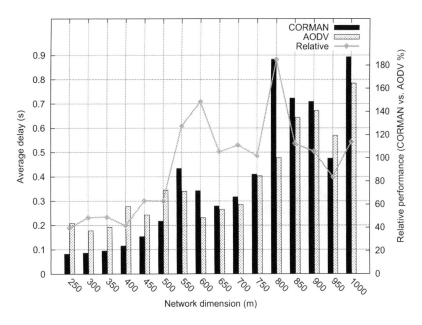

Figure 13.33 Packet delay vs. network dimension.

diversity at different receivers and allows them to cooperate with each other with a minimal overhead. Consequently, CORMAN has a strong resilience to link quality fluctuation and node mobility. Second, the route search of AODV does not function well with unreliable links. Recall that, in AODV, when a node finds that it does not have a next hop available for a given data packet, it broadcasts a RREQ (route request) to find one. Both the destination and any intermediate node that has a valid cached route can reply with a RREP (route reply). When links were perfectly symmetric, the RREP packet would take the inverse path leading

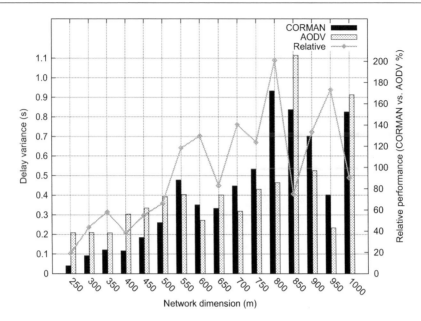

Figure 13.34 Delay jitter vs. network dimension.

to the initiating node of the route search. However, when links suffer from fading, the RREP packets may not be able to propagate back to the initiator because of transient low link quality in the reverse direction. As a result, it takes AODV much longer to obtain stable routes. In fact, this performance loss has been observed in earlier studies of AODV when large numbers of unidirectional links are present in the network [46]. In our NS-2 simulation using the Nakagami propagation model, the independent link quality fluctuation in both directions essentially produces temporary unidirectional links. This negative effect of fading links on AODV can be mediated to a degree when the node density increases. Therefore, in our next set of tests (Section 13.5.3.3), all simulation scenarios have a fairly higher node density for AODV to function reasonably well.

We are also interested in the end-to-end delay and its variance of CORMAN and AODV. Figure 13.33 presents the end-to-end delay of these protocols in different dimension scenarios. We see that, when the node density is higher (i.e., $250 \leq l \leq 500$ m), CORMAN has a shorter delay than AODV. For sparser scenarios (i.e., $550 \leq l \leq 1000$ m), CORMAN's delay is slightly longer than AODV but comparable. In CORMAN's implementation, a node determines that it can no longer contribute to a batch's progression if it has not seen an update of the map after 10 retransmits of its fragment. This is similar to 10 retries at the link layer in IP-style forwarding. (The default retry limit in 802.11 is 7.) Thus, the not-so-short end-to-end delay of CORMAN is caused by the larger number of data retransmits. This is a relatively small cost to pay for a significantly higher PDR (Figure 13.32). The delay jitter (standard deviation) measured for CORMAN and AODV has similar relative performance for these scenarios as in Figure 13.34. This is also because of the larger retry limit of CORMAN (10 times) compared to AODV over 802.11 (7 times).

13.5.3.3 Performance versus Velocity

We also study CORMAN's performance and compare it to AODV in different rates of node velocity. We conduct another set of tests in a network of 100 nodes deployed in a 300 × 300 (m²) space with a varying v_{max}, where $v_{max} = 0, 2, 4, 6, ..., 20$ (m/s). For each velocity scenario, we test CORMAN and AODV's performance in transporting CBR data flows again between a randomly selected source-destination pair. We repeat this process five times for a given scenario. We collect the PDR, end-to-end delay, and delay jitter for both protocols averaged over each scenario, and plot them in Figures 13.35 through 13.37, respectively. Note that these are fairly dense networks, so that AODV has a reasonably high PDR. Since the network diameter is rather small in this case, the measurements are fairly consistent across these different velocity scenarios.

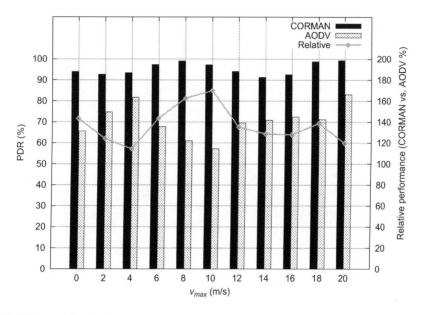

Figure 13.35 PDR vs. node velocity.

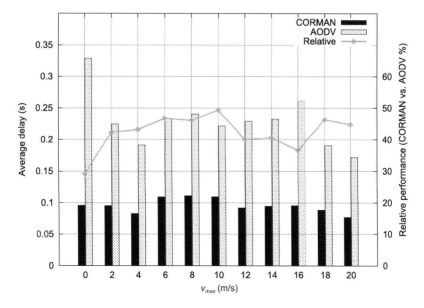

Figure 13.36 Packet delay vs. node velocity.

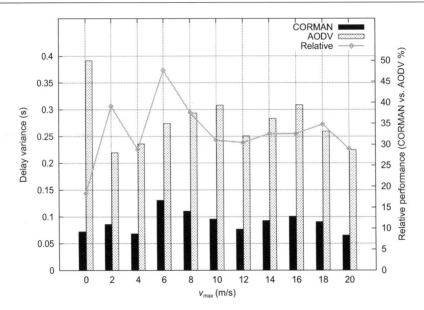

Figure 13.37 Delay jitter vs. node velocity.

From Figure 13.35, we observe that CORMAN's PDR is constantly around 95%, while that of AODV varies between 57% and 82%. With this very high network density, AODV's route search succeeds in a majority of the cases. Yet, it still does not have the same level of robustness against link quality changes as CORMAN. Compared to AODV, CORMAN has only a fraction of the end-to-end delay and variance (Figures 13.36 and 13.37) for two reasons. First, the opportunistic data forwarding scheme in CORMAN allows some packets to reach the destination in fewer hops than AODV. Second, the proactive routing (PSR) in CORMAN maintains full-on route information, whereas AODV still has to search for them if a route is broken. Although route search in AODV usually succeeds in dense networks with fluctuating link quality, the delay introduced by this process is inevitable.

Based on these observations, CORMAN can maintain its performance despite a high rate of node velocity with realistic channels emulated by the Nakagami model. Hence, it is very suitable for many mobile ad hoc network applications, e.g., vehicular ad hoc network (VANET) applications.

13.6 CONCLUDING REMARKS

13.6.1 Conclusions

In this chapter, we have proposed CORMAN as an opportunistic routing scheme for mobile ad hoc networks. CORMAN is composed of three components. (1) PSR – a proactive source routing protocol, (2) large-scale live update of forwarder list, and (3) small-scale retransmission of missing packets. All of these explicitly utilize the broadcasting nature of wireless channels and are achieved via efficient cooperation among participating nodes in the network. Essentially, when packets of the same flow are forwarded, they can take different paths to the destination. For example, in Figure 13.38, the route between nodes X and Y as determined by the routing module is indicated by the yellow band. The solid circles represent the listed forwarders, and the hollow ones are the small-scale retransmitters. In actual operations of CORMAN, packets p_1, p_2, and p_3 can take separate routes around this band

Figure 13.38 Packet trajectories.

depending on the transient link quality in the network. Such a decision is made on a per-hop and per-packet basis. Through computer simulation, CORMAN is shown to have superior performance measured in PDR, delay, and delay jitter.

13.6.2 Discussion

13.6.2.1 Proactive Source Routing Related

In particular, the PSR is motivated by the need of supporting opportunistic data forwarding in *mobile* ad hoc networks. In order to generalize the milestone work of ExOR for it to function in such networks, we needed a proactive source routing protocol. Such a protocol should provide more topology information than just distance vectors but has a significantly smaller overhead than link-state routing protocols; even the MPR technique in OLSR would not suffice this need. Thus, we put forward a tree-based routing protocol, PSR, inspired by PFA and WRP. Its routing overhead per time unit per node is in the order of the number of the nodes in the network as with DSDV, but each node has the full-path information to reach all other nodes. For it to have a very small footprint, PSR's route messaging is designed to be very concise. First, it uses only one type of message, i.e., the periodic route update, both to exchange routing information and as hello beacon messages. Second, rather than packaging a set of discrete tree edges in the routing messages, we package a converted binary tree to reduce the size of the payload by about half. Third, we interleave full dump messages with differential updates so that, in relatively stable networks, the differential updates are much shorter than the full dump messages. To further reduce the size of the differential updates, when a node maintains its routing tree as the network changes, it tries to minimize the alteration of the tree. As a result, the routing overhead of PSR is only a fraction or less compared to DSDV, OLSR, and DSR as evidenced by our experiments. Yet, it still has similar or better performance in transporting TCP and UDP data flows in mobile networks of different velocity rates and densities.

In the simulation in this work, we used PSR to support traditional IP forwarding for a closer comparison with DSDV and OLSR, while DSR still carried source-routed messages. In our simultaneous work, CORMAN [38], we tested PSR's capability in transporting source-routed packets for opportunistic data forwarding, where we also found that PSR's small overhead met our initial goal. That being said, as indicated in Section 13.5.1.2 earlier, while alleviating forwarding nodes from table lookup, DSR's source routing is especially vulnerable in rapidly changing networks. The reason is that, as a source-routed packet progresses further from its source, the path carried by the packet can become obsolete, forcing an intermediate node that cannot find the next hop of the path to drop the packet. This is fundamentally different from traditional IP forwarding in proactive routing with more built-in adaptivity, where the routing information maintained at nodes closer to the destination is often more updated than the source node. Although out of the scope of this research, it would be interesting to allow intermediate nodes running DSR to modify the path carried by a source-routed packet for it to use its more updated knowledge to route data to the destination. This is in fact exactly what PSR does when we used it to carry source-routed data in CORMAN. Granted, this opens up an array of security issues, which themselves are part of a vast research area.

As with many protocol designs, in many situations working on PSR, we faced tradeoffs of sorts. Striking such balances not only gave us the opportunity to think about our design twice, but also made us understand the problem at hand better. One particular example is related to trading computational power for data transfer performance. During one route exchange interval, a node receives a number of routing messages from its neighbors. It needs to incorporate the updated information to its knowledge base and share it with its neighbors. The question is when should these two events happen. Although incorporating multiple trees at one time is computationally more efficient, we chose to do that immediately after receiving an update from a neighbor. As such, the more accurate information takes effect without any delay. Otherwise, when a data packet is forwarded to a neighbor that no longer exists, it causes link layer retrial, backlogging of subsequent packets, and TCP congestion avoidance and retransmission. With the broadcast and shared nature of the wireless channel, the afore-mentioned effects are contrary to all other data flows in the area. Therefore, in research on multi-hop wireless networking, it almost always makes sense for us to minimize any impact on the network's communication resources even if there is penalty in other aspects. When it comes to when a node should share its updated route information with its neighbors, we chose to delay it until the end of the cycle so that only one update is broadcast in each period. If a node were to transmit it immediately when there is any change to its routing tree, it would trigger an explosive chain reaction, and the route updates would overwhelm the network. As we found out in our preliminary tests, this is the primary reason that WRP's overhead was significantly higher than the other protocols under study. PSR has just opened the box for us, and there will be many more things that we would like to investigate about it.

13.6.2.2 About Large-Scale Live Update and Small-Scale Retransmission

The large-scale live update is another way to utilize the broadcast nature in the wireless communication network, especially for the opportunistic data forwarding with proactive routing protocols in MANETs. In particular, to avoid too much routing overhead that is injected into the network, the proactive routing protocols can update the topology changes not in an event-driven manner but in a timer-driven one, which means the routing update should propagate from destination to source in a periodical way as we did in PSR (Section 13.3). As a result, the further from the source node to the destination node, the less accurate routing infor-mation maintained on the source node. Coherently, the Route-Prioritized Contention-based opportunistic data forwarding (Section 13.2.1.1.2) always wants the intermediate node that is nearest to the destination to forward the packets first, where much fresher routing infor-mation is maintained than upstream nodes which will forward the remaining packets later. Hence, large-scale live update uses the broadcast nature further in the wireless opportunistic forwarding network, and when a node takes itself as a frontier (Section 13.4.1) of the batch, it will update the forwarder list and pack the new one in the data packet. Therefore, when the data packets are delivered towards the destination, the fresher routing information can be propagated towards the source node more quickly with no additional overhead.

Furthermore, in the small-scale retransmission we proposed in Section 13.5, we broaden the opportunistic data forwarding further. In wireless communication networks, the tradi-tional IP forwarding, opportunistic data forwarding, and opportunistic data forwarding with small-scale retransmission compose an evolution process. In particular, the most fundamental approach is the IP forwarding, which is initially proposed for the Internet and only one next hop address marks the intended receiver in the forwarding process. Then, ExOR [9] makes a group of receiver to be intended by including a forwarder list in the data packet, and it makes all receivers forward the packet in a prioritized order. Now, our small-scale retransmission not only grants the forwarding ability to the receivers included in the forwarder list, but also lets the nodes which are not shown in the forwarder list participate in data forwarding between two of them. The broadened opportunistic data forwarding takes the broadcast nature one step further.

13.6.3 Future Work

Research based on CORMAN can be extended in the following interesting ways.

1. It would be informative to further test CORMAN. For example, we can compare CORMAN to ExOR and IP forwarding in static multi-hop networks with varying link quality to study their relative capabilities in data transfer. We will also test these data transfer techniques when multiple simultaneous flows are present, to study how well they share the network resources. Through these tests, we will be able to further optimize some parameters of CORMAN, such as the retry limit.
2. The coordination among multiple qualified small-scale retransmitters can be achieved with better measures than RSSI. In particular, if the "suitability" score is based on transient link quality rather than historic information, we will be able to better utilize the receiver diversity. Apparently, this would require non-trivial coordination among these retransmitters, which could challenging especially when we aim for zero extra overhead.
3. Nodes running CORMAN forward data packets in fragments. When the source and destination nodes are separated by many hops, it should allow nodes at different segments of the route to operate simultaneously. That is, a pipeline of data transportation could be achieved by better spatial channel reuse. The design of CORMAN can be further improved to address this explicitly. This may involve making the timing node back off more precisely and tightly, or even devising a completely different coordination scheme.

The potential of cooperative communication in multi-hop wireless networks is yet to be unleashed at higher layers, and CORMAN is only an example. PSR has just opened the box for us, and there will be many more aspects of it that we would like to investigate.

REFERENCES

[1] Imrich Chlamtac, Marco Conti, and J.-N. Liu. Mobile Ad Hoc Networking: Imperatives and Challenges. *Ad Hoc Networks*, 1(1):13–64, 2003.
[2] Rajmohan Rajaraman. Topology Control and Routing in Ad Hoc Networks: A Survey. *SIGACT News*, 33:60–73, 2002.
[3] Charles E. Perkins and Pravin Bhagwat. Highly Dynamic Destination-Sequenced Distance-Vector Routing (DSDV) for Mobile Computers. *Computer Communication Review*, pp. 234–244, 1994.
[4] Thomas Clausen and Philippe Jacquet. Optimized Link State Routing Protocol (OLSR). *RFC 3626*, October 2003.
[5] David B. Johnson, Yih-Chun Hu, and David A. Maltz. On the Dynamic Source Routing Protocol (DSR) for Mobile Ad Hoc Networks for IPv4. *RFC 4728*, February 2007.
[6] Charles E. Perkins and Elizabeth M. Royer. Ad hoc On-Demand Distance Vector (AODV) Routing. *RFC 3561*, July 2003.

[7]   Thomas M. Cover and Abbas A. El Gamal. Capacity Theorems for the Relay Channel. *IEEE Transactions on Information Theory*, 25(5):572–584, 1979.

[8]   Aria Nosratinia, Todd E. Hunter, and Ahmadreza Hedayat. Cooperative Communication in Wireless Networks. *IEEE Communications Magazine*, 42(10):74–80, 2004.

[9]   Sanjit Biswas and Robert Morris. ExOR: Opportunistic Multi-Hop Routing for Wireless Networks. In *Proceedings of ACM Conference of the Special Interest Group on Data Communication (SIGCOMM)*, pp. 133–144, Philadelphia, PA, August 2005.

[10] Peter Larsson. Selection Diversity Forwarding in a Multihop Packet Radio Network with Fading Channel and Capture. *ACM Mobile Computing and Communications Review*, 5(4):47–54, 2001.

[11] Eric Rozner, Jayesh Seshadri, Yogita Mehta, and Lili Qiu. *Simple Opportunistic Routing Protocol for Wireless Mesh Networks*. In *Proceedings of the 2nd IEEE Workshop on Wireless Mesh Networks (WiMesh)*, pp. 48–54, September 2006.

[12] Mathias Kurth, Anatolij Zubow, and Jens-Peter Redlich. *Cooperative Opportunistic Routing Using Transmit Diversity in Wireless Mesh Networks*. In *Proceedings of the 27th IEEE International Conference on Computer Communication (INFOCOM)*, pp. 1310–1318, 2008.

[13] Mohammad Naghshvar and Tara Javidi. *Opportunistic Routing with Congestion Diversity in Wireless Multi-Hop Networks*. In *Proceedings of the 29th IEEE International Conference on Computer Communication (INFOCOM)*, pp. 496–500, Piscataway, NJ, 2010. IEEE Press.

[14] Kai Zeng, Zhenyu Yang, and Wenjing Lou. *Opportunistic Routing in Multi-Radio Multi-Channel Multi-Hop Wireless Networks*. In *Proceedings of the 29th IEEE International Conference on Computer Communication (INFOCOM)*, pp. 476–480, Piscataway, NJ, 2010. IEEE Press.

[15] Shengbo Yang, Feng Zhong, Chai Kiat Yeo, Bu Sung Lee, and Jeff Boleng. *Position Based Opportunistic Routing for Robust Data Delivery in MANETs*. In *Proceedings of the 2009 IEEE Conference on Global Telecommunications (GLOBECOM)*, pp. 1325–1330, Honolulu, Hawaii, December 2009.

[16] Zifei Zhong and Srihari Nelakuditit. *On the Efficacy of Opportunistic Routing*. In *Proceedings of the 4th IEEE Communications Society Conference on Sensor, Mesh and Ad Hoc Communications and Networks (SECON)*, pp. 441–450, June 2007.

[17] Szymon Chachulski, Michael Jennings, Sachin Katti, and Dina Katabi. *Trading Structure for Randomness in Wireless Opportunistic Routing*. In *Proceedings of ACM Conference of the Special Interest Group on Data Communication (SIGCOMM)*, pp. 169–180, Kyoto, Japan, August 2007.

[18] Douglas S. J. De Couto, Daniel Aguayo, John Bicket, and Robert Morris. *A High-Throughput Path Metric for Multi-Hop Wireless Routing*. In *Proceedings of the 9th Annual International Conference on Mobile Computing and Networking (MobiCom)*, pp. 134–146, San Diego, CA, 2003.

[19] Jian Ma, Qian Zhang, Chen Qian, and Lionel M. Ni. *Energy-Efficient Opportunistic Topology Control in Wireless Sensor Networks*. In *Proceedings of the 1st International MobiSys Workshop on Mobile Opportunistic Networking (MobiOpp)*, pp. 33–38, New York, NY, 2007. ACM.

[20] Ilias Leontiadis and Cecilia Mascolo. *GeOpps: Geographical Opportunistic Routing for Vehicular Networks*. In *Proceedings of the IEEE International Symposium on a World of Wireless Mobile and Multimedia Networks (WoWMoM)*, pp. 1–6, Helsinki, Finland, June 2007.

[21] BoZidar Radunović, Christos Gkantsidis, Peter Key, and Pablo Rodriguez. *An Optimization Framework for Opportunistic Multipath Routing in Wireless Mesh Networks*. In *Proceedings of the 27th IEEE International Conference on Computer Communication (INFOCOM)*, pp. 2252–2260, April 2008.

[22] Dimitrios Koutsonikolas, Chih-Chun Wang, and Y. Charlie Hu. *CCACK: Efficient Network Coding Based Opportunistic Routing through Cumulative Coded Acknowledgments*. In *Proceedings of the 29th IEEE International Conference on Computer Communication (INFOCOM)*, pp. 2919–2927, Piscataway, NJ, 2010. IEEE Press.

[23] Rahul C. Shah, Sven Wiethölter, Adam Wolisz, and Jan M. Rabaey. *When Does Opportunistic Routing Make Sense?* In *Proceedings of the 3rd Annual IEEE International Conference on Pervasive Computing and Communications (PerCom)*, pp. 350–356, March 2005.

[24] Jonghyun Kim and Stephan Bohacek. *A Comparison of Opportunistic and Deterministic Forwarding in Mobile Multihop Wireless Networks*. In *Proceedings of the 1st International MobiSys Workshop on Mobile Opportunistic Networking (MobiOpp)*, pp. 9–16, New York, NY, 2007. ACM.

[25] Kai Zeng, Wenjing Lou, and Hongqiang Zhai. *On End-to-End Throughput of Opportunistic Routing in Multirate and Multihop Wireless Networks*. In *Proceedings of the 27th IEEE International Conference on Computer Communication (INFOCOM)*, pp. 816–824, April 2008.

[26] Llorenc Cerdà-Alabern, Vicent Pla, and Amir Darehshoorzadeh. On the Performance Modeling of Opportunistic Routing. In *Proceedings of the 2nd International Workshop on Mobile Opportunistic Networking (MobiOpp)*, pp. 15–21, New York, NY, 2010. ACM.

[27] Mei-Hsuan Lu, Peter Steenkiste, and Tsuhan Chen. *Video Transmission over Wireless Multihop Networks Using Opportunistic Routing*. In *Proceedings of the 16th IEEE International Packet Video Workshop (PV)*, pp. 52–61, November 2007.

[28] Fan Wu, Tingting Chen, Sheng Zhong, Li Erran Li, and Yang Richard Yang. *Incentive-Compatible Opportunistic Routing for Wireless Networks*. In *Proceedings of the 14th ACM International Conference on Mobile Computing and Networking (MobiCom)*, pp. 303–314, New York, NY, 2008. ACM.

[29] J. J. Garcia-Luna-Aceves and Shree Murthy. A Path-Finding Algorithm for Loop-Free Routing. *IEEE/ACM Transactions on Networking*, 5:148160, 1997.

[30] Jochen Behrens and J. J. Garcia-Luna-Aceves. *Distributed, Scalable Routing Based on Link-State Vectors*. In *Proceedings of ACM SIGCOMM*, pp. 136–147, 1994.

[31] Kirill Levchenko, Geoffrey M. Voelker, Ramamohan Paturi, and Stefan Savage. *XL: An Efficient Network Routing Algorithm*. In *Proceedings of ACM SIGCOMM*, pp. 15–26, 2008.

[32] Shree Murthy and J. J. Garcia-Luna-Aceves. An Efficient Routing Protocol for Wireless Networks. *Mobile Networks and Applications*, 1(2):183–197, 1996.

[33] J. J. Garcia-Luna-Aceves and Marcelo Spohn. *Source-Tree Routing in Wireless Networks*. In *Proceedings of the 7th Annual International Conference on Network Protocols (ICNP'99)*, pp. 273–282, Toronto, Canada, October 1999.

[34] Xin Yu. Distributed Cache Updating for the Dynamic Source Routing Protocol. *IEEE Transactions on Mobile Computing*, 5(6):609–626, 2006.

[35] Yih-Chun Hu and David B. Johnson. *Implicit Source Routes for On-Demand Ad Hoc Network Routing*. In *Proceedings of the 3rd ACM International Symposium on Mobile Ad Hoc Networking and Computing (MobiHoc '01)*, pp. 1–10, Long Beach, CA, October 2001.

[36] Bin Hu and Hamid Gharavi. *DSR-Based Directional Routing Protocol for Ad Hoc Networks*. In *Proceedings of the 2007 IEEE Conference on Global Telecommunications (GLOBECOM)*, pp. 4936–4940, Washington, DC, November 2007.

[37] Douglas West. *Introduction to Graph Theory* (2nd Edition). Prentice Hall, Upper Saddle River, NJ, August 2000.

[38] Zehua Wang, Yuanzhu Chen, and Cheng Li. *CORMAN: A Novel Cooperative Opportunistic Routing Scheme in Mobile Ad Hoc Networks. IEEE Journal on Selected Areas in Communications*, to appear in 2012.

[39] Zehua Wang, Cheng Li, and Yuanzhu Chen. *PSR: Proactive Source Routing in Mobile Ad Hoc Networks*. In *Proceedings of the 2011 IEEE Conference on Global Telecommunications (GLOBECOM)*, Houston, TX, December 2011.

[40] Zehua Wang, Yuanzhu Chen, and Cheng Li. *A New Loop-Free Proactive Source Routing Scheme for Opportunistic Data Forwarding in Wireless Networks. IEEE Communications Letters*, 15(11):1184–1186, 2011.

[41] Zehua Wang, Yuanzhu Chen, and Cheng Li. *PSR: A Lightweight Proactive Source Routing Protocol for Mobile Ad Hoc Networks*. In *2012 IEEE Transactions on Vehicular Technology*, 63(2):859–868, 2014 (February), doi: 10.1109/TVT.2013.2279111.

[42] Zehua Wang, Cheng Li, and Yuanzhu Chen. *Local Cooperative Retransmission in Opportunistic Data Forwarding in Mobile Ad-hoc Networks*. In *2012 IEEE International Conference on Communications (ICC)*, Ottawa, ON, 2012, pp. 51–62, doi: 10.1109/ICC.2012.6364129.

[43] Charles Reis, Ratul Mahajan, Maya Rodrig, David Wetherall, and John Zahorjan. *Measurement-Based Models of Delivery and Interference in Static Wireless Networks*. In *Proceedings of ACM SIGCOMM*, pp. 51–62, August 2006.

[44] Mei-Hsuan Lu, Peter Steenkiste, and Tsuhan Chen. *Design, Implementation and Evaluation of an Efficient Opportunistic Retransmission Protocol*. In *Proceedings of the 15th Annual International Conference on Mobile Computing and Networking (MobiCom)*, pp. 73–84, New York, NY, 2009. ACM.

[45] White Paper Wild Packets Inc. Converting Signal Strength Percentage to dBm Values, November 2002.

[46] Mahesh K. Marina and Samir R. Das. *Routing Performance in the Presence of Unidirectional Links in Multihop Wireless Networks*. In *The Third ACM International Symposium on Mobile Ad Hoc Networking and Computing (MobiHoc '02)*, pp. 12–23, Lausanne, Switzerland, June 2002.

Chapter 14

Security in Opportunistic Networks

Mohini Singh and Anshul Verma
Banaras Hindu University, Varanasi, India

Pradeepika Verma
Indian Institute of Technology (BHU), Varanasi, India

CONTENTS

14.1 INTRODUCTION

The opportunistic network does not require an end-to-end connection between source and destination nodes for communication. It is the best way for providing communication in an intermittent mobile ad hoc network. It may consist of innumerable nodes that are frequently connected or disconnected due to the mobility of the nodes. The messages are exchanged or forwarded by exploiting expedient contacts between nodes without relying on pre-existing network infrastructure [1]. The delay tolerant network is a superclass of the opportunistic network [2].

The opportunistic network works on the store-carry-and-forward mechanism [3, 4]. As shown in Figure 14.1, the source sends a message to the destination, but there is no direct communication link between them. Therefore, some intermediate nodes are utilized to forward the message towards the destination. The intermediate nodes store and carry the message until they find the destination node or better intermediate node to pass the message to the destination. The routing protocols also exploit much of the users' context information to optimize the routing task [5–7].

The opportunistic network poses various unique features, such as high node mobility, intermittent communication links, and long communication delays. Therefore, designing opportunistic forwarding protocols are very difficult and vulnerable to various security issues. The design of opportunistic networks faces serious problems, such as how to effectively implement node authentication and access control, confidentiality, and data integrity and how to ensure routing security, privacy protection, cooperation, and trust management. In other words, systematic research on security solutions for opportunistic networks is still open and challenging [8]. Security is a process or technique of ensuring the integrity, confidentiality,

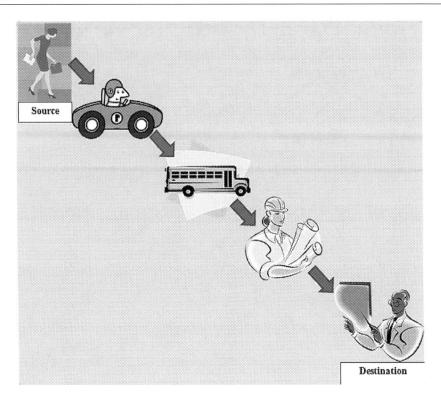

Figure 14.1 Opportunistic network.

non-repudiation, availability, authorization, and authentication of a network system. Despite the various challenges, many real-life applications have been developed based on opportunistic networks [9].

The rest of the chapter is organized as follows. Section 14.2 describes characteristics of opportunistic networks. Sections 14.3 and 14.4 discuss general issues and security issues in opportunistic networks, respectively. A study of main thrust areas of security as well as research works that have been done in these areas are presented in Section 14.5. Section 14.6 describes the general architecture of the security in opportunistic networks. Section 14.7 presents various security attacks in opportunistic networks and their defense mechanisms. Finally, we conclude the chapter and give some future research directions.

14.2 CHARACTERISTICS OF OPPORTUNISTIC NETWORKS

Many developing and potential communication circumstances do not accept the web's underlying presumptions. In a lack of the end-to-end path between source and destination, the TCP/IP protocol does not work properly. Some drawback occurs, such as intermittent connectivity, high delay, and high error rate. However, the opportunistic network is suitable for an environment that is characterized by intermittent connectivity, high delay, and error rate.

Mobility: Mainly, an opportunistic network is formed by handheld portable mobile devices/ nodes carried by users. Therefore, nodes are highly mobile and frequently change their locations that are the main reasons for intermittent connectivity. However, mobility is treated as an advantage rather than a drawback because the mobility of nodes creates an opportunity to meet with the destination or better forwarding nodes.

Intermittent connectivity: Opportunistic networks improve network communication where the connectivity of the network is intermittent/irregular. An opportunistic network compensates for intermittent connectivity via the "store-carry-forward" techniques. Military applications in opportunistic network regions are allowing critical information retrieval in mobile battlefield circumstances using only intermittent connectivity network communication. Intermittent connectivity might be scheduled or opportunistic.

Long delay: The dynamic nature of opportunistic networks affects the connectivity behavior in topology. Packet loss is one of the challenges for the wireless network. But opportunistic networks use packet loss challenge as an opportunity with the help of "store-carry-forward" techniques. In the opportunistic network, nodes store the message until the better forwarding node comes into the communication range. These techniques increase the delay of the network.

High error rate: The higher error rate may be less trustworthy for communication or data transmission. A high error rate is the outcome of intermittent connectivity. The degree of errors encountered during data transmission over a communication or network connection is responsible for corrections in bit errors. So, it requires more bit processing or packet retransmission as well as increases the traffic in the network. On the other hand, the link error rate requires low retransmission. Hop-to-hop transmission reduces the number of packets retransmitted in comparison to Internet-type transmission.

14.3 ISSUES IN OPPORTUNISTIC NETWORKS

Bandwidth: An opportunistic network has a limited bandwidth in comparison to a wired or traditional wireless network. The opportunistic network has a low capacity as compared to infrastructure networks. The opportunistic network has deficient interference conditions and multiple accesses and fading effects.

Connectivity: Due to the limitations of the opportunistic network, connectivity is a challenge. Short-range communication and mobility of nodes decrease the connection between the nodes in the opportunistic network. Low connectivity might affect the topology as well as end-to-end service for forwarding data in the opportunistic network. High connectivity increases the efficiency of the performance.

Congestion: An opportunistic network supports intermittent connectivity. It overcomes delays as well as communication disruptions, which are drawbacks of a traditional network. Data transmission is allowed between the source and the destination, where the end-to-end path does not exist. In the opportunistic network, nodes' connectivity is limited, but some nodes are well connected in the network. Hence, these nodes' connections are exploited by most of the ongoing transmissions; as a consequence, these connections become more vulnerable to congestion. It also increases the unusable disconnection and low delivery rate. Congestion control [10, 11] is a more challenging issue in the opportunistic network because of intermittent connectivity. The opportunistic network mainly targets the management of buffer for congestion control by eliminating unrequired data from the buffer.

Routing: An opportunistic network observes issues related to end-to-end services, where the pre-existing path does not exist. The dynamic behavior of nodes increases the challenge for route establishment. It is necessary to overcome main issues such as mobility, dynamic topology, and routing table. So, an opportunistic network is forwarding data using the "store-carry-forward" technique [8–10].

a. *Mobility*: Mobility is a reason for the dynamic behavior of the network [12, 13]. Due to mobility, nodes frequently change their locations in the network and might increase the coverage area of the network.

b. *Topology*: Due to the frequent mobility of nodes, the topology between nodes is also dynamic. Sometimes a temporary topology is established between the nodes, but it is not stable due to the nodes' mobility.

c. *Routing table*: In the opportunistic network, unlike traditional wireless networks, pre-existing path information is not saved in routing tables because nodes are highly mobile; they change their positions frequently. If a node changes its position, that affects the routing table. The routing table frequently changes its records, and updating routing tables at each node frequently generates extra overhead.

Security: An opportunistic network observes security issues that are concerned with end-to-end path services as well as where the end-to-end path may not exist for data forwarding. It is necessary to have some authorized, authentication, privacy, and access control mechanism to handle the security-related issues in the network. Various objectives have been created for the security component of the opportunistic network. The first objective is immediately avoiding unapproved applications from having their information brought through the opportunistic network. The second objective is avoiding unapproved applications from claiming control over the opportunistic network. The third objective is to instantly dispose of groups that are harmful or inappropriate for the network. In conclusion, the objective is to recognize expeditiously and de-approve bargained elements.

14.4 SECURITY ISSUES IN OPPORTUNISTIC NETWORKS

Mostly, security of the network techniques is a challenge to collective message integrity and authorized identities of users. These techniques do not challenge routers' authorization, which processes the information for end-to-end services. This segment examines a portion of the issues that occur in the security of an opportunistic network.

Key management: A significant open issue in opportunistic network security is the absence of a delay-tolerant technique for key administration [14]. At this phase, we are just truly realizing how to utilize existing plans, which at last require an online status testing service or key dissemination service, which isn't viable in a high delay or exceptionally interrupting condition. The main, for the most part, appropriate plans we have right now are virtually identical to shared confidential facts or, in all likelihood, irreversible public key (or authentication-based) plans. These are the regions where more research work could create fascinating outcomes. The key management is based on the typically traditional cryptography. It does not sustain router authentication.

Replay management: Handling replay is one of the significant issues in the opportunistic network. It became a challenge to discourse replay issues in the opportunistic network because of the intermitted connectivity in the network. In major networking circumstances, we either try to remove or else drastically minimize the possibility of a message being replayed. The key management based on the handling replays depends mostly on the message possibly being replayed. In some opportunistic network circumstances, this determination also is mostly the case, as replaying (for example, authorized or authenticated) data can be a reasonably straightforward method to consume rare network resources. Delay elements in the opportunistic network also make handling the replay a difficult task. Therefore, replay management remains an open security issue in the opportunistic network.

Security of routing protocols: A major issue in the opportunistic network is routing protocol security. On the other hand, if any opportunistic network routing protocol was supposed to use either Licklider Transmission Protocol (LTP) [15–17], or Bundle Protocol, it could utilize their available security features. The mechanism of security is established for the block of metadata. This mechanism has been used for another block of non-payload data. For some issues, it might provide solutions, but most of the issues are still unaddressed.

Traffic investigation: A general traffic investigation assurance plot is not a reasonable objective for the opportunistic network, and there has been no requirement of a non-exclusive way to deal with this issue. For some disturbance lenient systems, concealing traffic might be a significant security necessity. Furthermore, the leading issue is the degree to which there is a genuine requirement for a nonexclusive plan for security against traffic investigation. The subsequent issue is how to characterize such a plan to be postponement- and interruption-lenient and which additionally doesn't devour an excessive number of assets. At last, traffic investigation insurance might be left as a nearby issue for the fundamental system layers.

Multicast security: In the opportunistic network, there is a lack of techniques that define which nodes register for anycast or multicast end-point services. Therefore, there is a need for techniques that will clearly define which node is registered for anycast or multicast for a particular service. For example, the signup for a mailing list is similar to multicast end-point registration.

Evaluation of performance: The security arrangement for the opportunistic network will expend extra bandwidth utilization cost and computation cost for the network's nodes. Also, if more than one security management techniques are utilized, greater bandwidth amount and computation power will probably be utilized. Therefore, security techniques should be developed according to the need of the applications, bandwidth, and computation power available for the security features; it should not generate extra overhead on the network.

14.5 THRUST AREAS OF SECURITY

Security is an important issue in opportunistic networks; therefore, several research works have been done to resolve various aspects of security. Here we discuss research works that have been done related to bundle security protocols, key and trust management, security initialization, fragment authentication, etc.

Bundle security protocols: Bundle security protocols [18] provide a guarantee for hop-by-hop authentication, end-to-end authentication, and end-to-end confidentiality of messages, and they are described in the RFC 6257 standard. The Bundle Protocol specification [19] has four extension blocks that are added to a message, namely the Bundle Authentication Block (BAB), Payload Integrity Block (PIB), Payload Confidentiality Block (PCB), and the Extension Security Block (ESB). However, the protocol does not include a key management module and assumes that all nodes must have some provision to perform cryptographic operations. BAB is responsible for the hop-to-hop authentication of a message between two adjacent nodes. PIB ensures end-to-end authentication of a message between the source and destination. PCB ensures end-to-end confidentiality by encrypting messages between the source and destination. Whereas, ESB protects the metadata portion (non-payload part) of a message. The standard also defines the way of using cryptographic approaches to process the security blocks on the intermediate nodes or final destination node. The standard also has the provision of a security-zone, security-source, and security-destination, which divides the security control over multiple network organizations. This also provides security to the devices that do not support cryptographic operations.

Key management: Secure key distribution is necessary for a network to ensure message integrity, authenticity, and confidentiality. In the traditional Internet, secure key distribution is implemented by using online servers. Whereas, online servers cannot be assumed in opportunistic networks for key distribution and certificate verification [20]. To solve this problem, Identity Based Cryptography (IBC) [21] was proposed in [20] for secure key distribution in intermittently connected networks. Time-based keys [22] were used for controlling the rights of compromised nodes and malicious users, in which a signing key is valid for a short period. To check the applicability of the IBC approach, a comparative study between IBC and traditional PKI approaches was performed in [23] over message integrity, authenticity, and end-to-end confidentiality. It was concluded that IBC does not provide any advantage in opportunistic networks over the traditional PKI approach in terms of message integrity and authenticity. However, it was shown that IBC performs better than the PKI approach in terms of end-to-end confidentiality. A security architecture was proposed in [24] that introduces physical and cryptographic approaches to protect Internet kiosks. Based on a similar concept, Internet kiosks were used in [25] as PKGs to generate users' private keys. However, this approach does not guarantee user authentication.

Security initialization: Bootstrapping security in opportunistic networks is a challenging issue. An existing mobile network infrastructure-based approach was proposed in [23] that is assumed to be a widely available authentication scheme. A social context information-based approach was presented in [26] that uses social context information of nodes such as knowledge of present and past affiliation or encounter history of nodes to create an initial security context between two nodes that do not have any security relationship. They used the distribution of keys within some affiliated entities that are common for both nodes, and through those keys, they established secure communication between these two nodes.

Trust management: The main goal of trust management is preventing malicious nodes from conducting Sybil and black hole attacks. In the Sybil attack [27], a malicious node creates multiple identities in the network to get a larger influence in the network. The black hole attack [28] is a type of denial of service (DoS) attack in which messages are dropped by malicious nodes or other compromised nodes, malicious nodes may pretend to be a destination and may attract messages. An approach was developed in [29] to initializing the trust by building certificate chains (such as the Pretty Good Privacy system) that assume unconditional transitivity of trust along the chain path. In [30], social trust is used to detect Sybil attacks. Two types of social trusts (Explicit and Implicit social trusts) were used to detect Sybil attacks. The result of this approach is used to create an integrated framework for the secure content dissemination known as PodNetSec [31], which is built on top of PodNet [32]. To prevent black hole attacks, a reputation-based approach was proposed in [33] to predict the compatibility of an encountered node to deliver a message to a destination. In their extended work [34], the authors have applied it on PRoPHET to investigate its effectiveness against black hole attacks. A *data-centric trust*-based framework was presented in [35] to prevent black hole attacks, whose validity is inferred based on the Dempster-Shafer theory [36].

Fragment authentication: A secure message fragmentation scheme is proposed in [37] that is named toilet paper, in which checkpoints are inserted into fragments by using a cryptographic hash. A signature is also inserted in the message along with hashes that help any intermediate node to authenticate the fragment by using them. An enhanced version of this approach is proposed in [23]. A special authentication function is proposed in [37] for fragments authentication; however, how to construct such an efficient function is not described.

14.6 SECURITY ARCHITECTURE FOR OPPORTUNISTIC NETWORKS

The opportunistic network utilizes interactions of intermediate nodes to forward messages towards a destination. Intermediate nodes may be unreliable, and that raises various serious issues related to authentication, authorization, security, and privacy. To overcome these issues, a general security architecture is proposed in [8] (see Figure 14.2) that includes five modules: authentication, secure routing, access control, trust management and cooperation, and application/user-specific privacy protection. However, all modules may not be necessary for all opportunistic networks, and one or more modules may be eliminated depending upon user demands, network requirements, or threat scenarios.

First, the *authentication* security module verifies the identity of the nodes to prevent unauthorized nodes from joining the opportunistic networks. Second, the *access control* module verifies whether a particular node or user is authorized to access a network resource or to perform an operation. Third, the *secure routing* module ensures the message contents' confidentiality and avoids message tampering as well as delivers the message to the destination eventually. Fourth, the *trust management and cooperation* module monitors the behavior of the intermediate nodes to detect any malicious activity as well as implements a trust evaluation mechanism to promote nodes' cooperation and prevent selfish behavior of the nodes. Finally, the *application/user-specific privacy protection* module provides customization of privacy features to the nodes or applications according to their needs. This means a node or application may express or hide some information related to them selectively.

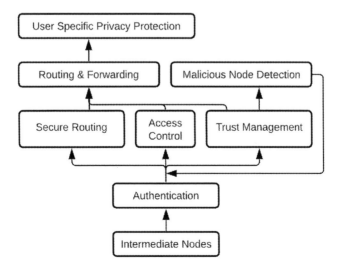

Figure 14.2 Security Architecture for Opportunistic Networks.

14.7 SECURITY ATTACKS AND THEIR DEFENSE MECHANISMS

In opportunistic networks, attacks can be classified into two classes: internal and external attacks [38–41]. Internal attacks occur due to the internal nodes that are part of the network and therefore are more harmful for the network because internal nodes have access rights over various network resources. Whereas, external attacks occur due to the external nodes those are not part of the network therefore are less harmful in comparison internal attacks because they do not have access rights over the network resources. Opportunistic networks are vulnerable to the various attacks such as wormhole attacks, black hole attacks, selfish attacks, Sybil attacks, and selective dropping attacks. These attacks and their defense mechanisms are discussed as follows.

Wormhole attacks: In wormhole attacks [42], malicious nodes pretend to be the source node of some packets in the network; those are actually sent by another node. Malicious nodes may have shortest path for some destinations and therefore may attract traffic or other nodes. The packet leash technique was proposed in [43] to protect from wormhole attacks in which geographical leashes and temporal leashes information can be added to packets to protect them from the attacks. This technique requires secure and tight time synchronization. Geographical leashes restrict the distance between the source and destination of a packet. Whereas, temporal leashes restrict the lifetime of a packet, and as a consequence, it restricts the maximum distance travelled by a packet. However, this kind of technique needs secure and tight time synchronization. An ad hoc on-demand distance vector (AODV) protocol-based wormhole attack detection and isolation technique was proposed in [44]. The source node sends a route request to the destination and in response receives all available routes with the number of hops. These routes are used to detect malicious nodes by applying route redundancy and routes aggregation and calculating the round-trip time (RTT) for all listed routes. However, this approach is not feasible for the opportunistic networks because end-to-end connections between source and destination may not be present, and it is very hard to find out more than one route to the destination.

Black hole attacks: In black hole attacks, malicious nodes discard some or all of the received packets and advertise that they have route information of the most demanding destinations on the network. An "encounter ticket" (ET)-based defense technique was proposed in [45] against a black hole attack in which ET represents evidence of the encounters of nodes. However, this technique does not work in all cases, and it detects an attacker only when claiming non-existent encounters and does not handle packet discarding due to black hole attack. A watchdog-based approach was proposed in [34, 46] to detect black hole attacks in which a trust/reputation is estimated for each node in the network to detect whether it is a malicious node or not. Trust/reputation is estimated for each node by other nodes based on its message forwarding history. Say a node A forwards a message to node B to deliver the message to node C. Node C sends a positive feedback message (PFM) to node A on successfully reception of the message from node B; otherwise, node A registers node B as a suspicious node until PFM arrives. In this way, the trust/reputation system is built. However, PFM is sent using epidemic routing that generates network overhead. An approach based on maintaining secure packets exchanging history between nodes is proposed in [47]. When two nodes encounter, they record the number of exchanged packets after encryption using a private key to prevent black hole attacks. However, initializing keys to the nodes manually or using a key distribution scheme will be difficult in opportunistic networks.

Dropping and selective dropping attacks: In a normal situation, packets are dropped according to the predefine policies of the network, such as resource limitation or expiration of time to live (TTL). A packet weight-based dropping policy and mechanism is proposed in [48] where weight is estimated using inter-contact time between nodes. However, it is very difficult to detect dropping or selective dropping attacks because malicious nodes drop few or all received packets, and source and destination nodes do not know where and when dropping takes place. Authenticated end-to-end or link-to-link acknowledgment mechanisms can be used for detecting packet dropping attacks [49, 50]. A network coding-based mechanism is proposed in [51] to assess the effect of a packet selective dropping attack. A destination node calculates delivery ratio and sends it to the source node. The source node starts dynamically adjusting the redundancy factor to cover the degradation in the delivery ratio caused by the attack. The effect of packet dropping on routing performance is evaluated by theoretical analysis as well as simulation. In [52], a packet dropping attack detection mechanism is proposed in which intermediate nodes send an acknowledgment to the source node on safe reception of a packet. Source node constructs a Markle tree on the basis of received acknowledgments and then compares the value of the tree root with precalculated value. If both the values are equal, then no packet dropping occurred on this path; otherwise, there is packet dropping. This mechanism can detect the path with the malicious node and can suggest alternative paths for retransmission. However, this mechanism generates network overhead and does not detect the exact malicious node on the path. A cooperative participation-based packet dropping detection mechanism is proposed in [53] in which cooperative participation occurs at the network-bootstrapping phase. Alternative routes are chosen to avoid the malicious or non-trusted path. This mechanism generates network overhead. A data provenance-based packet dropping attack detection technique is proposed in [54]. The watermarking-based secure provenance transmission mechanism and the inter-packet timing are used to identify malicious nodes. The technique performs three main tasks: detecting lost packets, identifying the presence of the attacks, and identifying and isolating the malicious paths. However, the technique does not identify the exact malicious node on the path.

Watchdog and pathrater techniques are used in [55] to improve throughput. The watchdog technique is used to detect the malicious node by monitoring the packets' transmission pattern. Whereas, the pathrater technique is used to predict the best path by using the information from the watchdog and the link reliability data. The watchdog technique is not efficient in case of ambiguous collisions, receiver collisions, false misbehavior, and limited transmission power. An ExWatchdog technique is proposed in [56] to overcome the drawbacks of the watchdog technique and to enhance the malicious node detection process. ExWatchdog is capable of detecting those malicious nodes who partition the network by falsely declaring other nodes as malicious. A reputation-based technique for detecting packet dropping attacks is presented in [57], in which direct and indirect observation information is used to estimate full reputation weight. Nodes are included or excluded from the network on the basis of reputation weights. Moreover, historical reputation and fuzzy logic are used to improve the performance of the fault tolerance. A packet-dropping-attack detection and malicious node detection algorithm is presented in [58]. The algorithm uses an indicative field in the packet header that is used to investigate whether a node received the complete original number of packets from the previous node or not. An attack detection mechanism is proposed in [59] to detect a special type of packet dropping attack where malicious nodes inject fake packets in place of dropped packets. The creation time of the packets are used to predict whether they are original or fake. The results show the mechanism achieves high accuracy and detection rate.

Selfish and Sybil attacks: Selfish nodes use network resources and services but deny to offer services for other nodes due to some constraints, such as, selfish nodes do not forward packets due to low battery or buffer overflow. Defense techniques to protect from selfish attack are classified into two classes: barter based and credit based. A barter-based technique to encourage selfish nodes to cooperate is proposed in [60]. The technique has two major parts: a reputation system and a virtual payment or rewarding system. When two nodes encounter, they exchange summaries of the messages currently in their buffer. Thereafter, they decide which messages have to be exchanged and in which order. Messages are exchanged one by one from both sides. If any node cheats, the transmission is canceled. After each successful message transmission, the nodes receive a score that is accumulated to calculate their total score at the end. However, this technique assumes both that the encountered nodes have the same number of messages to be exchanged, and that the connection time will be enough to exchange all the decided messages that are not practical in real scenarios. A credit-based incentive technique named MobiCent is proposed in [61], in which each node pays the relay nodes for forwarding messages. The payment scheme uses two algorithms: first is a payment set selection algorithm that decides which relays to be paid, and second is a payment calculation algorithm that decides how much has to be paid to each selected relay. Due to rewards, each relay node cooperates in the forwarding of messages without hesitation unless the reward is not sufficient. This technique will not effectively work if the network has selfish nodes in the majority.

In a Sybil attack, malicious nodes drop received packets as well as create fake identities. Therefore, it is very difficult to detect an actual malicious node because it uses different identities to communicate with neighbors. The definition and taxonomy of the Sybil attack is described in [62]. Various defense techniques are classified as old and new techniques. For example, the resource testing is described as an old technique, and radio resource testing, verification of key sets for random key predistribution, registration, position verification, and code attestation are shown as new defense techniques for Sybil attacks. A reputation-based defense technique to detect Sybil attacks is described in [63], in which explicit and implicit social trust establishment mechanisms are used. The trust is established between nodes by the contact quality between nodes and the trustworthiness of the nodes' opinions.

Anti-localization techniques: A node location can be predicted by tracking the path and movement of the node. The node location information can be misused by malicious nodes; thus, it is a serious concern in opportunistic networks. ALAR [64] is one of the techniques used to implement location privacy. ALAR divides a message into many encrypted segments and sends them to destination through different paths. The decryption key is stored in the last segment; therefore, a node cannot interpret the message contents unless it receives all the segments. The technique maximizes the sender's location privacy as well as message delivery. However, delivery ratio and delivery latency are degraded due to overhead of setting number of segments and multiple paths. In [65], nodes of the network are divided into groups; each node must be a member of a group. Each node is assigned its public/private key pair as well as its group's public/private key pair. Each node also maintains a copy of the public keys of all other nodes and groups. The source node divides a message into multiple parts and encrypts these parts using the destination node's public key. The source inserts the destination address into each encrypted message part and sends all the parts through multiple paths. Each node checks whether it is the destination of the message; if it is, it decrypts the message, otherwise it is forwarded to the neighbors. However, this technique is vulnerable to Sybil attacks. The application of k-anonymity for location-based services (LBS) and its recent advancements are reviewed in [66]. The applicability of k-anonymity for LBS is mainly classified on the basis of architecture, algorithms for anonymization, and the types of k-anonymity (according to the different query processing techniques).

14.8 CONCLUSION

The aim of this chapter was to discuss main security issues in opportunistic networks, and the research works have been done to address these issues. The chapter described the opportunistic network and its characteristics. The chapter discussed general issues as well as security issues of opportunistic networks. Moreover, various thrust areas of security were discussed along with the research works that have been done to address issues of these thrust areas. A general security architecture for opportunistic networks was also described. Finally, various security attacks and their defense techniques were discussed. As a new networking paradigm, opportunistic networks have very potential real-life applications in diverse domains such as wildlife monitoring, underwater networks, mobile social networking, pervasive computing, Internet communication in rural areas, and intelligent transportation. Therefore, security, privacy, and trust issues that are considered serious challenges for opportunistic networks must be addressed.

For future research, developing cooperative incentive mechanisms and fairness-aware incentive schemes for trust management may be good research directions. Moreover, the data-centric trust mechanism rather than the entity trust mechanism should be deeply studied in opportunistic networks. Privacy preservation is also a main concern for opportunistic networks that should be further investigated. Mostly, the correctness of security and trust protocols are validated through the simulators; that is not sufficient. Therefore, formal verification of security and trust protocols by using formal methods will be an emerging research direction in opportunistic networks.

ACKNOWLEDGMENT

This chapter is a part of the research work funded by "Seed Grant to Faculty Members under IoE Scheme (under Dev. Scheme No. 6031)"[1] and "DST-Science and Engineering Research Board (SERB), Government of India (File no. PDF/2020/001646)"[2].

NOTES

1 Awarded to Anshul Verma at Banaras Hindu University, Varanasi, India.
2 Awarded to Pradeepika Verma at Indian Institute of Technology (BHU), Varanasi, India.

REFERENCES

[1] Wahid, A., Kumar, G., & Ahmad, K. (2014). *Opportunistic networks: Opportunity versus challenges-survey*. In *National Conference on Information Security Challenges* (pp. 14–24).
[2] Boldrini, C., Conti, M., & Passarella, A. (2008). Autonomic behaviour of opportunistic network routing. *International Journal of Autonomous and Adaptive Communications Systems*, 1(1), 122–147.

[3]   Ciobanu, R. I., & Dobre, C. (2013). Opportunistic networks: A taxonomy of data dissemination techniques. *International Journal of Virtual Communities and Social Networking (IJVCSN)*, 5(2), 11–26.

[4]   Bharamagoudar, S. R., & Saboji, S. V. (2017, December). *Routing in opportunistic networks: taxonomy, survey*. In 2017 International Conference on Electrical, Electronics, Communication, Computer, and Optimization Techniques (ICEECCOT) (pp. 300–305). IEEE.

[5]   Verma, A., & Srivastava, A. (2011). Integrated routing protocol for opportunistic networks. *International Journal of Advance Computer Science and Application*, 2(3), 85–91.

[6]   Verma, A., Pattanaik, K. K., & Ingavale, A. (2013). Context-based routing protocols for oppnets. In *Routing in Opportunistic Networks* (pp. 69–97). Springer, New York, NY.

[7]   Verma, A., & Pattanaik, K. K. (2017). Routing protocols in opportunistic networks. *Opportunistic Networking: Vehicular, D2D and Cognitive Radio Networks*, 125–166.

[8]   Wu, Y., Zhao, Y., Riguidel, M., Wang, G., & Yi, P. (2015). Security and trust management in opportunistic networks: a survey. *Security and Communication Networks*, 8(9), 1812–1827.

[9]   Verma, A., Singh, M., Pattanaik, K. K., & Singh, B. K. (2018). Future Networks Inspired by Opportunistic Networks. *Opportunistic Networks: Mobility Models, Protocols, Security, and Privacy*, 104, 229–246.

[10] Grundy, A. (2012). *Congestion control framework for delay-tolerant communications* (Doctoral dissertation, University of Nottingham).

[11] Coe, E., & Raghavendra, C. (2010, March). *Token based congestion control for DTNs*. In 2010 IEEE Aerospace Conference (pp. 1–7). IEEE.

[12] Fathima, M., Ahmad, K., & Fathima, A. (2017). Mobility models and routing protocols in opportunistic networks: a survey. *International Journal of Advanced Research in Computer Science*, 8(5), 354–371.

[13] Mota, V. F., Cunha, F. D., Macedo, D. F., Nogueira, J. M., & Loureiro, A. A. (2014). Protocols, mobility models and tools in opportunistic networks: a survey. *Computer Communications*, 48, 5–19.

[14] Sahadevaiah, K., & Reddy, P. P. (2011). Impact of security attacks on a new security protocol for mobile ad hoc networks. *Network Protocols and Algorithms*, 3(4), 122–140.

[15] Ramadas, M., Burleigh, S., & Farrell, S. (2008). Licklider transmission protocol-specification. *IETF Request for Comments RFC*, 5326.

[16] Burleigh, S., Ramadas, M., & Farrell, S. (2008). Licklider transmission protocol-motivation. *IETF Request for Comments RFC*, 5325.

[17] Wang, R., Burleigh, S. C., Parikh, P., Lin, C. J., & Sun, B. (2010). Licklider transmission protocol (LTP)-based DTN for cislunar communications. *IEEE/ACM Transactions on Networking*, 19(2), 359–368.

[18] Symington, S., Farrell, S., Weiss, H., & Lovell, P. (2007). Bundle security protocol specification. *Work Progress*, October, 28.

[19] Scott, K., & Burleigh, S. (2007). Bundle protocol specification, https://tools.ietf.org/html/rfc5050.

[20] Seth, A., & Keshav, S. (2005, November). *Practical security for disconnected nodes*. In 1st IEEE ICNP Workshop on Secure Network Protocols, 2005 (NPSec). (pp. 31–36). IEEE.

[21] Boneh, D., & Franklin, M. (2001, August). *Identity-based encryption from the Weil pairing*. In Annual International Cryptology Conference (pp. 213–229). Springer, Berlin, Heidelberg.

[22] Fall, K. (2003). Identity based cryptosystem for secure delay tolerant networking. *Manuscript*.

[23] Asokan, N., Kostiainen, K., Ginzboorg, P., Ott, J., & Luo, C. (2007). Towards securing disruption-tolerant networking. *Nokia Research Center, Tech. Rep. NRC-TR-2007-007*.

[24] Ur Rahman, S., Hengartner, U., Ismail, U., & Keshav, S. (2008, August). *Practical security for rural Internet kiosks*. In Proceedings of the Second ACM SIGCOMM Workshop on Networked Systems for Developing Regions (pp. 13–18).

[25] Ma, D., & Tsudik, G. (2010). Security and privacy in emerging wireless networks. *IEEE Wireless Communications*, 17(5), 12–21.

[26] El Defrawy, K., Solis, J., & Tsudik, G. (2009, July). *Leveraging social contacts for message confidentiality in delay tolerant networks*. In 2009 33rd Annual IEEE International Computer Software and Applications Conference (Vol. 1, pp. 271–279). IEEE.

[27] Douceur, J. R. (2002, March). *The Sybil attack*. In *International Workshop on Peer-to-Peer Systems* (pp. 251–260). Springer, Berlin, Heidelberg.

[28] Raymond, D. R., & Midkiff, S. F. (2008). Denial-of-service in wireless sensor networks: Attacks and defenses. *IEEE Pervasive Computing*, 7(1), 74–81.

[29] Capkun, S., Buttyan, L., & Hubaux, J. P. (2003). Self-organized public-key management for mobile ad hoc networks. *IEEE Transactions on Mobile Computing*, 2(1), 52–64.

[30] Trifunovic, S., Legendre, F., & Anastasiades, C. (2010, March). *Social trust in opportunistic networks*. In *2010 INFOCOM IEEE Conference on Computer Communications Workshops* (pp. 1–6). IEEE.

[31] Trifunovic, S., Anastasiades, C., Distl, B., & Legendre, F. (2010, January). *PodNetSec: Secure opportunistic content dissemination*. In *Proceedings of ACM Mobisys* (p. 1).

[32] May, M., Karlsson, G., Helgason, O., & Lenders, V. (2007, October). *A system architecture for delay-tolerant content distribution*. In *IEEE Conference on Wireless Rural and Emergency Communications (WreCom)*.

[33] Li, N., & Das, S. K. (2010, February). *RADON: reputation-assisted data forwarding in opportunistic networks*. In *Proceedings of the Second International Workshop on Mobile Opportunistic Networking* (pp. 8–14).

[34] Li, N., & Das, S. K. (2013). A trust-based framework for data forwarding in opportunistic networks. *Ad Hoc Networks*, 11(4), 1497–1509.

[35] Raya, M., Papadimitratos, P., Gligor, V. D., & Hubaux, J. P. (2008, April). *On data-centric trust establishment in ephemeral ad hoc networks*. In *IEEE INFOCOM 2008-The 27th Conference on Computer Communications* (pp. 1238–1246). IEEE.

[36] Shafer, G. (1976). *A Mathematical Theory of Evidence*. Princeton University Press.

[37] Partridge, C. (2005, November). *Authentication for fragments*. In *Proceedings of ACM SIGCOMM HotNets-IV Workshop*.

[38] Gupta, S., Dhurandher, S. K., Woungang, I., Kumar, A., & Obaidat, M. S. (2013, October). *Trust-based security protocol against blackhole attacks in opportunistic networks*. In *2013 IEEE 9th International Conference on Wireless and Mobile Computing, Networking and Communications (WiMob)* (pp. 724–729). IEEE.

[39] Gonçalves, M. R. P., dos Santos Moreira, E., & Martimiano, L. A. F. (2010, April). *Trust management in opportunistic networks*. In *2010 Ninth International Conference on Networks* (pp. 209–214). IEEE.

[40] Wu, B., Chen, J., Wu, J., & Cardei, M. (2007). A survey of attacks and countermeasures in mobile ad hoc networks. In *Wireless Network Security* (pp. 103–135). Springer, Boston, MA.

[41] De Fuentes, J. M., González-Manzano, L., González-Tablas, A. I., & Blasco, J. (2014). Security models in vehicular ad-hoc networks: A survey. *IETE Technical Review*, 31(1), 47–64.

[42] Hu, Y. C., Perrig, A., & Johnson, D. B. (2006). Wormhole attacks in wireless networks. *IEEE Journal on Selected Areas in Communications*, 24(2), 370–380.

[43] Hu, Y. C., Perrig, A., & Johnson, D. B. (2003, March). *Packet leashes: a defense against wormhole attacks in wireless networks*. In *IEEE INFOCOM 2003. Twenty-second Annual Joint Conference of the IEEE Computer and Communications Societies (IEEE Cat. No. 03CH37428)* (Vol. 3, pp. 1976–1986). IEEE.

[44] Shin, S. Y., & Halim, E. H. (2012, October). *Wormhole attacks detection in MANETs using routes redundancy and time-based hop calculation*. In *2012 International Conference on ICT Convergence (ICTC)* (pp. 781–786). IEEE.

[45] Li, F., Wu, J., & Srinivasan, A. (2009, April). *Thwarting blackhole attacks in disruption-tolerant networks using encounter tickets*. In *IEEE INFOCOM 2009* (pp. 2428–2436). IEEE.

[46] Singh, U. P., & Chauhan, N. (2017, July). *Authentication using trust framework in opportunistic networks*. In *2017 8th International Conference on Computing, Communication and Networking Technologies (ICCCNT)* (pp. 1–7). IEEE.

[47] Ren, Y., Chuah, M. C., Yang, J., & Chen, Y. (2010, June). *Detecting blackhole attacks in disruption-tolerant networks through packet exchange recording*. In *2010 IEEE International Symposium on "A World of Wireless, Mobile and Multimedia Networks" (WoWMoM)* (pp. 1–6). IEEE.

[48]   Ke, M., Nenghai, Y., & Bin, L. (2010, November). *A new packet dropping policy in delay tolerant network*. In *2010 IEEE 12th International Conference on Communication Technology* (pp. 377–380). IEEE.

[49]   Zhang, X., Jain, A., & Perrig, A. (2008, December). *Packet-dropping adversary identification for data plane security*. In *Proceedings of the 2008 ACM CoNEXT Conference* (pp. 1–12).

[50]   Carbunar, B., Ioannidis, I., & Nita-Rotaru, C. (2004, October). *JANUS: towards robust and malicious resilient routing in hybrid wireless networks*. In *Proceedings of the 3rd ACM Workshop on Wireless Security* (pp. 11–20).

[51]   Chuah, M., & Yang, P. (2009, August). *Impact of selective dropping attacks on network coding performance in DTNs and a potential mitigation scheme*. In *2009 Proceedings of 18th International Conference on Computer Communications and Networks* (pp. 1–6). IEEE.

[52]   Baadache, A., & Belmehdi, A. (2012). Fighting against packet dropping misbehavior in multi-hop wireless ad hoc networks. *Journal of Network and Computer Applications*, 35(3), 1130–1139.

[53]   Sen, J., Chandra, M. G., Balamuralidhar, P., Harihara, S. G., & Reddy, H. (2007). *A distributed protocol for detection of packet dropping attack in mobile ad hoc networks*. In *2007 IEEE International Conference on Telecommunications and Malaysia International Conference on Communications* (pp. 75–80). IEEE.

[54]   Sultana, S., Bertino, E., & Shehab, M. (2011, June). *A provenance based mechanism to identify malicious packet dropping adversaries in sensor networks*. In *2011 31st International Conference on Distributed Computing Systems Workshops* (pp. 332–338). IEEE.

[55]   Marti, S., Giuli, T. J., Lai, K., & Baker, M. (2000, August). *Mitigating routing misbehavior in mobile ad hoc networks*. In *Proceedings of the 6th Annual International Conference on Mobile Computing and Networking* (pp. 255–265).

[56]   Nasser, N., & Chen, Y. (2007, June). *Enhanced intrusion detection system for discovering malicious nodes in mobile ad hoc networks*. In *2007 IEEE International Conference on Communications* (pp. 1154–1159). IEEE.

[57]   Song, J., & Ma, C. (2009, November). *A reputation-based scheme against malicious packet dropping for mobile ad hoc networks*. In *2009 IEEE International Conference on Intelligent Computing and Intelligent Systems* (Vol. 3, pp. 113–117). IEEE.

[58]   Alajeely, M., & Doss, R. (2014, September). *Defense against packet dropping attacks in opportunistic networks*. In *2014 International Conference on Advances in Computing, Communications and Informatics (ICACCI)* (pp. 1608–1613). IEEE.

[59]   Alajeely, M., Doss, R., & Mak-Hau, V. (2014, November). *Packet faking attack: a novel attack and detection mechanism in oppnets*. In *2014 Tenth International Conference on Computational Intelligence and Security* (pp. 638–642). IEEE.

[60]   Buttyan, L., Dora, L., Felegyhazi, M., & Vajda, I. (2007, June). *Barter-based cooperation in delay-tolerant personal wireless networks*. In *2007 IEEE International Symposium on a World of Wireless, Mobile and Multimedia Networks* (pp. 1–6). IEEE.

[61]   Chen, B. B., & Chan, M. C. (2010, March). *Mobicent: a credit-based incentive system for disruption tolerant network*. In *2010 Proceedings IEEE INFOCOM* (pp. 1–9). IEEE.

[62]   Newsome, J., Shi, E., Song, D., & Perrig, A. (2004, April). *The sybil attack in sensor networks: analysis & defenses*. In *Third International Symposium on Information Processing in Sensor Networks, 2004. IPSN 2004* (pp. 259–268). IEEE.

[63]   Trifunovic, S., Legendre, F., & Anastasiades, C. (2010, March). *Social trust in opportunistic networks*. In *2010 INFOCOM IEEE Conference on Computer Communications Workshops* (pp. 1–6). IEEE.

[64]   Lu, X., Hui, P., Towsley, D., Pu, J., & Xiong, Z. (2010). Anti-localization anonymous routing for delay tolerant network. *Computer Networks*, 54(11), 1899–1910.

[65]   Jansen, R., & Beverly, R. (2010, October). *Toward anonymity in delay tolerant networks: threshold pivot scheme*. In *2010-Milcom 2010 Military Communications Conference* (pp. 587–592). IEEE.

[66]   Zuberi, R. S., Lall, B., & Ahmad, S. N. (2012). Privacy protection through k. anonymity in location based services. *IETE Technical Review*, 29(3), 196–201.

Index

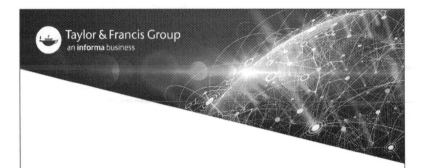